AF831837

Karl Leonhard

Classification of Endogenous Psychoses and their Differentiated Etiology

Second, revised and enlarged edition

Edited by Helmut Beckmann

Springer-Verlag Wien GmbH

Prof. Dr. med. Dr. h. c. Helmut Beckmann
Psychiatrische Klinik und Poliklinik, Universitäts-Nervenklinik,
Würzburg, Deutschland

Prof. Dr. Karl Leonhard†, Berlin

Translated from German by Charles H. Cahn

Originally published *as Aufteilung der endogenen Psychosen und ihre differenzierte Ätiologie, 7. neubearbeitete und ergänzte Auflage*
© 1995 Thieme, Stuttgart

The first English edition was published by Irvington Publishers, Inc., © 1979

This work is subject to copyright.
All rights are reserved, whether the whole or part of the material is concerned, specifically those of translation, reprinting, re-use of illustrations, broadcasting, reproduction by photocopying machines or similar means, and storage in data banks.

© 1999 Springer-Verlag Wien
Originally published by Springer-Verlag Wien New York in 1999

Typesetting: Bernhard Computertext KG, A-1030 Wien

Graphic design: Ecke Bonk

Printed on acid-free and chlorine-free bleached paper
SPIN: 10711564

Library of Congress Cataloging-in-Publication Data

Leonhard, Karl, 1904–
[Aufteilung der endogenen Psychosen und ihre differenzierte Ätiologie, English]
Classification of endogenous psychoses and their differentiated etiology /
Karl Leonhard. – – 2nd, rev. and enlarged ed. / edited by Helmut Beckmann. p. cm.
Previously published: Classification of endogenous psychoses. New York : Irvington Publishers, 1979.
"Originally published as: Aufteilung der endogenen Psychosen und ihre differenzierte Ätiologie, 7. neubearbeitete und ergänzte Auflage. © 1995 Thieme, Stuttgart" – – CIP t.p. verso. Includes bibliographical references and index.
ISBN 978-3-211-83259-2 ISBN 978-3-7091-6371-9 (eBook)
DOI 10.1007/978-3-7091-6371-9
1. Psychoses Classification. I. Beckmann, Helmut, 1940– .
II. Leonhard, Karl, 1904– Aufteilung der endogenen Psychosen. English. III. Title.
RC512.L413 1999
616.89'001'2– –dc21

ISBN 978-3-211-83259-2

Editor's Comment

Half a century ago KOLLE spoke about the "Oracle of Delphi of the Endogenous Psychoses". Since then in spite of all innovations and technological achievements in the past decades this pronouncement has not lost anything of its significance and its application to the present day research situation.

Mental and emotional disorders belong to the large group of "diseases of the people". From the public health point of view they are of extraordinary import for every community. Accordingly, some countries have generously supported research efforts in the neurosciences as well as in the physical sciences with the aim of achieving tangible progress in prevention and precise diagnosis, as well as rehabilitation. Such efforts have amongst others led to the declaration of the "Decade of the Brain" in the United States and more recently to that of the "World Decade of the Brain" in which all international societies concerned with the problem of mental health and rehabilitation have set the objective of introducing and carrying out more effective measures.

However public support should not be overemphazised. Compared with cardiovascular diseases, rheumatic disorders, neurological conditions such as multiple sclerosis etc. the support for research in the psychoses is comparably scanty and in many countries of the world completely absent.

Psychiatry in the true sense of the word has achieved objective progress only since, beginning with the Age of Enlightenment, it has become more closely allied with the natural sciences, and through the influence of some prominent representatives, such as PINEL in France and GRIESINGER in Germany, become associated with faculties of medicine. This desirable situation has only been attained in some parts of the world; setbacks are continually experienced and peculiar amalgamations take place between the natural sciences and shamanism.

With the 7th new edition of Karl LEONHARD'S "Classification of Endogenous Psychoses and their Differentiated Etiology" the "World Decade of the Brain", proclaimed in 1990 approaches its middle. Results from the so-called endogenous psychoses have so far been sparse. It is true that some symptoms of these disorders may be relieved with neuroleptic drug therapy; cures however have not been achieved. Nothing is known about effective prevention. For the affective disorders the introduction of lithium as a prophylactic measure is noteworthy, but even this treatment is not entirely reliable and may be associated with undesirable side effects placing an undue burden on the patient. The treatment of endogenous psychoses has almost stood still in the last three decades. It is still necessary to use partly effective and insufficiently reliable chemical substances in our treatment. In spite of all the enticements of modern marketing strategies better medications have not yet been found.

It is worth remembering that already in the 19th century (for instance KAHLBAUM and several French authors) a number of nosological entities had been described among the endogenous psychoses. In the same century, however, researchers such as Heinrich NEUMANN and Wilhelm GRIESINGER had postulated a "unitary psychosis".

Emil KRAEPELIN (1856–1925) proposed a partial compromise by separating the large field of dementia praecox with unfavorable prognosis from the field of manic-depressive conditions with favorable prognosis. This creation of a bimodal concept led to a dichotomy which for research has not proved to be very fruitful up to the present time. In spite of this subdivision KRAEPELIN described masterfully many subcategories, taking into acount cross-sectional and longitudinal symptomatology which even today may be considered to be quite valid, but towards the end of his life he lost the strength to develop this classification further. Eugen BLEULER (1857–1939) took over the concept of dementia praecox/manic-depressive illness, but completely ignored the prognostic aspects which KRAEPELIN (apart from a few exceptions) had held to be of greatest value. Thus he combined a large proportion of psychoses, which KRAEPELIN had included under the manic-depressive illnesses, with the conditions which he now called "schizophrenias" or "schizophrenia". He himself was convinced that he was dealing with several nosological entities; nevertheless it is a tragic aspect of his creative research that all his life he was looking for so-called "basic disturbances" of "schizophrenia", which of course he did not find, because they do not exist. BLEULER did not bluntly reject FREUD'S psychoanalysis; his positive attitude helped by Adolf MAYER resulted in his being made to feel welcome in Anglo-American psychiatry. Furthermore his cross-sectional symptomatological description, leaving aside prognosis, was accepted there more easily, for example by MAYER-GROSS.

During the last several decades at certain intervals of time and with good intention classification systems have been worked out by psychologists and psychiatrists by means of voting and consensus; descriptions of these systems come into the hands of mental health professionals in great regularity in the form of authoritative guidelines. All too easily it is being overlooked that behind it all no progress is made in research, merely restructurization. The latter has not been derived from lifelong observations of patients. Therefore from the scientific point of view this remains at least questionable. Interrater reliability keeps on being stressed at the expense of clinical validity. It even happened that clinical investigations requiring a high degree of knowledge and experience possessed by only the most skilful were delegated to "trained" students, psychologists, and scientific assistants in their first years of apprenticeship. Of course such mathematically attractive results cannot provide satisfaction. Witness the fruitlessness of our decades long research.

Contemporaneously with KRAEPELIN'S efforts Carl WERNICKE (1848–1905) in Berlin, Breslau and later Halle worked in the field of central neurology (particularly of aphasia) as well as in the field of descriptive psychiatry. From his findings in psychopathology he repeatedly postulated a "theory of disjunction" ("Sejunktionstheorie"), that is, an interruption of the connections between neural systems, leading either to a loss of functions, an excess of function or a faulty function. This disjunction may, for example in the area of psychomotility, lead to akinesia, hyperkinesia, or parakinesia. He suggested that a similar process occurred in thinking and in disturbances of the will. In so doing he found a powerful opponent in Karl JASPERS, who labelled him as a "brain mythologist". JASPERS had evidently overlooked the essential element in Wernicke's research, that is the careful elaboration of psychopathological conditions in cross-section as well as in their longitudinal course; even today in their precision WERNICKE'S descriptions are of great value. Among other surviving concepts WERNICKE coined the terms "akinesia", "hyperkinesia", "anxiety psychosis".

His pupil Karl KLEIST (1879–1960) followed WERNICKE in neurology as well as in psychiatry and psychopathology; by means of extensive studies of patients with brain trauma, KLEIST thoroughly confirmed and expanded WERNICKE'S observations (TEICHMANN 1990).

Differences between WERNICKE and KRAEPELIN became more apparent leading to a certain antagonism between them. This was continued with KLEIST, who never accepted such a gross division into two as KRAEPELIN and his pupils had proposed. As well he doubted the untiy of manic-depressive illness and asked his pupil Edda NEELE to investigate monopolar/bipolar depressions (1949). Furthermore he separated the cycloid psychoses from the field of manic-depressive illness and from a part of the schizophrenias. From his observations are derived masterful psychopathological descriptions which can hardly be further improved. Unfortunately he lacked the time to summarize all his psychopathological investigations so that now they have to be read in various original papers.

Karl LEONHARD was a pupil of KLEIST. In 1936 he came from the Gabersee psychiatric hospital to Frankfurt/Main, and brought with him the concept of "defect schizophrenic clinical pictures". He thus became a recognized lecturer (in German "Habilitation") at Frankfurt University with the full support of KLEIST. Here was found for the time a clear separation of a large portion of "schizophrenic" diseases, which at first KLEIST but later also LEONHARD did not include under the group of schizophrenias, but were considered as "system diseases" of the brain. These "systematic" schizophrenias were thought to have originated in a weakness either caused by a constitutional hereditary disposition or by environmental factors. They have a chronic insidious onset and their course is progressive with poor prognosis. In fact investigations for several decades, already introduced at KLEIST'S clinic, revealed that evidently these conditions with a deleterious course but little genetic predisposition showed nosologically a sharply limited characterization. After early admixtures of accessory symptoms such as hallucinations and delusions they become stable some years later, and may be identified again and again whatever therapeutic procedures are used. These hebephrenias, paraphrenias, and catatonias show entirely different clinical pictures which can only be learned and recognized by applying a high degree of intellectual effort. Nevertheless the latter is worthwhile, since here we do not have hereditary forms of mental illnesses before us and thus can avoid a source of error in therapy and research.

Of course Karl LEONHARD'S etiological considerations which in this book he points out in a highly differentiated manner for each of his diagnostic entities may be held to be somewhat imaginative speculations. For all that, they are based on his conscientious observations and passionately ingenious reflections throughout the decades. Nevertheless he himself accepts that there may be other ways of explaining his findings. Comparisons of his views on early childhood catatonia with those in the literature must be considered to be highly weighted in his favor and may turn out to be very relevant in the education of future generations.

The so-called "unsystematic schizophrenias" (periodic catatonia, affective paraphrenia, cataphasia) are of great scientific significance in so far as in them heredity plays a striking role, thus leading themselves to modern genetic research par excellence. Their course at the beginning is usually stormy, later there are often phases with the development of more or less clear-cut defects. When strong affective elements are present symptoms may be well controlled by means of modern neuroleptic therapy and thus offer a good therapeutic field for modern pharmacotherapy. Nevertheless it must be mentioned critically that true cures are hardly possible. These illnesses show, even with few symptoms, the formation of typical defects (periodic catatonia: apathy; affective paraphrenia: mistrust, suspicion; cataphasia: perplexity, poverty of affect). Even here LEONHARD, beside the striking genetic findings, expressed important thoughts concerning the role of parents and siblings during the development of the patient's childhood; even if these thoughts are not convincing they are at least noteworthy.

With regard to the cycloid psychoses (anxiety-happiness psychosis, excited-inhibited confusion psychosis, hyperkinetic-akinetic motility psychosis) LEONHARD in building on the research of his predecessors WERNICKE and KLEIST sees things rather differently. He deserves the credit of having differentiated these from the other forms of psychoses. He describes precise psychopathological clinical pictures which at times temporarily overlap with other cycloid psychoses, so that they are not always easily recognized in cross-section even by experienced observers. Occasionally there are transitions, at least of short duration, with unsystematic schizophrenias or even manic-depressive illness. This diagnostic problem may however be solved in most cases if the longitudinal development of the condition is carefully analysed. What is important for therapy is that medications in practice produce only symptomatic improvement and do not have any real influence on the course of the psychosis. Cycloid psychoses also show spontaneous remissions and may not leave any defect behind. Continued medication with neuroleptics during the healthy periods are more a hindrance than a help and may lead to toxic side effects which may disable the patient in his every day life and brand him as mentally ill. Because of this LEONHARD has made a passionate appeal for the careful differentiation of the cycloid from the other psychoses. The prognosis, which is exceedingly important not only for the patient but also for his family may be of great value, keeping in mind how much the burden of mental illness is carried by the whole family. To be sure, great caution is indicated with regard to predicting the duration of a phase, since there are cycloid psychoses which last from a few days to several years. Nevertheless there is good reason in the last analysis to retain a favorable prognosis. The familiar incidence in contradiction to the unsystematic psychoses is low (about 4%). Here too his etiological reflections in terms of family and siblings are interesting, and can only be understood on the basis of LEONHARD'S collective works (Biopsychology of Endogenous Psychoses, 1970; Biological Psychology, 1993). They will be rejected by many. But this does not in the least diminish the value of the exact descriptions and consideration of hereditary factors offered here.

The concept of manic-depressive illness with its usual bipolar course was taken over by KLEIST and by LEONHARD from KRAEPELIN, and required no essential additions. However LEONHARD differentiated it clearly from pure melancholia and pure depression. This had always been surmised by KLEIST, but was confirmed by Edda NEELE (1949), and then proved by LEONHARD and his co-workers from their investigations of families through observing the clinical and genetic characteristics. There is thus no doubt that Karl LEONHARD has conclusively proved that the discovery of the monopolar/bipolar dichotomy of manic-depressive illness in its phenomenology and genetics was valid. ANGST (1966), PERRIS (1966) and WINOKUR (1969) in their investigations have always confirmed this even without having undertaken a phenomenological cross-sectional and longitudinal analysis. On the basis of his investigations Karl LEONHARD assumes for manic-depressive illness as differentiated by him an important hereditary flaw, equivalent to genetic dominance. No connection with the X-chromosome was found by him in his extensive material despite such proposals by several authors. Even the discrepancy found in monozygotic twins with regard to the incidence of these illnesses he explains as due to a different affective lability, which does not have to be genetically inherited, but which may eventually promote or prevent the onset of the illness in one or the other of the twin partners. This he mentions primarily as a hypothesis in order to stimulate discussion.

The separation of mania occurring in phases has likewise been recognized all over the world. These are rather rare diseases to which particular attention ought to be paid; this happens all too infrequently. LEONHARD has quite independently separated five pure depressions: agitated (also translated as "harried"), hypochondriacal, self-tortured, suspicious, apathetic, as well as five pure euphorias: unproductive, hypochondriacal, exalted, confabulatory, indifferent. These clinical pictures occur infrequently, but are being seen again and again by experienced clinicians, sometimes misdiagnosed as atypical depression or as neuroses. But Karl LEONHARD has described them with such sharp circumscription that the diagnoses cannot be missed from observation. It is quite incomprehensible that these characteristic clinical pictures continue to be pushed this way or that way diagnostically. Here he postulated in each case an affliction of a particular emotional layer in which pathological thinking is occasioned by the type of pathological affect. This he underscores in his book "Biological Psychology" where he gains valuable insights from the pure depressions, the pure euphorias, as well as the systematic schizophrenias as derived from his observations of the symptoms. These insights ought to help modern scientific psychology emerge from the blind alley of fruitless research.

The question has often been asked why the "WERNICKE–KLEIST–LEONHARD–SCHOOL" of psychiatry has not prevailed internationally. There are several answers to this question: first the fact that WERNICKE died early and thus was unable to defend his point of view from that of the influential KRAEPELIN; secondly, critics such as JASPERS (who worked in psychiatry for only a few months), and many others who branded him and his pupil KLEIST as "brain mythologists" and ignored him as well.

Against the differentiated nosology of the "WERNICKE–KLEIST–LEONHARD–SCHOOL" numerous other objections have been raised by essentially uninformed critics. Many modern psychiatrists reproached LEONHARD saying that his nosological entities were too subjective, they had been elaborated from clinical observations, and lacked "objective" confirmation. He was also criticized in that he did not develop any "rating scales" by means of which individual clinical pictures could have been identified in "operationalized" form by third parties. To counter these criticisms is the fact that LEONHARD had personally with his co-workers examined several hundreds, even thousands of patients before each nosological entity was elaborated. His investigations spanning many decades, were conducted by him personally and were meticulously described by him in writing; in his "Frankfurt series" they consisted of several hundred cases, in his "Berlin series" of 1465 cases. Of course the figures of the differentiated clinical pictures vary greatly, from the very frequent manic-depressive illness and the cycloid psychoses to the single forms of unsystematic and systematic schizophrenias. Single subforms he observed only occasionally, a fact which he mentions every time. In rare cases (e.g. indifferent euphoria) he presented only a few case histories. Complicated combined systematic forms likewise were seen by him in relatively few cases, because these were almost always being cared for in psychiatric institutions because of their severity. This has led to criticisms that he had modelled individual categories, as if on a drawing board. From observations that my co-workers and I were able to make regarding his explorations of many years I cannot remember a single case in which he was unable to describe a type of illness in all its psychopathological differentiations in the way he had suggested. In this way it became understandable again and again why his method of diagnosing remained closed indefinitely to so many psychiatrists. The diagnosis of his nosological categories is complete only if all symptoms found in his descriptions can be diagnosed in the patient. Individual mainly qualitative changes exclude the correct diagnosis. Therefore in his descriptions one is not dealing with only clinical impressions, artful circumscriptions, reminiscences or suppositions, but with psychopathological descriptions made and modified in their empirical exactitude for decades, representing the most important elements in a most striking manner. Thus the most careful operationalizations were practised, in which firm associations of symptoms were described. It is unthinkable that symptoms the way he sees them can be removed from the groupings to form other clinical entities. Each disturbance of thinking, for instance in systematic paraphrenia, has its own characteristic form, and can be found only with this condition. The same applies to auditory hallucinations which may occur in practically all psychiatric conditions, but which for LEONHARD acquired their nosological significance only in their definite firm legitimate symptomatic connection, that is, a syndrome. Deviation from this firm conformity or inattentiveness will always lead to the wrong diagnosis. Numerous scientists have tried to "operationalize" LEONHARD and have always failed, because by doing so less exact or less appropriate pictures arose. Of course this increases the difficulty of accepting and transmitting his point of view. In comparison with neurology in which one is essentially dealing with motor and sensory phenomena whose differential diagnosis presents enormous problems, why should these be any less when dealing with the highest human functions? In neurology additional somatic findings (imaging and laboratory techniques) considerably facilitate diagnosis. Here in psychiatry it is to be hoped that soon with the help of modern scientific procedures additional diagnostic techniques may bring more certainty to this field. In the meantime we have to fall back on the painstaking road of psychopathological differentiation in order to obtain the most homogeneous groups for investigation.

Early results in clinical genetics have already been shown to be impressive by means of such a nosological differentiation (FRANZEK and BECKMANN 1991). Differential methods of therapy as well have found a more sensible application than if they had been used in an undifferentiated manner for a supposedly unitary psychotic continuum (BECKMANN et al. 1992).

With regard to prognosis LEONHARD'S classification of endogenous psychoses brings enormous advantages, thereby avoiding many misjudgements which have contributed to giving our profession such bad publicity. Furthermore it protects us from deceptive illusions, encourages us in our therapeutic endeavours, and keeps in check exaggerations which occur here and there.

In research concerning the etiology of endogenous psychoses the type of differentiated psychopathology according to LEONHARD is indispensable. Here too some early encouraging results have been obtained, which have been described partially by LEONHARD himself (UNGVARI 1993).

The possibilities of danger because of early childhood isolation in single child families or distinct position in the order of siblings are shown impressively. Other investigators found distinct environmental influences in the prenatal period in differential nosological subdivision (STÖBER et al. 1993a, 1994). Imaging techniques have revealed quite significant findings in certain nosological subdivisions (BECKER et al. 1993). Research in electroencephalography has revealed similar findings (STRIK et al. 1993, WARKENTIN et al. 1992).

The undersigned came to know the validity of LEONHARD'S concepts through his work for several years with Karl LEONHARD himself. Eventually he and his co-workers Ernst FRANZEK and Gerald STÖBER undertook lengthy studies to test the validity of LEONHARD'S concepts; by means of a series of investigations independent from one another they achieved high coefficients of reliability (Cohen's kappa 0.90).

Since a number of other experienced clinicians have been able to validate the classification system of the WERNICKE–KLEIST–LEONHARD–SCHOOL, there now exists the wellfounded prospect that this unfortunately rather difficult but singularly relevant concept may succeed in the future despite of the prevailing international classifications (ICD and DSM). In favor of this concept should be mentioned the confirmation by ANGST, PERRIS and WINOKUR of the monopolar/bipolar dichotomy, the confirmation of the concept of the cycloid psychoses (PERRIS 1974, BROCKINGTON et al. 1982, BECKMANN et al. 1990) and of periodic catatonia, an unsystematic form of schizophrenia described by GJESSING (1974); furthermore the supporting contribution by ASTRUP (1979) on the systematic schizophrenias. Individual clinical descriptions have been published by STÖBER et al. (1993b) (self-tortured depression) or STÖBER et al. (1993c) (proskinetic catatonia).

The preponderance of Anglo-American psychiatry after World War II, based on KRAEPELIN, Kurt SCHNEIDER and BLEULER, offered in a greatly simplified form a two-diagnoses-system and was therefore accepted much more gratefully than the highly differentiated nosological system of Karl LEONHARD. The continued changes proposed in each revision of ICD or DSM seem to promote the feeling that progress is being made, although the direction is further and further away from the cross-sectional/longitudinal diagnostic system of the most experienced clinicians. One may suspect that this path may lead to many decades of fruitless research, and will not serve well the "World Decade of the Brain". Even today critical voices have been heard with regard to the present day diagnostic scheme (VAN PRAAG 1993, BROCKINGTON 1992, FRANZEK and BECKMANN 1991).

During LEONHARD'S lifetime there was no lack of international attention. His principal work has been translated into several languages and published repeatedly. His other books as well have been published in several editions, which due to the fact that they were published in the German Democratic Republic had very limited numbers and now are no longer available in second-hand book stores. The scientific climate has resulted in a considerable number of publications of which several are listed at the end of this chapter. The findings from research with twins which had been omitted from the last edition of the book have now been included because of their importance; in the international literature they had been totally neglected, but are relevant as a point of departure for LEONHARD'S reflections on differentiated etiology. Besides even without these findings they emphasize the hereditary factors in the various nosological entities.

In conclusion cordial thanks are due to Ernst FRANZEK and Gerald STÖBER for their collaboration and review of the manuscript. They deserve great credit for continuing the scientific work of Karl LEONHARD. Bruno PFUHLMANN and SABINE VOSS gave very valuable assistance in the correction and typing of the manuscript.

Würzburg, April 1995 Helmut Beckmann

References

Angst J (1966) Zur Ätiologie und Nosologie endogener Psychosen. Monogr. Gesamtgeb. Neurol. Psychiat., H. 112. Springer, Berlin

Astrup C (1979) The chronic schizophrenias. Universitetsforlaget, Oslo

Becker T, Stöber G, Lanczik M, Hofmann E, Franzek E (1994) Cranial computed tomography and differentiated psychopathology – are there patterns of abnormal CT findings? In: Beckmann H, Neumärker KJ (Eds) Endogenous psychoses. Leonhard's impact on modern psychiatry. Ullstein Mosby, Berlin

Beckmann H, Fritze J, Franzek E (1992) The influence of neuroleptics in specific syndromes and symptoms in schizophrenics with unfavorable long-term course. Neuropsychobiol 26: 50–58

Beckmann H, Fritze J, Lanczik M (1990) Prognostic validity of the cycloid psychoses. Psychopathol 23: 205–212

Bleuler E (1911) Dementia praecox oder die Gruppe der Schizophrenien. In: Aschaffenburg, G (Ed) Handbuch der Psychiatrie. Deuticke, Leipzig Wien

Brockington IF, Perris C, Kendell RE, Hillier VE, Wainwright S (1982) The course and outcome of cycloid psychoses. Psychol Med 12: 97–105

Brockington IF: Schizophrenia: yesterday's concept. Eur Psychiat 7: 203–207

Franzek E, Beckmann H (1991) Syndrom- und Symptomentwicklung schizophrener Langzeitverläufe. Nervenarzt 62: 549–556

Gjessing LR (1974) A review of periodic catatonia. Biol Psychiat 8: 23–45

Kleist K (1947) Fortschritte der Psychiatrie. Kramer, Frankfurt

Kraepelin E (1923) Psychiatrie. Ein Lehrbuch für Studierende und Ärzte, 8. Aufl. Barth, Leipzig

Neele E (1949) Die phasischen Psychosen nach ihrem Erscheinungs- und Erbbild. Barth, Leipzig

Perris C (1974) A study of cycloid psychoses. Acta Psychiat Scand 50: 7-75 (Suppl. 253)

Perris C (1966) A study of bipolar (manic-depressive) and unipolar recurrent depressive psychoses. Acta Psychiat Scand 42 (Suppl. 194)

van Praag H (1993) "Make-Believes" in psychiatry or the perils of progress. Brunner/ Mazel, New York

Stöber G, Franzek E, Beckmann H (1993a) Obstetric complications in distinct schizophrenic subgroups. Eur Psychiat 8: 293–299

Stöber G, Franzek, E, Beckmann H (1993b) Die selbstquälerische Depression. Eine Form monopolarer endogener Depressionen. Nervenheilkunde 12: 166–169

Stöber G, Franzek E, Beckmann H (1993c) Die "Proskinetische Katatonie". Ein kasuistischer Beitrag zur Psychopathologie chronisch schizophrener Psychosen. Krankenhauspsychiatrie 4: 70–73

Stöber G, Franzek E, Beckmann H (1994) Schwangerschaftsinfektionen bei Müttern von chronisch Schizophrenen. Nervenarzt 65: 175–182

Strik WK, Dierks T, Franzek E, Maurer K, Beckmann H (1993) Differences in P300 amplitudes and topography between cycloid psychoses and schizophrenia in Leonhard's classification. Acta Psychiat Scand 87: 179–183

Teichmann G (1990) The influence of Karl Kleist on the nosology of Karl Leonhard. Psychopathol 13: 267–276

Ungvari GS (1993) The Wernicke-Kleist-Leonhard school of psychiatry. Biol Psychiat 34: 749–752

Warkentin S, Nilsson A, Karlson S, Risberg G, Franze'n L, Gustafson L, Wernicke C (1992) Cycloid psychosis: regional blood flow correlates of a psychotic episode. Acta Psychiat Scand 85: 23–29

Wernicke C (1900) Grundriß der Psychiatrie in klinischen Vorlesungen. Thieme, Leipzig

Winokur G, Clayton PJD (1969) Family history studies. I. Two types of affective disorders separated according to genetic and clinical factors. In: Wortis J (Ed) Recent advances in biological psychiatry. Plenum, New York (pp. 35–50)

Preface to the 6th Edition

No new edition of the "Classification of Endogenous Psychoses" required on my part so many modifications as this 6th edition. Extensive investigations carried out in recent years resulted in unexpected new knowledge regarding the etiology of endogenous psychoses, leading me to change the title of this book to "Classification of Endogenous Psychoses and their Differentiated Etiology". The new findings might have necessitated a much longer text, but this was avoided by shortening the clinical section. Previously I had presented a large number of case histories in order to make the descriptions more meaningful and concrete. In this edition many of these have been omitted. Interested readers may find them in earlier editions. Thus space has been provided for describing in detail etiological observations beside the clinical descriptions. Having previously recognized independent diseases in the various syndromes which are described under endogenous psychoses, my concepts have been given a much sounder base by demonstrating distinct etiologies. Some forms depend primarily on heredity, others on different kinds of psychosocial factors. The influence of siblings on one another, as well as absence of siblings, has become of special significance. A totally distinct etiology is found in early childhood schizophrenia; this disorder has not been described in earlier editions, as I learned it only in recent years.

Karl Leonhard

Translator's Notes

Karl Leonhard's division of schizophrenia into 3 major groups which he called systematic schizophrenias, unsystematic schizophrenias, and cycloid psychoses has found relatively little acceptance in international psychiatry, in which "operational and atheoretical diagnostic systems" prevail (DSMs and ICDs)[1].

There are probably several reasons why Leonhard's concepts have been all but ignored in the by now huge literature – Leonhard's relative isolation during the latter part of his life in East Germany, his use of complex German diagnostic terminology, and his almost stubborn resistance to compromise with the efforts of most other experts in psychiatric diagnostic classifications. But in recent years Helmut Beckmann and his collaborators, Eli Robins, Frank Fish, Thomas Ban, Christian Astrup, Carlos, Perris, Ian Brockington and others, have started to revive interest in Leonhard's viewpoints, as the evidence from neurodevelopmental, brain-imaging, genetic and psychosocial research has mounted, to show more and more convincingly that "schizophrenia" (singular) is not one entity but consists of several quite different disorders to be considered as "schizophrenias" (plural) or "schizophrenia spectrum disorders"[2].

The previous translation into English of the 5th edition of Leonhard's "Classification of Endogenous Psychoses" was published in 1979. Before his death in 1988 Leonhard added two new chapters, one on the significance of psychosocial circumstances, and the other on childhood catatonia, which appeared in the 6th edition in 1986; it was subsequently worked over and re-edited by Professor Beckmann, giving the translator the opportunity to present a new translation (and at the same time to correct a considerable number of mistakes in the 5th edition).

Leonhard took great pains in his examination of the patients whom he observed, often conducting "psychic-experimental tests" (not to be confused with "intelligence tests") in which he asked patients to explain the meaning of proverbs. Some of the latter have English equivalents, others do not.

---

[1] DSM means Diagnostic and Statistical Manual (of the American Psychiatric Association); ICD means International Classification of Diseases (of the World Health Organization)

[2] For review see: Beckmann H, Neumärker KJ (eds) (1995) Endogenous psychoses – Leonhard's impact on modern psychiatry. Ullstein Mosby, Wiesbaden

Explanatory footnotes are given in the text. Since patients' responses may include play on certain words contained in the German proverbs but not in the English equivalents, on some occasions the exact translation from the German had to be given.

Apart from the translation of Leonhard's differentiated diagnostic terms (which in English might not always reflect accurately what the German terms meant), the translator also preferred to refer to and retain "Nervenklinik"as "Psychiatric Hospital", and "kranke Eltern" or "kranke Geschwister" as "affected parents" or "affected siblings."

The Würzburg School of Psychiatry, led by Helmut Beckmann, has been conducting extensive research on schizophrenia spectrum disorders for the last 15 years, testing many of Leonhard's concepts and finding them valid in many respects. Their work has been published for the most part in European and some American psychiatric journals, but one important paper on a twin study was published in the January 1998 issue of the American Journal of Psychiatry[3] which should be of particular interest, not only to North American psychiatrists, but also to all those conducting research on the genetics of endogenous psychoses.

Montreal, September 1998 Charles H. Cahn

[3] Franzek E and Beckmann H (1998) "Different genetic background of schizophrenic spectrum psychoses: a twin study". Am J Psychiatry 155: 76–83

Table of Contents

Introduction

Psychiatry has been endeavouring for many decades to clarify the endogenous psychoses. Overemphasis on heredity has diminished; instead psychosocial factors have been moved into the foreground without however knowing what form they should take. Theories which have been suggested have not been confirmed. Because of this the wheel is now being turned backwards: hereditary causative factors are again becoming more prominent. In consideration of the lack of true progress many authors have adopted an attitude of resignation, stating that clinical research in etiology should be abandoned until bodily causes have been found. My opinion is quite different. Lack of success in clinical etiological research in my opinion is the result of the fact that endogenous psychoses have been roughly divided into only two forms. In this way a majority or even a multitude of disorders have been lumped together and therefore no uniform etiology could of course be found. Above all so-called schizophrenia (singular!), taken as a unit, is completely unsuited for etiological research. Clinically I see here the most significant differences. How can fantastic paraphrenia be combined as a unit with foolish hebephrenia or negativistic catatonia? KRAEPELIN'S division into only two forms of endogenous psychoses in this respect has been a great disadvantage. It is true that KRAEPELIN himself beyond the division into two undertook many more fine separations, but his successors did not pay any attention to this, they saw only the gross division of endogenous psychoses into Dementia praecox or schizophrenia and manic-depressive illness. The teaching of mental illnesses in which no external cause was found was in this way shockingly oversimplified. While the sister discipline of neurology has identified perhaps hundred endogenous disorders and continues to describe others, psychiatry recognized only two, perhaps three or four. And while neurologists in the hereditary forms of endogenous diseases are seeking the heredity of each single form and even in conditions which clinically seem to be unitary find several types of heredity, psychiatrists are still disagreeing about one hereditary form of schizophrenia, as if underlying the so colorful and even extremely different clinical pictures there was only one disease with only one uniform heredity.

The development of psychiatry would certainly have taken a different course if WERNICKE, KRAEPELIN'S great opponent, had not died prematurely. WERNICKE, like no other, understood how to pick out and describe from the multitude of endogenous psychoses characteristic distinct clinical pictures; it is true that without KRAEPELIN'S etiological considerations he would have remained very one-sided, but by means of his classical descriptions he would certainly have succeeded in maintaining the interest of clinical psychiatrists with regard to the many distinct forms. Today there are probably very few psychiatrists who have read his "Grundriss". Those who have are impressed with how much more the different endogenous psychoses meant for him than for the psychiatrists of today who seem to be interested only in schizophrenia and manic-depressive disease, perhaps now also "schizoaffective" psychosis. Perhaps not too much attention would be paid to WERNICKE'S theoretical concepts derived from the theory of sejunction, but the focus would be on his descriptions of the various disorders. WERNICKE who had many students of cerebral pathology unfortunately had none in psychiatry. Only KLEIST took up and pursued WERNICKE'S ideas. The spirit of these two great psychiatrists will be apparent in my own presentations.

If not a few but many endogenous psychoses are to be differentiated psychiatry becomes a complex science. But the differentiation is unavoidable if psychiatrists wish to emerge from the blind alley in which the doctrine of endogenous psychoses has stagnated for a long time. In the case of the phasic psychoses my concepts have at last been accepted through the investigations of ANGST, PERRIS, WINOKUR and others, specifically that monopolar and bipolar endogenous depressions are genetically separate disorders; but only a small beginning has been made.

In considering schizophrenia for decades a peculiar contradiction has existed. Most psychiatrists are in agreement with EUGEN BLEULER, who originally did not speak of a single schizophrenia, but a "group of schizophrenias", and thereby expressed doubts as to the unity concept; but most scientific publications sound as if there was only a single schizophrenia. In maintaining such a contradictory attitude a solution of the many problems which we meet when dealing with the schizophrenias will never be found. If there is no desire to consider each single form of endogenous psychosis separately, at least a separation should be made between the groups in which a loose unity may be found. In this way five groups may be differentiated: **Monopolar phasic psychoses, manic-depressive illness, cycloid psychoses, unsystematic schizophrenias, and systematic schizophrenias.**

The last named group is of quite special significance because it contains the most severe psychoses with relatively little hereditary loading. Fortunately, just in this area I was recently able to make some important etiological observations. Investigation of twins using the differentiated diagnosis produced important etiological findings. Follow-up investigations of schizophrenic children brought about further progress. Having examined patients supposedly considered severely mentally retarded, whom even KRAEPELIN had partially considered as childhood schizophrenics, gave me clinically and etiologically important results quite recently. I may mention here in advance that the findings to be reported are quite concrete and in so far as they are essential possess statistical significance.

KRAEPELIN at one time had considered his classification as etiological, because he correctly assumed a different etiology in conditions with different prognoses. But his prognostic concept could not be maintained. It is true that on one side there was the benign manic-depressive illness, but among the cases which are nowadays frequently labeled schizophrenic there are forms that remit and forms ending with defect. Therefore today the diagnosis of schizophrenia has nothing to say about prognosis. However in making a differentiated diagnosis a prognosis is obtained at the same time. Patients and their relatives may be informed that if the diagnosis is motility psychosis and not catatonia, if it is a happiness psychosis and not a paranoid schizophrenia, if the diagnosis is anxiety psychosis or confusion psychosis, i.e. in any of these psychoses the present phase will remit completely. Moreover **therapy** is different according to whether the psychosis results in a spontaneous remission or whether there is the risk of a defect developing if appropriate therapy is not provided. Unfortunately these days I see many patients with cycloid psychoses who because of prolonged medication show toxic disabling symptoms, whereas without medication they would be perfectly well. The damage is aggravated by the fact that after prolonged medication, as ALBERT (1986) has reported, habituation occurs so that even in the case of phasic psychoses discontinuation of medication produces a relapse. In my opinion a patient with a cycloid psychosis should only be treated psychopharmacologically during the acute phase. After the latter has abated the medication should be rapidly discontinued by gradual lowering of the dose so that no habituation may occur. In order to make sure that this was properly done, while I was still directing the department, I saw to it that patients were discharged only after the medication had been discontinued. When this was done no relapses occurred. Habituation occurs in all likelihood because metabolism becomes adjusted to the addition of the medication and after discontinuation of the medication loses the previously acquired equilibrium.

In discharging patients only after they were free of medications I went against the modern challenge of early discharge. I did not worry about this, for on the one hand early discharge of patients whose disorders were in phases had no great significance; on the other hand it was important to prevent permanent damage from medical interventions. I was unable to make sure that medication after discharge was discontinued in good time. Such omission produces much damage. Toxic lowering of drive, of affectivity, and the presence of extrapyramidal symptoms represent a severe toxic disorder. This is a risk for many patients, because about a third of all cases nowadays diagnosed in the wide concept of schizophrenia in reality have cycloid psychoses with good remissions. This shows how important differentiated diagnosis is in therapy as well.

If by means of prognostic delimitations and differentiation of many special forms psychiatry becomes a very difficult subject in my opinion this is merely an advantage. In today's situation the ease with which a psychiatrist can make his diagnosis – he could hardly be wrong if he chooses only one of two possibilities – has led to the fact that many psychiatrists have altogether stopped learning how to describe and evaluate a clinical picture correctly.

Instead inroads have been made into psychiatric concepts with which the disorder – usually a form of schizophrenia – is characterized and judged.

A demonstration of the fact that an endogenous mental disorder belongs to an independent disease entity may be given when observations are made of its repeated appearance. If a psychosis is manifested several times in similar forms, either in patients with recurrent mental disorders, or in relatives who have become psychotic, this indicates the existence of an independent disorder. It is true that allowance must be made for the breadth of symptoms in the same disorder. Some mental illnesses always recur in the same form, while others show a considerable variety of symptoms. In order to present the confirmatory findings in earlier editions of the book I described many case histories where other patients in the family were available for comparison.

The disorders of the patients to which I refer had to be evaluated statistically; therefore I could not choose them at will, but had to conform to predetermined conditions. With regard to the phasic psychoses I took cases seen between 1938 and 1942 in the Psychiatric Hospital of Frankfurt.

Together with NEELE I have evaluated and later completed these cases **statistically.** NEELE (1949) has published some of these findings. In the cases of schizophrenias we had to take patients whose disorders had begun a long time ago, thereby enabling us to obtain a good view of the course of the disorder. I have not evaluated statistically the findings from my investigations in the Institution Gabersee (LEONHARD 1936). But the fact that KLEIST with his coworkers including myself undertook many prolonged examinations gave me the opportunity to collect my cases. This included all the patients whom I personally examined, as well as 55 patients who in the years 1937–1940 were in hospital and had already been affected by the illness for ten years or longer. Furthermore I was able through the kindness of BRUNO SCHULZ to examine a considerable number of chronic schizophrenics in Bavarian institutions. Altogether my early cases included 526 patients with phasic psychoses among which were also cycloid forms, and 337 patients with schizophrenias. In Berlin I was able to greatly increase these numbers. In the last 15 years I was able to undertake examinations in psychiatric hospitals and together with Frau VON TROSTORFF to evaluate statistically 1465 cases (691 men and 774 women). The total number of schizophrenics mounted to 1450. The number of phasic including cycloid psychoses with the help of my coworkers rose to 1163. Added to these were 352 patients (150 men, 202 women) whom I reexamined in the last 15 years, giving me a total of 1515 cases of phasic psychoses that were evaluated statistically. Of course the cases that had been observed clinically greatly exceed this number, for I hardly need to add that the cases of endogenous psychoses seen by me daily were not just part of the routine but allowed me to observe them with clinical interest. The **case histories** to be presented are derived principally from my time in Frankfurt, that is, these were older cases. This is an advantage because in those days psychopharmacological treatments were not yet provided. The clinical pictures therefore are not artificially changed. The fact that patients' utterances belonged in a different world, especially during the time between the two world wars, does not diminish our understanding of the clinical pictures.

In some endogenous psychoses heredity is the chief etiological criterion, in others psychosocial factors stand in the foreground. In spite of this I have no hesitation in continuing to speak of "**endogenous**" psychoses and omit the "so-called" often mentioned these days. Psychosocial causes influence the disorder not immediately before its onset but in childhood and adolescence; during that time they generate an inner predisposition from which the disorder later emerges. When this occurs the underlying endogenous disposition makes the difference.

Clinical Pictures of Phasic Psychoses (without Cycloid Psychoses)

Until very recently nearly all psychiatrists were united in the opinion that manic and depressive disease pictures were all part of manic-depressive illness. It was the work of ANGST (1966) and PERRIS (1966) that helped spread my theory that unipolar and bipolar diseases were separable. I carried out my first investigations together with NEELE. KLEIST (1943), with whom we were working, shared my opinion. Previously, KLEIST had claimed that there was no independent manic-depressive illness, but rather only melancholia and mania with a certain reciprocal affinity. Thus he had already postulated the independence of the unipolar forms, but had gone too far by denying the independent existence of manic-depressive illness. The genetic difference between the unipolar and bipolar forms was seen in that the manic-depressive form had a significantly higher rate of psychoses among relatives than did unipolar forms. WINOKUR (1969) and others followed in this line of thinking. However the two disease forms also have different clinical pictures. The bipolar form displays a considerably more colorful appearance; it varies not only between the two poles, but in each phase offers different pictures. The unipolar forms, of which there are several (ANGST, PERRIS, WINOKUR do not treat this) return, in a periodic course, with the same symptomatology.

Every individual form is characterized by a syndrome associated with no other form. On the other hand, in bipolar cases no clear syndromes can be described since there are many transitions between various formations and the picture may even shift during one phase. **Thus, one can generally recognize a bipolar form during the first phase.** In the same sense, bipolar forms may be recognized as such when there is only an accidental swing toward one pole but where the potential toward the other pole exists. Consequently the differentiation is more easily made between **polymorphic** (bipolar) and **pure** (unipolar) forms. The number of polymorphic disease forms with no tendency at all to the opposite phase is very small, insofar as attention is paid to minor indications. Not infrequently the opposite phase will merely be hinted at, though it cannot fail to be recognized. Manic patients may have a depressive mood for hours, depressive patients as their depression disappears may show excitement with gaiety and overactivity. Both do not have to be regarded as the expression of separate phases, but these variations do show the disease potential toward the other pole. Furthermore, depressive patients may often be excited by encouragement and led out of their depression. They become lively, talkative, with hardly a sign of the depressed mood.

In fact a patient during such an encouraging discussion may become so lively that eventually he might be described as hypomanic. Afterwards the depressive mood returns. In the unipolar forms there are no signs of lability toward the other pole; a unipolar melancholic never shows a manic trait – no matter how long it lasts or how often it returns – and a unipolar manic never has a trace of depression. One should expect reactive depressions only when a patient realizes after the end of the psychosis what mischief he or she has done during the manic phase. Such depressive reactions can generally be differentiated from depressive phases of manic-depressive illness by the presence of normal motivation.

Manic-Depressive Illness

The clinical picture of manic-depressive insanity described by KRAEPELIN is well known. But since usually attempts have only been made to differentiate it from the schizophrenias, and other phasic psychoses, i.e. pure forms and cycloid psychoses, have been disregarded, it behooves us to review our concepts about the broad range of symptoms.

In one group of cases the symptomatic picture is restricted to the manic and the melancholic basic syndrome. The first is composed of euphoria, that can easily develop into annoyance, elevated self-consciousness, flight of ideas, pressure of speech, and overactivity, while the second includes a depressive mood with ennui, feelings of insufficiency, thought blocks, psychomotor retardation, difficulty in making decisions and depressive idea construction. We will also find these syndromes in pure mania and pure melancholia – there they are in fact necessary forms, while in manic-depressive disease these syndromes rarely appear in their pure form.

More frequent in manic-depressive illness are cases in which the basic syndrome of mania and of melancholia, or of both, has changed. Essential single symptoms may be absent. In mania there may merely be a euphoric mood without flight of ideas and without overactivity, as is the case in "unproductive mania". If overactivity is absent, on stimulation the patients may develop flight of ideas and pressure of speech but show no other motor restlessness. Conversely, the euphoric mood may be absent, while flight of ideas and overactivity are present. Similarly in melancholia, depressive mood and ideas of inferiority may be the only symptoms, while thought inhibition and psychomotor retardation are absent. If on the other hand depressed mood is absent and only thought and psychomotor inhibitions are present, or perhaps only one of these, then simple inhibited states arise.

The incompleteness of mania and melancholia – I want to speak of manic and melancholic **"partial states"** – can go further. With the pure euphorias and depressions we will see that there are different forms of affect and thus different forms of affective disorders. Agitated depression, hypochondriacal depression, exalted euphoria, confabulatory euphoria, and so on, apparently arise when only one type of emotion is affected by the disorder. Thus there may be a dissociation within the affective sphere which may at times be imitated by the manic-depressive illness, so to that its partial states resemble agitated depression, unproductive euphoria, and so on. This is rarely fully developed, but more often in one of the phases of manic-depressive illness one or another type of emotion may predominate, so that there will at least be hints of one of the pure euphorias or depressions.

More rare than the partial states are the **mixed states** in which in the place of the missing symptom a symptom of the opposite pole appears. KRAEPELIN described this extensively. Euphoria may be combined with psychomotor inhibition. The latter may in fact be extreme and manic stupor may arise. Unproductive mania, already mentioned as a partial state, may also be mixed. When thought inhibition joins euphoric mood and overactivity, then unproductivity of thought and speech replaces manic flight of ideas. Conversely, manic thought disorder may be combined with depression, creating depression with flight of ideas accompanied by pressure of speech. Other forms of excitement need not be present; the excitement in thinking alone may lead to pressure of speech. If, on the other hand, psychomotor excitement joins the depression, then it is difficult to decide whether the affect is merely looking for discharge leading to anxious agitation, or whether there is a manic component in the excitement. The latter is ascertained if overactivity without connection with depressive mood is recognizable.

Other variations of manic-depressive illness arise when events go beyond the normal borderline and create symptoms otherwise associated with **cycloid, i. e. different forms of bipolar disorder.** Best known of the manic-depressive variations is **confused mania** in which the thought disorder has a character otherwise associated with confusion psychosis. There is not necessarily a disorder of consciousness. In hyperkinetic forms there are traces of motility psychosis, and in ecstatic forms there are traces of happiness psychosis in the otherwise manic picture. In stuporous depression the inhibition is similar to that of the mutism of confusion psychosis or the akinesia of motility psychosis. Severe anxiety corresponds to the affective disorder of anxiety psychosis.

The overall picture of manic-depressive illness with its wide range of symptoms will appear when **I describe single cases.** The most informative cases are those in which other cases of the illness have appeared in the same family. Such cases are presented below. They come mostly from my **Frankfurt investigations.** The fact that these are older cases is actually advantageous since neuroleptic treatment was not yet carried out at that time. First I list a patient who herself and in her family showed only the **typical symptoms** of manic-depressive illness. In some of these cases atypical traits were apparently absent because the phases appeared in a particularly mild form. A greater intensity of the disease process may easily produce transitions to the features of cycloid psychoses.

Case 1: Elisabeth Kol, born in 1908, became depressed in September 1937, made self-accusations, thought she was incapable, was afraid her father would die, and nearly stopped speaking. She appeared normal from January to March 1938, then she became excited with pressure of speech and activity. She was admitted to hospital, had flight of ideas, was partially happy, partially irritable. After she had calmed down, she was discharged. In 1921, the mother's sister had spent ten days in Institution T. due to a depressive mood. She was obsessed with religious questions, made self-accusations, and complained about a lack of feeling toward her children.

This patient showed nothing really atypical. If both poles had not been obvious in the clinical picture, then pure mania or pure melancholia might have been suspected. However, that is not very common. In rare cases it might happen that a manic-depressive illness was imitated by the accidental combination in one person of a pure mania and a pure melancholia. The family history in the next case points to such an interpretation.

Case 2: Stefan Sud, born in 1900, who had entered hospital only for evaluation, went through repeated episodes of mild mania with heightened self-consciousness, wealth of ideas, industriousness, and mild depression, with loss of energy, some self-accusations and inhibitions. There were no psychoses in the family, but we found out the following about the parents: The **father** was always very industrious, co-founder of many societies, administrator in many honorary positions, and ten times an honorary member. The **mother** is quiet, sensitive, and religious. After marriage she went through a depression that subsided after hydrotherapy.

In most cases of manic-depressive psychosis, the atypical symptomatology shows that we are not dealing with a mixture of pure mania and pure melancholia. This will become clearer in the following case.

Case 3: Maria Ma, born in 1901, had experienced periods of depression since the age of 25. In 1942 she was despairing for two days, then exhibited pressure of speech in a mood at times depressed and at times cheerful. This state continued for a few days in the hospital, then severe inhibition followed with nearly complete mutism. After a few days, provocative jocularity was observed. Irritable, whining, agitated and stuporous days followed. Thereafter a mixed state began in which the patient lay in bed the whole day hardly moving, but displaying flight of ideas or incoherence in her broken verbalizations. Following further changes the disturbance ended four months after it had started.

In atypical forms various emotions may be affected. The clinical picture resembles "pure depression", not "pure melancholia".

Case 4: Dorothea Drus, born in 1890, experienced "nervous exhaustion" in 1926/1927. In 1938 she spent four weeks in Hospital G. for a neurasthenic-depressive condition with complaintiveness. In 1940 she showed a similar picture in the hospital in Frankfurt; her mood was more labile than depressive. Suicide attempts had a somewhat demonstrative character. Then strong fear replaced the lability, and it became even stronger in Institution G., to which she had been transferred. The patient cried, complained, begged, was full of despair, and tried to tie her bed jacket tightly around her neck. After her fears receded she demanded repeatedly and urgently to be discharged. In May she recovered rapidly and was discharged in a happy mood. In 1918 the **father** had had a mania that was often interrupted by a crying mood but which had no other atypical characteristics.

If this patient had only been observed during her excitement, her illness would probably have been regarded as **agitated depression,** because of her constant expressions of fear. But from the very different picture in the rest of the course, this diagnosis must be rejected. The very varied picture shows that this was a case of manic-depressive illness, as confirmed by the father's mania.

Case 5: Anna Hell, born in 1914, in 1931 experienced anxious excitement, hallucinated ("the walls say I am a swindler"), and was difficult to diagnose. Then she became increasingly stuporous and was finally fully motionless. There was no longer depressive affect, and at times the patient would even smile. After three months she recovered. In 1937 she returned to hospital and was now hypomanic, somewhat forward, without distance. In 1939 she was again depressive and complained in hospital particularly of alienation of feeling: she had no more heart and no more feelings, the greenness and blossoming of nature meant nothing more to her, everything was so colorless. Her child and her husband seemed strange to her. She felt inwardly changed, that she was somebody else. There was also a feeling of inability to accomplish anything. After six weeks the depression disappeared, and she was then hypomanic again.

The patient, after short hallucinatory fear, went through an akinetic state, no longer recognizable as depressive. Later I will speak of similar states. Here I am interested in the second phase which was shaped quite differently and which revealed the multifacetedness of manic-depressive illness. In this phase the alienation phenomena were in the foreground. In pure melancholia too, the complaint about a subjectively felt cooling of the emotions is often heard. With the patient, however, the complaint that she had lost her normal emotions was practically the only depressive symptom, except that there was also a feeling of inadequacy. According to the picture in this phase of the illness, one might assume that this was a case of **"apathetic depression"**, which disorder will be described under "pure depressions". The first phase and the hypomanic interphases, show, however, that the "alienation depression" was only one episode within manic-depressive illness.

Case 6: Otto Wil, born in 1907, developed feelings of anxiety in 1938, and was admitted to hospital unconscious after a suicide attempt with sleeping pills. Here he was very inhibited, indecisive, made self-accusations, complained of loss of interest and claimed he would never get well. There were also hypochondriacal complaints: his body was out of line, everything hung to the right, hands and feet would fall asleep, the legs jerked, there was something wrong in the lower abdomen, the intestine had ceased functioning. After a period of six months, the above phenomena disappeared. In 1926 a **sister** made a reactive suicide attempt, but became depressive again in 1934 for no reason and was brought to Institution M. She displayed certain hypochondriacal traits and was partially inhibited and partially anxiously excited. After one year she became manic with pressure of speech and flight of ideas. During the decline of the mania she displayed temporary muteness with compulsive movements. Then she became depressive again for a short period of time. **A second sister** arrived in 1943 in the Frankfurt Hospital with depression, which at first appeared with inhibition, dislike of work, hopelessness, anxious fears, and indeterminate bodily complaints. After a few weeks it turned into a severe agitation, during which the patient cried inarticulately, struck out aimlessly with her fists, and spoke of poisoning. The mood was fearful with elements of irritation. She died in this excited state of bronchopneumonia. The **father** of the three patients entered Institution M. at the age of 53 years due to depression. With mild fluctuations, he remained stuporous for two years, but despite severe inhibition moaned, rubbing his hands together, and kept on moaning. Finally the psychosis was arrested. A quiet, shy manner remained. He died in 1933 without any further mental illness.

The last mentioned patient is of interest: at the age of 53 he showed the picture I have described as characteristic for "involutional depression". This often develops into "hardened[1] regressive depression" (LEONHARD 1937), that is, anxious-stuporous behaviour, in which, apart from severe inhibition, there is restlessness of the hands, often accompanied by moaning. In my sample of patients with phasic psychoses, I found my position upheld that this form of depression almost only occurs in persons of climacteric age. However, the development into a hardened regressive depression need not follow. In the few cases from other age groups, it was a question of complications from arteriosclerosis or fever. The fingering movements here could be a sign of delirium. I could not decide if the climacterium could have a similar effect in the sense of an exogenous noxious influence. With the patient described above, the psychoses of his three children speak for his illness also being manic-depressive, and not involutional depression.

The mixed state of the first sister of the subject is of interest. In this state she showed mute restlessness, most likely excitation with thought inhibition. Immediately beforehand, she had experienced excitement and flight of ideas. A turn into depression followed. The second sister, too, seemed to have a mixed state during her excitement, but with a sudden switch from a previously inhibited condition, it is more probable that the excitement was of a manic nature, while the affect still remained depressive. Both sisters had only mild hypochondriacal traits. Because they were very marked in the subject, I have described the family relations here.

The peculiar complaints that have been presented here will occupy us later with **hypochondriacal depression.** If there were not the simultaneous presence of melancholic symptoms in the form of inhibition, indecisiveness, self-accusations and hopelessness, one would have to suspect hypochondriacal depression. There were no fears about bodily well-being – that too is called hypochondriacal – but actual bodily misperceptions, which are directly reminiscent of the hypochondriacal hallucinations of schizophrenics. The total picture of the depression also speaks for its really being a manic-depressive illness. Confirmation is also found in the relatives who display manic traits. Depression had, therefore, once again produced a certain type of feeling and thus showed similarity with one of the pure forms.

Case 7: Marta Emm, born in 1907, years before her first severe illness had depressions in which she spoke of death. In 1936 she entered hospital, was somewhat fearful, complained, and expressed self-accusations that she was a bad mother. Further, peculiar bodily complaints appeared – she had the feeling that her head was too large and the hands too long; the stomach had changed; the intestines no longer worked properly. But not only her body, the whole environment seemed changed, everything seemed larger. She also had the feeling that people were laughing at her. After four weeks she was improved and discharged. In 1940 she became ill again. This time feelings of inadequacy appeared, a feeling of incapability of dealing with anything. There was a hopelessness that she would never get well. She only lacked the courage for suicide. She lacked motivation, was weak

[1] „erstarrend" in German

and inhibited. There were also misperceptions and feelings of estrangement; she had stomach cramps as if she had to vomit; a feeling in her head that it was growing larger. Everything external seemed strange and far away from her, she felt as if she no longer belonged to humanity. She could no longer grasp ideas correctly. After four months she was improved and discharged. A **sister** of hers was in hospital six times. She generally showed a melancholic picture with inhibition, occasionally she had periods of anxious excitement, and alienation phenomena were sometimes recognizable. She had short manic excitement twice, one a manic stupor during which she spoke no word but laughed silently and tried to put on jewelry. The **father** had a labile mood and was in the Frankfurt Hospital for a short period in 1930 with a reactive depression.

In Marta Emm we find a connection between phenomena of hypochondriasis and alienation. The latter are different, however, from those in case 5. There we found a subjectively perceived cooling of feeling while here there is a feeling of strangeness toward the patient's own body and external events. We will see in the pure forms that this type of alienation is a phenomenon frequently seen in hypochondriacal depression. This can also be imitated in manic-depressive illness. Melancholic symptoms, inhibition, and lack of energy with feelings of inadequacy all also point to its not being true hypochondriacal depression in which agitation is a much more characteristic symptom. The illness of the sister confirms the presence of manic-depressive illness. Her mixed state of a manic stupor is worthy of interest.

Case 8: Hedwig Schab, born in 1905, became melancholic in 1932, believing that she was being spied on, that the newspapers were writing about her, that people were playing games with her. She was admitted to hospital after a suicide attempt. Here she was inhibited, laconic, spoke in a monotonous voice about her suicide attempt, "I was so stupid, no one dies at all; it just never happens that people die." She continued to express ideas of reference of having been observed, but complained more about hypochondriacal problems. She felt as if worms were crawling around in her body, she must have a child in her abdomen, she no longer had a real body, her stomach was missing. She was discharged seven months later in a happy equable mood. In 1936 she became ill again, was more tearful than deeply depressed and often had a smiling expression. In the intelligence test, she easily fell into flight of ideas. This time no hypochondriacal ideas were present, but rather many ideas of reference. People were maligning her and, although she had killed no one, someone wanted to kill her and put poison in her food. After four months she was again happy and composed and could be discharged. In 1938 she entered hospital for a third time. She was very labile, "cried and laughed simultaneously", expressed ideas of reference, such as people in the street were singing about her, and so on. Her whole environment seemed to her to be changed. After three months this phase ended completely. Her **mother** had experienced several depressions, and, in Institution I., exhibited a manic state with paranoid-querulous traits.

In the first depression, which was the deepest, hypochondriacal ideas were present, which were again accompanied by alienation phenomena. Schab also believed her stomach was missing and that neither she nor other people could die. In the first depression and more markedly in the following two, ideas of reference became apparent and dominated the picture to a large extent. We will return to the relationship between anxiety and ideas of reference in connection with anxiety psychoses. When anxiety is minor and not particularly recognizable externally, it may be of a case of **suspicious depression** which we will examine under the pure forms. Here again a pure form is imitated by manic-depressive illness. The similarity is, of course, only approximate. In the first phase the nihilistic hypochondriacal coloration goes in quite a different direction, and in both of the other phases the fluctuating affect cannot be overlooked. On one occasion there was a definite mixed state with flight of ideas and a depressive basic mood. The bipolarity of the psychosis was even more clear in the **mother.**

Some patients show **anancastic coloration** of their depression. There are often hints of compulsive phenomena, particularly in inhibited forms with indecisiveness. A markedly anancastic coloration of depression probably only appears when the manic-depressive tendency is accompanied by a compulsive neurotic tendency. It is then magnified by depression and indecisiveness. It is, however, something altogether different in cases where compulsive neurotics become depressed when reacting to their disorder from which they try to escape in vain. In this form of "compulsive depression" – as with reactive depressions – the depression is clearly related to its cause. Depressed patients with compulsive neurotic coloration, on the other hand, show their depression quite independently from their compulsive thoughts.

After having seen several pictures of manic-depressive illness that resembled pure depressions, we also have euphoric phases that are reminiscent of **"pure euphoria".** The somewhat exuberant state of one subject resembled **"exalted euphoria",** as will be seen later. This patient claimed that she had fulfilled a secret mission, a task assigned to her by Hitler. The behaviour of the **mother** of the subject with her simple friendliness without content was reminiscent of **"unproductive euphoria",** which we will examine with the pure forms.

Now I shall mention a patient whose manic behaviour resembled that of **"confabulatory euphoria".** This patient demanded 100,000 men in order to occupy Offenbach, he claimed to be the richest man to have taken control of innumerable firms and to have written 25,000 business letters. He also claimed to have been in the Foreign Legion, to have been a pilot in the Spanish War, and to have crashed three times.

Manic-depressive patients whose psychosis resembles a pure form should be checked to see if in their personality any characteristic features can be found that might explain the special form.

In some cases of manic-depressive illness the clinical picture resembles that of one or other **cycloid psychosis.** This occurs often when the disorder is of a severe nature. **Confusion psychosis** may appear to be a bit like **confused mania.**

Case 9: Therese Fick, born in 1912, went through a period of confused mania, with false recognition of people and mood lability, followed by melancholia with strong inhibitions, particularly thought inhibitions. Her **mother's mother** had several episodes of melancholia with inhibition, indecisiveness, and confused mania. As well, two **sisters of this afflicted grandmother** were mentally ill temporarily. No medical record was available for one, the other had mania with severe confusion. Of the latter, two daughters were temporarily mentally ill: one had inhibited depression with some anxiety; the other had two periods of mania with totally incoherent thought process and false recognition of people.

Confused manias appear in marked form almost only in women. Here the incoherence of the thought process can be understood as the highest degree of flight of ideas. We will find it to be different in confusional psychosis. Generally there is at the same time a high degree of pressure of speech, as the latter is much more closely related to the thought disorder than to the other forms of excitement. The mood in confused manias is often no longer purely euphoric, but rather labile with a tendency to slip temporarily into depression. Perhaps this mood lability is not only an expression of manic-depressive bipolarity, but partially comparable to the lability of symptomatic psychoses; hence a sign that the manic-depressive psychosis has reached an endotoxic level. At times disturbances of the sensorium accompany the confusion. This appears to be an endotoxic effect of the gradually heightened disease process, particularly because consciousness is at times clouded at the height of the disease.

A female patient heard many voices, as she confirmed after her recovery; she perceived a smell of death and had visual hallucinations. She saw a golden beam on the wall, she saw her dead cousin dressed in white, and her father confessor sitting on her bed. As this patient had visual hallucinations, we must consider the possibility of clouding of consciousness.

The more severe the confusion becomes and the more false recognition of persons and hallucinations occur the greater becomes the question whether the diagnosis of confused mania should not be confusion psychosis.

Differential diagnosis is often more difficult in **stuporous states.** Perplexed stupor is characteristic for inhibited confusion; in manic-depressive illness, on the other hand, particularly strong inhibition can lead to depressive stupor. – In the following case the diagnosis might have been questionable if the patient's aunt had not experienced manic and melancholic phases.

Case 10: Emilie Räuch, born 1890, was in hospital eight times suffering from depression. In between there were short manic states with excitement and irritability. When depressed she had some ideas of reference, but no other disturbances of thinking. Inhibition was very marked, and became stronger in all phases, approaching stupor. The patient seemed somewhat confused, but her facial expression revealed her depressive mood. The **mother's sister** was hospitalized six times, four times with typical mania, and twice with depression accompanied by moderate inhibition and poverty of ideas.

Similar to the picture of confusion psychosis, manic-depressive illness may resemble the picture of **motility psychosis,** as the following case shows.

Case 11: Maria Bat, born in 1892, was in the Frankfurt Hospital eight times for mania, which was often followed by inhibited-depressive swings. In her excitement she was generally confused and displayed hyperkinetic traits. For example, she danced, threw herself on the bed, stood up again, ran through the room, shook her head with her hair flowing, hit the wall rhythmically, spread out her arms, and made many other expressive movements. A **brother** experienced manic and depressive phases at home, and in 1915 in the Frankfurt Hospital showed the picture of mania with expansive-confabulatory traits.

Sometimes the stupor accompanying the depression is not comparable to the perplexed stupor of confusion psychosis, but rather to the **akinesia** of **motility psychosis.** In the latter there is the tendency towards freezing of expressive movements. Here the disturbance is not due to thought inhibition, as in perplexed stupor, but rather to psychomotor inhibition. The following case shows again that we are dealing with a manic-depressive patient.

Case 12: Hildegard Mör, born in 1895, was in Institution I. ten times since 1922 and in the Frankfurt Hospital once in 1940. She had episodes of mania that were often somewhat confused; mixed states in which depressive mood was combined with confused excitement, or cheerfulness and flight of ideas was combined with psychomotor inhibition; also depressions with stuporous behaviour. Then she became totally rigid, resistive, had to be tube-fed, for hours she remained in the same position. In 1928 her **mother** had a manic episode for 14 days (there was no complete description). In 1933 she had a depression, which after admission to Institution I. rapidly passed through an irritable state into mania, associated with confusion and ideas of grandeur. A **sister of the mother** had a manic episode in 1927 in Institution W. (not fully described), and in 1928 in Institution M. a depressive episode with self-accusations, ideas of impoverishment, and mild anxious excitement.

Finally there are cases showing transition to the third cycloid psychosis, i.e. **anxiety-happiness psychosis** with both of its poles.

Case 13: Hedwig Schleuß, born in 1906, became ill in 1942, believed she was being followed by the Gestapo and claimed that her boarder was working with them. She believed the people on the street were observing her, and that a baby carriage in the street was only an observation place from which to spy on her. She reinterpreted every noise in the apartment and looked for the Gestapo's "bug". She also heard voices and believed to have heard her sister. In hospital she continued to express such ideas, but also produced self-accusations that she had falsely signed bills and that she was not worthy of her girl friend. At first she was rather motionless and perplexed, then she became anxiously excited, begged for help, and clung to others. After a few weeks, the anxiety decreased and was replaced by a hypomanic state. After relapse into anxiety, again with ideas of reference, a hypomanic state continued. The patient was lively, talkative, laughed a lot and became querulous when she was not immediately discharged. A **mother's sister** had been in the Frankfurt Hospital in 1917 where she was depressive, lachrymose, preoccupied with hypochondriacal ideas, claiming that a nerve was strained, her nerves were gone, her head felt as if it were bleeding inside, and her heart was burning. She was periodically anxious and excited. Eventually she was discharged free of symptoms.

In this patient ideas of reference, with anxious basic mood, played a large role. This might have been a case of anxiety psychosis; there were however also melancholic symptoms. The hypomanic state surely was the reason for the diagnosis of manic-depressive illness. The diagnosis can be considered confirmed by the depression of the mother's sister who did not have ideas of reference. – Features of **happiness psychosis** were observed in the following patient.

Case 14: Josef Göpf, born in 1871, was in the Frankfurt Hospital five times between 1923 and 1940. He had only mild depressive moods, but pronounced manic episodes, partially typical, partially accompanied by ideas of having been blessed. Thus in 1923 he wanted to go through the world as an evangelist; he claimed he was a spirit and did not need to eat. A **sister** in 1937 was in Institution L. with stuporous depression and later committed suicide.

Cases have here been selected to show clearly that features of other endogenous psychoses may be present and that therefore manic-depressive illness is polymorphous in its clinical pictures. In most cases patients could be observed for longer periods of time, and the benign course of the condition could be confirmed.

Summary

The symptomatology of manic-depressive illness with its range of symptoms has been demonstrated above, in all cases with **hereditary disposition.** The essential feature is the manic and melancholic **basic syndrome;** atypical forms may be **mixed states,** or **partial states,** which occasionally mimic pure euphoria and depression, or an extension of the symptomatology into that of **other bipolar forms.** Certain similarities to cycloid psychoses can be understood as increase in degree of severity – thus confused mania and stuporous depressive pictures. The picture will be completed by the statistical results.
Certain points of view, however, which could be of clinical importance, are not being discussed here, since I have nothing new to add to them. I have no new results on the duration of the separate phases which, for mania, is considered to be about 3 to 6 months, and for melancholia about 6 to 9 months. My results also confirmed the fact that phases may exceptionally last only hours or continue for years and, in fact, become chronic. I have ignored normal temperaments associated with manic-depressive disease, not because I saw no possibility of adding to our knowledge, but because **special investigations** were planned (LEONHARD 1963 a and b, LEONHARD et al. 1962, VON TROSTORFF 1970). It is clear that hypomanic, hypomelancholic, and cyclothymic temperaments are present in the families of the patients as forms of dilutions of mania, melancholia, and manic-depressive mixed state. Beyond this, it is of great interest to what extent there are similarities with and differences from the temperament associated with other bipolar psychoses. **Somatic findings** in manic-depressive illness should be checked, if investigations thereof have been carried out in other phasic disorders.

Pure Melancholia and Pure Mania

The full significance of the polymorphous nature of manic-depressive illness only becomes clear when the pure forms with their stable symptomatic pictures are discussed, starting with pure melancholia and pure mania. Some points have already been touched upon, since the variable syndromes of manic-depressive illness occasionally mimic the pure forms.

The "basic syndromes", with which the discussion of manic-depressive illness began, also represent the pictures of pure melancholia and pure mania.

Pure Melancholia

The most important symptom of melancholia is the **depressed mood.** There can be implicit anxiety, but more often apathy gives the depression its characteristic form. The indifferent behaviour could make the patients seem "sad", but this concept, reflecting a higher mental state of mind, does not correctly describe melancholic depression; it appears to be more a physical decline of the feelings, not corresponding to mental sadness. K. SCHNEIDER's concept of "vital" depression is very much like that of pure melancholia. Depression cannot be described in terms of normal psychology; only the somewhat general term of "depressedness" can be understood in the sense of this anergic (i.e. incomprehensible by normal mentality) mood. In mild cases, the patients appear almost more apathetic than actually depressive; in severe cases, however, deep suffering develops. We will later find anxiety to be a special component of the emotional disturbance of pure depression. It is certainly also included in melancholic depression but is generally covered up by other emotional components and by inhibition often declared subjectively. High degrees of anxiety, however, exclude the diagnosis of pure melancholia; anxiety leading to severe reactions does not occur here.

The second cardinal symptom is **psychomotor inhibition** which accompanies the depression of melancholia, and can be of higher or lower degree, generally running parallel to the depth of the disturbance. In pure melancholia, inhibition never reaches stupor. A slow speech pattern, a monotonous and low voice, slowing down of all movements, poverty of expressive movements, all these characterize melancholic inhibition. This does not restrict the ability to carry out daily functions. Stuporous depressions are common in extremes of manic-depressive illness; they also appear in other bipolar psychoses, but not in the pure forms.

Thought inhibition, the third cardinal symptom of pure melancholia, becomes recognizable when the patient is asked questions which require a certain, albeit small, amount of reflection. Reaction time is markedly longer than could be explained by psychomotor inhibition alone. While the simplest questions, about personal data for example, can still be answered quickly, it takes a disproportionately long time to receive answers about data not readily available or about more complicated information from the past. Difficult questions may not even be understood the first time. Simple questions in intelligence testing, which normally require hardly any thought, can be answered quickly. But depending on intelligence and educational level and the degree of severity, thought inhibition becomes apparent and leads to noticeably long reaction times. Thought inhibition does not reach high degrees in pure melancholia, and mutism from thought inhibition does not occur.

Other symptoms of pure melancholia can be deduced from the three cardinal symptoms. Thought inhibition and psychomotor inhibition make decisiveness more difficult. Thus **indecision** is an important symptom of pure melancholia. In milder cases it is often not easy to demonstrate the presence of inhibition, but at least the subjectively perceived greater difficulty in making decisions points out the inhibition. While patients may be able to operate fairly freely in an already moving process, they have a hard time beginning the process itself. Even when there is not much thought content patients cannot make up their minds. Thus inactivity may occur, more noticeably in the domestic milieu than in hospital. Sometimes indecision leads to manifestation of an **anancastic tendency,** as we have already seen in inhibited depression of manic-depressive illness. No compulsive activity develops opposing the inhibition, but obsessive brooding does occur.

The **feeling of inadequacy,** or insufficiency, is yet another very important symptom of pure melancholia. It is also derived from psychomotor and thought inhibition; it corresponds particularly to the feeling of lack of accomplishment resulting from the inhibition. The patients complain that they no longer complete anything or do not progress in their work. On the other hand the complaint of inadequacy can imply an objectively groundless self-denigration in the sense of **ideas of sinfulness.** The relationships are reciprocal, since the idea of sin can conversely be partially based on real lack of achievement. Difficulty in making decisions, and feelings of inadequacy are particularly important when making the differential diagnosis from the other pure forms of depression, for in the latter they are absent since inhibition is absent.

Furthermore, **various depressive ideas** may characterize the clinical picture. We will see later how, in pure depressions, pathological processes may produce abnormal ideas at different emotional levels. In pure melancholia all levels are affected, so all sorts of idea construction are present. **Self-accusations, ideas of inferiority, fear, anxiety, hypochondriacal ideas, ideas of alienation and ideas of reference** are all observable. They do not, however, all appear in the intrusive form, such as we find in pure depression. One or another pathological idea may predominate in pure melancholia, probably depending on the personality of the subject, but the total idea construction here is not obtrusive. Often one first discovers it when directly asking about it. Apparently, inhibition, particularly thought inhibition, works against abnormal elaboration of ideas. Thus one finds many pure melancholics who display the three cardinal symptoms in their characteristic form but without abnormal content. Patients' utterances may be limited to complaints about lack of enjoyment of life, difficulty in making decisions, and inability to achieve anything. Outright questions about this or that depressive idea often will be answered in the negative. Self-accusation is absent less frequently, probably because it is nurtured by feelings of inadequacy. Thus the characteristic picture of pure melancholia is one of quiet, moderately inhibited depression, with little content. Since obtrusive symptoms are absent and the depression itself is hardly noticeable in taciturn patients, one frequently has the impression of a milder disease.

But, in fact, this is basically a serious illness. The suicidal risk in pure melancholia is great. Because the whole range of emotions is changed by the depression, the capacity for normal emotional reactions is completely lost, so that the patients no longer have any feeling for the positive side of life. Thus they often commit suicide, astonishingly treating it simply as a matter of course, not driven by strong affect, in contrast to patients with anxiety psychosis. Even in cases of pure melancholia that seem mild on the basis of the total picture, suicide may occur. Suicide would probably be much more common if inhibition and indecision did not prevent it.

Deficient capacity for normal emotional reactions moreover leads to **impoverishment of interest;** this is recognizable subjectively as well as objectively. On the same basis arises a symptom, which we will begin to know more thoroughly in the description of "apathetic depression" that is, the patients' complaint that they have no feelings at all, that they **cannot even feel sad anymore.** The complaint of no longer being able to cry fits in here almost of necessity, for crying only occurs when there are fluctuations of mood, as I have described elsewhere (LEONHARD 1976 b).

With manic-depressive illness a greater number of cases had to be cited to gain an overview of the motley clinical picture. In pure melancholia there is not a great range of symptoms. The cases largely resemble one another; this is, of course, the meaning of the concept of a pure form. However, misdiagnosis may occur. It has already been emphasized that manic-depressive illness occasionally mimics a pure form, but the family picture usually helps to make the correct diagnosis, for instance in the following case.

Case 15: Anna Schei, born in 1867, became depressive for the first time in 1914 and complained of pressure in the head. In 1923 she entered the Frankfurt Hospital with a new depression. She could no longer run her home, could no longer work well, was interested in nothing anymore, and accused herself about that. In hospital she sat around, hardly moving, with an empty depressive facial expression. At first she expressed suicidal ideas, later she claimed she felt nothing anymore, not even sorrow, she was like a piece of wood, she could not cry anymore. She was discharged against medical advice, recovered at home, and remained well for 20 years. In 1942 she returned to hospital and was again depressive and tired of life. This time she complained that her head was tied up, as if a ring was wrapped around it; she had no more interest in her household. She could not think anymore or read. The depression disappeared after about three months, although at discharge, she was still somewhat quiet and unsure of herself. A **sister** became ill in 1909, was anxious, and explained that something must have changed in her head. There were also self-accusations that she had lied as a child, she did not want to eat, but would rather die. At home she was allegedly periodically perplexed and restless. In the Frankfurt Hospital she sat hunched over and answered sparsely and slowly. She maintained her depressive ideas. Once she claimed she heard voices, but nothing particular about their content could be determined; perhaps she only meant the voice of her conscience. After three months she was discharged recovered. She did not become ill again. The mother of the subject also had depressions and hanged herself while suffering from melancholia.

In the **family histories** of probands with pure melancholia we find only depressive elements, often accentuated by hypo-melancholic temperament. Manic episodes are completely absent. The psychoses of the relatives present the recognizable typical picture of pure melancholia, in so far as they are adequately described. Only a few peculiarities are worth mentioning. The sister of case 15 said something vague about hearing voices. It seems that in pure melancholia – as we will find with the other pure depressions – the ideas may assume a pseudo-hallucinatory character. Against assuming that these are genuine hallucinations is the fact that the patients complain about the voices in the same way as they do about their subjective symptoms, for example, those of a hypochondriacal nature; they have at most the fear but not the conviction that these experiences have external origin. Sometimes an anxious coloration is reported. Subject 15 and her sister were said to have been anxiously excited at home. In hospital both were inhibited and depressive, and showed no evidence of anxiety. It often happens, and not only with melancholia, that many more obtrusive symptoms are reported at home than appear in hospital. That can readily be explained by the fact that at home inappropriate handling by the family may easily precipitate emotional reactions absent in hospital, because they are not expressions of the disease itself.

Even in cases of pure melancholia where there were no psychoses in the family, the relatives appeared different from those of manic-depressive patients. Among the relatives of pure melancholics we find many quiet, serious natures, i. e. hypomelancholic personalities, and occasional suicides, without definite evidence of a psychosis. The temperamental type of the subjects themselves is also often dysthymic. Our investigations in this connection have been described above under manic-depressive illness.

A flowing transition from dysthymic temperament to a phase of pure melancholia is not infrequently found. Sometimes it is reported that, at a certain point in time, the mood worsens, although not to the actual level of disease, but enough so that general unhappiness and often suicidal thoughts occur. Such forms of hypomelancholia can continue for years and then develop into an acute phase. Chronic bad moods may also develop in the wake of depressive phases. An actual phase of pure melancholia may occasionally last for an extremely long time. More usually it tends to recede after six to twelve months as do the depressions of manic-depressive illness. The tendency toward a chronic course is found in pure depression, but in pure melancholia this applies more to dysthymia than to the actual melancholic phase. As we will see in the statistical section, the number of phases in pure melancholia on the average is considerably less than in manic-depressive illness.

Summary

I believe that I have demonstrated the independence of pure melancholia from manic-depressive illness by the clinical descriptions and the family histories. It is a **unipolar** disease, without manic phases. Clinical differentiation from manic-depressive illness is not possible in every case, but evidence from family histories has shown misdiagnosis to be rare. The clinical picture is characterized by depression, psychomotor inhibition, thought inhibition, indecision, feelings of insufficiency, and varying depressive ideas, not particularly obtrusive in any one direction. The **danger of suicide** is great. Somatic symptoms add to the psychic picture, as with the melancholia of manic-depressive illness. Comparative investigations, however, would be desirable.

Pure Mania

In all its traits pure mania is the opposite of melancholia. Corresponding to depression, we find an **elevation of mood.** Patients are often described as cheerful, but this concept of a higher emotional state does not reflect the elevation of mood in mania which is partially a bodily phenomenon. On the other hand, the concept of the notion of euphoria may be used in this sense. Just as the notion of "vital" depression the notion of "vital" euphoria is justified, characterizing the basic manic mood.

In manic-depressives, euphoria often turns into irritability. The pure manic may also become irritated when he meets resistance, but this is usually a momentary reaction. Irritability as an essential component of the basic mood includes a depressed component and is not present in pure mania, while it is common in the mania of manic-depressive illness.

Thought in pure mania is characterized by **flight of ideas.** In mild cases the patient may find the way back to a theme after a digression, while in more severe cases he completely loses the logical thread and continually makes new superficial associations. This internal distractibility is paralleled by distractibility from external impressions. Flight of ideas in pure mania never reaches confusion; mania with confusion belongs to manic-depressive illness. The excitement of thought is accompanied by **psychomotor excitement,** as exemplified by overactivity. The patients involve themselves in everything going on around them. Many gestures and facial expressions accompany their restless activity. Motivation is still retained, and the activities still serve some purpose, even if only a superficial momentary one; expressive movements, however, jump about considerably, although generally they still correspond to the mental state at the moment. Mania lacks more primitive reactive movements which have lost their purpose, and expressive movements developing autonomously; we will find these in the description of motility psychosis. **Pressure of speech,** which is always found in pure mania, is partially precipitated by overstimulation of thinking which leads to a wealth of ideas. Psychomotor excitement increases it further.

Decision-making is easier due to thought and psychomotor excitement. Beyond this the euphoria obscures difficulties. Thus **short-circuiting** appears in the making of plans and their execution. Failed business attempts and marriage proposals, and the like, are the consequences. The ease with which everything subjectively flows is above all responsible for the **feeling of competence,** found in all patients with pure mania. One consequence is that they never feel sick, instead they often feel healthier than ever. **Elevated self-consciousness** is partially dependent on this feeling of competence and is partially the immediate result of euphoria, which allows only positive aspects of the self to be seen. **Ideas of grandeur** consequently are very characteristic of pure mania, even more so than in the mania of manic-depressive illness. This is certainly an expression of the fact that to pure mania belongs a pure form of euphoria. Other ideas recede considerably because flightiness of thought hinders the appearance of delusional ideas. But intimations of development of ideas from the disturbed affect seen in other pure euphorias may appear in pure mania. Fleeting ideas of happiness and confabulations may occur. Hints of hypochondriacal ideas, despite the clearly euphoric mood, may be found as well. We will see the reason for this in "hypochondriacal" euphoria. Ideas of grandeur, too, do not have a character of fixed delusions, but change from day to day, and fit very easily into the fluid total picture of pure mania. I present a case of pure mania.

Case 16: Gustav Roh, born in 1883, in 1926 suddenly developed an elevated mood while on a trip and spent a lot of money. The mood soon returned to normal. In 1934 he became excited again, expressed ideas of grandeur with political content, saying that he was waiting for the position of von Papen, and addressed his wife as "Excellency". He had many other plans as well, for example, he wanted to have several children from blonde women. He was admitted to the psychiatric division of a hospital, where he expansively developed his plans with flight of ideas and pressure of speech. When he came to speak of the "stealing of his freedom", he became irritated, but he soon forgot it in his elevated mood in which he felt "healthier than ever". Three weeks later he became calmer and could be discharged. In 1937 he became ill again and entered Sanatorium K. He again was full of plans: he wanted to open a hotel in Switzerland, to take over the firm of Kosmos, and to win a gold medal in sports. He felt well in the sanatorium, joked with the girls, was amiable, witty, and loud-mouthed everywhere. He showed pressure of speech and writing, displayed flight of ideas, and felt "stronger and more capable than ever". Four weeks later he became calmer, but was still hypomanic when discharged. In 1939 a new phase occurred. Roh bought some real estate this time, an antique living room, as well as pictures of naked women. He wanted to publish a book on gymnastics with these pictures and make a lot of money with it. He made his payments with bad cheques. He was admitted to the psychiatric ward of a hospital, where he developed his plans and praised his own bodily and mental capacities. After one day he was transferred to the Frankfurt Hospital, where he was in the best of moods, and spoke about all his erotic projects. He developed pressure of writing and speaking, flight of ideas, and was continuously busy. He displayed lively facial expressions and gestures. This time the phase lasted three months. When Roh was only mildly hypomanic, similar to when he was well, he could be discharged. There are no psychoses in the family, but all his siblings are allegedly somewhat "exalted".

Roh offered the picture of pure mania in an impressive form. Depressive traits were totally absent, while his temperament and that of his siblings tended towards the manic. Symptoms of pure mania were all marked: euphoric mood, overactivity, pressure of speech, and flight of ideas. The feeling of great competence was particularly visible and included the sexual sphere. The first phases passed very quickly, while the last one continued for three months, i. e. the usual length of time for mania. If the post-psychotic hypomania was included the duration of the phases would have been much longer, but differentiation from the normal temperament would have been impossible. In the following case, the continuous state of mania can no longer be considered a peculiarity of temperament, rather it was pathological to such an extent that we must call it **chronic mania.**

Case 17: Franz Misch, born in 1875, had always been somewhat excited. In 1928 the excitement increased. He harassed his alleged fiancée who did not want him, and he behaved shamelessly. He was admitted to the Frankfurt Hospital, where he was very high spirited, scolded other people, demanded justice, was excessively self-assertive, and developed pressure of speech. He was transferred to Institution H.; after six months he calmed down and was discharged. In 1934 he again was excited. The admitting doctor found him to be arrogant, jovial, proud of his qualities, and easily distracted. After admission to hospital he was self-important, kept no distance from others, displayed pressure of speech with flight of ideas and excessive gesticulation. He was again transferred to Institution H., and in April 1935 he was discharged, somewhat calmer. In July 1935 he was readmitted to the Frankfurt Hospital after having molested women while at home. The admitting doctor found that he spoke without interruption, made long digressions, and was easily distracted. In hospital he again showed heightened self-importance, he took the other "poor patients" under his wing, and he blusteringly scolded the attendants without being deeply irritated; he proudly presented himself as an accordion player. After three months he was transferred to Institution W., from which he was discharged after a further three months, somewhat calmed down. At home he remained moderately excited, but became worse in 1938, disturbing others at home, swearing in vile language and exhibiting himself to children. He was again admitted to hospital, where he made light of everything with a self-righteous posture. He was supercilious to the attendants. An urge to speak, flight of ideas, and distractibility were again present. Transferred to Institution H., he remained excited and absconded in 1939. He was readmitted to the Frankfurt Hospital, and, as before, was in an elevated mood, displayed flight of ideas, superciliousness, and loquacity. Back in Institution H. he remained manic with no changes. In 1943 he was given a trial discharge. In 1944 he re-entered the hospital, having performed all kinds of practical jokes at home, playing the accordion by night, and knocking on windows. His clinical picture was unchanged. He ranted about his readmission, but was easily distracted. He laughed and joked, still displayed flight of ideas, praised himself as a milliner, and wanted to perform on the accordion. He was again transferred to Institution H., where he remained manic.

Misch basically was a primitive personality. Hence his manic superciliousness and overactivity took on a somewhat impoverished character of fantasy. All the manic symptoms are clear in Misch: elevated mood with heightened self-importance, overactivity, flight of ideas, distractibility, and pressure of speech. That Misch felt healthy and competent was not directly mentioned, but was clear from his total behaviour.

I present the case due to the peculiarity of its course. At first manic attacks were always followed by periods of quietness, albeit incomplete. Later the condition remained unchanged in mania of moderate degree. At the end of the observation period the patient had been manic for 10 years. Chronic mania also occurs in manic-depressive illness, but I found it more often among the pure manias. Similar to what we saw with pure melancholia, pure mania may show gradual transition to a milder temperamental disturbance. Hypomanic temperament, in my patients as well in their relatives, often characterize entire lives. Thus occasionally a clear cut manic constitution may appear to be a stable quality of a person. In another case of pure mania, the condition had lasted ten years before admission to hospital and has not yet ended. In several other cases, mania appeared to be only the climax of a chronically hypomanic state.

Summary

Pure mania is characterized by euphoria, restless activity, wealth of expressive motions, overtalkativeness, flight of ideas, heightened self-importance, and rapidly changing ideas of grandeur. The excitement does not change into confusion or hyperkinesia. Thus pure mania appears milder than many cases of mania of manic-depressive illness. But more than mania of manic-depression it tends towards a **chronic course,** at times in the form of hypomania. The **accompanying bodily phenomena** are similar to those of mania of manic-depressive illness, but the subjective feeling of great bodily strength is more pronounced, whereas overexertion from excitement which could cause bodily harm is absent.

Pure Depressions and Pure Euphorias

Pure melancholia and pure mania do not represent purely affective diseases; thought and will are also disturbed. There are, however, psychoses in which only the emotional side becomes diseased, although not in its totality, but only on a certain level. In any case, that is how I interpret pure depressions and pure euphorias. Pathological construction of ideas appears to be determined only by pathological affect, as we shall see. The number of forms of affective syndromes is greatly increased by pure depressions and pure euphorias. This may make the overview more difficult, but because of their pure forms, the pictures can be differentiated easily. They are also of great importance for the psychology of normal emotional life, as I shall discuss elsewhere. Practical experience creates the necessity of noting all these separate forms.

Pure Depressions

Pure depressions are common enough to be repeatedly observed. Once familiar with their picture it will seldom be difficult to make the correct diagnosis.

Agitated Depression[2]

What is called "agitated melancholia" can be a paranoid anxiety psychosis, but more often is a case of agitated depression, which in my earlier publications was called "pure anxiety psychosis" (LEONHARD 1937, 1939). Anxiety accompanied by agitated excitement completely dominates the picture; it is the only essential symptom of the illness.

One certainly cannot fully equate the emotional state of the patient with normal anxiety; this is shown by ideas which develop from the diseased affect and which certainly are not based on fears only. As we shall see, self-accusations or hypochondriacal ideas may occur. It is apparently a dysphoria, most closely related to anxiety but not entirely congruous with it. From the remarks of patients it is possible to repeatedly detect a deep feeling being tortured, with repercussions throughout the whole personality as well as the body. The feeling of being threatened is not always found. The immediate feeling of being tortured is so much in the foreground that the formation of thoughts generally seems parenthetical and is occasionally absent. No doubt this is quite different from normal mood.

A suicidal tendency is frequently present in patients with agitated depression, but less in the sense of a primary negation of life and more as a flight from the painful state. Suicide can also occur despite direct fear of death and punishment of hell. That too shows the difference from normal anxiety. This difference is also recognizable in that agitation, which is often but not necessarily precipitated by normal anxiety, inevitably accompanies this pathological anxiety. Agitated depression, as the name implies, is driven by anxiety into constant restlessness. If external stimuli are absent, restlessness may stop for a short time, but the least cause will revive it.

In particular, agitation is expressed in the patient's moaning, complaining, crying, screaming, begging, clinging, running around, wringing of hands, and more. In the beginning various expressions of anxiety can be mixed together, but later almost always one or the other form predominates and repeats itself in a uniform manner. Spoken expressions, in particular, may become iteratively uniform. Thus a patient may repeatedly moan the same words for days, or even weeks or months, that she is lost, or that someone should help her. Schizophrenic stereotypes may be thought of when confronted with this uniformity, but they have nothing to do with it. On the contrary, the underlying anxiety confirms that it is a more benign form, i. e. agitated depression. Rarely, as we shall see, does anxiety psychosis imitate this picture.

Attempts to influence the patient in cases of agitated restlessness are totally in vain. Whether one tries it with friendliness or severity, with eloquence or threats, the restlessness will remain unchanged. On the contrary, the complaining and moaning generally increases then, probably since the attempts at intervention act as new stimuli. The most that is achieved is when the patients

[2] May also be translated as "harried depression".

are avoided and no attention paid to their complaints. Subjectively they probably suffer more when one tries to become involved, since they are more agitated by the attempts than calmed. It is, of course, often impossible for the other patients and the staff to ignore the constant complaining because it can be extraordinarily disturbing. This is particularly true when it continues all night. The temptation then may arise to overlook the anxiety, which appears to have an uncommonly compelling quality for the patient, and to see the disruptive behaviour as obstinacy or even maliciousness. But the regularity with which this phenomenon occurs in all patients of this type shows that it is fully independent of character traits. If the agitation should stop, and not immediately reappear when provoked by external stimuli, then the patients may seem to be more apathetically exhausted than truly pacified. Occasionally they avoid stimuli in order to maintain some sort of rest. Not infrequently, however, this labile period of quiet may lead to a more permanent one.

In other cases the agitation assumes another, more annoying form when the anxiety recedes. Out of the constant complaining and begging, a **constant sense of oppression** may develop. With the same tenacity with which the patients previously complained about their horrible inner pains, they may now demand their immediate discharge. They may become more obviously querulous. Yet this sense of oppression, which still has a tortured component, differentiates them from other querulous patients. It, too, cannot be influenced, and here it is even more tempting than with anxiety to consider it as obstinacy. The correct interpretation surely is that the restlessness is neither accidental nor psychologically reducible to anxiety, but that the pathological affect, even if of little depth, contains this immediate driving element within itself. Presumably in agitated depression a deep emotional level is affected in which the patient's behaviour is instinctive and is barely influenced by the overlying personality.

Correspondingly, the anxiety of a patient with agitated depression need not have any mental content. The patient's expressions may be restricted to constant complaining about the terrible inner suffering; loud moaning may frequently be the only outer expression of anxiety. Of course the anxiety does seek some content, particularly at the onset, when uniformity has not yet developed; the content may be multifaceted. Fears may be expressed of being martyred, dismembered, or burned, alone or with the whole family. There may also be ideas of sin; but these ideas are derived from the fear of threatening punishment in this world or in hell. Hyochondriacal ideas are common too, and are related to the anxiety, for they either represent fears about bodily well-being, although genuine misperceptions are absent; or they may be felt as tightness in the neck or chest, i. e. they represent a symptom of anxiety. All forms of the depressive ideas in agitated depression are rather external, and meaningless for the psychosis itself. The total picture remains basically the same whether the patient complains that he is suffering greatly or has committed a terrible sin. The picture of agitated depression does not even change if no idea at all is present. How essential this circumstance is will become clear in the discussion of "self-tortured depression".

Occasionally symptoms take on a hallucinatory, or better perhaps, a pseudohallucinatory character. Patients may say that they see future horrors awaiting them or they believe they hear threatening noises or voices in this connection. Such delusions, too, maintain a peripheral character; the patients draw no conclusions from them and do not become more anxious than they already are. The affect appears here only to be looking for a content, perhaps hardly any different from what a normal person who is terribly upset expects might happen. A normal person, too, may misinterpret events. In anxiety psychosis, we shall find quite different hallucinations, which mean much more to the not only anxious but also distrustful patient, and which are experienced with much more independent affect. However, hallucinations are not common in agitated depression and probably are derived from individual predisposition.

Regardless of whether the anxiety has much or little content, it is very difficult to fixate the patient on anything besides what he or she is preoccupied with. Often not even the simplest data about patients' lives may be obtained, since instead of giving rational answers, they just begin their complaints again. That they would like to answer can be seen from the fact that after repeated questioning they will often give a fleeting answer, almost as an aside, as if they had to exploit short moments in which the pressure of the anxiety had become somewhat milder. Again the longer no answer is forthcoming the greater is the temptation to assume an obstinate refusal to cooperate. However it is beyond doubt that it is the anxiety which so fully dominates thinking and acting that there is no room for anything else.

In our present therapeutic era of **neuroleptic treatment** such severe forms of agitation are no longer found. However, the constant complaining may not be totally suppressed by medication. It is nevertheless advantageous that I present some patients as they behaved without medication.

Case 18: Irma Krein, born in 1912, suddenly became ill in 1941 with severe anxiety and restlessness. She was admitted to hospital; here she got out of bed, wrung her hands, clutched her breast, clung to others, and wailed loudly and continually. Occasionally she hastily exclaimed that she was innocent, but mostly she only uttered unarticulated sounds. This severe state of agitation, hardly modified by opium, lasted eight weeks. Then she rapidly became quiet, and was discharged in a normal mood. A **sister** was in Institution W. due to a mental disorder, and later remained healthy. Unfortunately no medical history was available.

Another patient wailed, cried, tried to strangle herself, and called out: "What crime did I commit that I must suffer so terribly?" These two patients showed agitation with very little content; the agitation remained objectless. Other patients showed agitation with varying content. One patient called herself a criminal who wanted to remain innocent but could not do so. At the same time she shouted: "Shoot me dead, don't torture me any longer." Still another patient moaned continually, claimed her brain was poisoned, she was going to suffocate. Still another patient declared that she was reduced to beggary, for her and her child there was no salvation. When told that she was a wealthy woman with an estate of 400,000 Mark she always answered with the complaint: "Dear God, it is terrible." Such a delusion may have a psychological explanation. It would have been particularly painful for the patient, who had built her whole life on her wealth, to feel impoverished. Her pathological affect, which in agitated depression selects contents more by accident, for external reasons led to this specific delusion. – Another case is presented here, in which hallucinations developed.

Case 19: Robert Schmelz, born in 1888, became ill in 1939, and immediately was admitted to hospital. He ran around wringing his hands, distorted his face, screamed that he was under arrest and that his relatives must die with him. Then he cried out repeatedly "this cannot be made better". He believed he had contracted syphilis and had passed it on to others. His anxiety still increasing, he screamed "it cannot be made better in a thousand years". He heard tortures being prepared for him, how other patients worked with tools under their blankets, and also how someone said "tonight his testicles will be twisted off". He jumped off a table with suicidal intent. After six weeks he was transferred to Institution N., where he remained, unable to be influenced, in a state of anxious excitement. Not until six months later did increasing calmness set in. Schmelz finally made a full recovery and could be discharged. A **sister** was in an institution in 1889–1900 at the age of 24 for a mental disorder, but later remained healthy. The name of the institution unfortunately could not be ascertained.

The hallucinations of this patient revealed the severe anxiety he felt when hearing the threat that his testicles would be twisted off in the night. In this form hallucinations do not really represent a complication of the illness, they were produced by the severe anxiety and belong to the illness itself. As well, it is significant that in a suicide attempt the patient jumped off a table. The goal of such an attempt is not to die, but the affect of agitated depression contains a powerful urge to put an end to the unbearable suffering. Suicide attempts in agitated depression generally have a quality different from pure melancholia. Patients with agitated depression seek death not as a negation of life, but because they can no longer tolerate the suffering. A female patient who strangled herself did this in the presence of nurses, thus again without particular purpose. Such suicides contain a convulsive element. These patients may express fear of death, yet in desperation jump out of the window.

The following patient was greatly afraid of punishment in hell, nevertheless she begged to be killed. – There was a similar case in her family; moreover her agitated depression took a **chronic course.**

Case 20: Kath. Schürg, born in 1900, became ill for the first time in 1932, and was admitted to Institution I. She was anxious and wailed without interruption. She claimed she was a sinner, a lost soul, and must burn in hell forever. Her wailing could not be stopped, she clung to others, asked to be put to death, and tried to cut her wrists. She remained restless throughout the following year and made several serious suicide attempts. She repeatedly wailed that she was damned forever, there was nothing left to do, and she was already a living corpse. Some improvement occurred a year and a half after admission, but a mild anxious depressive mood remained and she was irritable during this period of time. In 1939 she became worse. She was so anxious that she could no longer work. She returned to the Frankfurt Hospital where she was observed to be fearful and restless. Most of the time she held back, but in reaction to external stimuli she began to complain. After four weeks she was discharged, but was readmitted six months later. Now she was more labile than actually depressive. After ten days she was transferred to Institution W. and then discharged five weeks later. The **mother's mother** became ill in 1889 during her menopause, continually wailing "I am sick. I can no longer perform my duties. I am incurable. I must die." After this condition had continued at home for eight months she was admitted to the Frankfurt Psychiatric Hospital. Here she cried out loudly "Oh God, Oh God! Be merciful, am I really to be condemned today?" She clapped her hands over her head in desperation and cried further "Now they are taking me, now they are burning me." Then hypochondriacal ideas appeared: her body would collapse and she had been poisoned. The howling, screaming, and complaining continued steadily, always the same "it is frightful, horrible, terrible. This has never happened before and will never happen again." She was afraid she would fall into a boiling pot. The complaints became gradually more constant. When more occupied with something on the ward, she kept on lamenting without interruption "Oh God, Oh God, I have no bowel movements." After ten months she was discharged against medical advice, then returned ten days later. The medical record skips to 1892, stating that she wailed loudly and that she was being burned and quartered. Thereafter she became quiet, and finally recovered. Four weeks later the patient was discharged, and remained healthy for twenty years. She died in 1915 of old age.

In this patient a chronically depressive picture apparently never ending followed her agitated depression. It was not an actual psychosis, but agitated depression was still recognizable when the patient began to wail as soon as she was irritated. The diagnosis is confirmed by the family picture; the mother's mother also showed a clear picture of agitated depression. The period of disease of three years corresponds to the daughter's protracted course, but complete recovery set in and lasted until death. In other cases with psychoses among relatives there were no exact descriptions of these conditions. Altogether there was no hint of mania or hypomania. On the other hand several suicides occurred but not as often as in pure melancholia.

Summary

Agitated depression is characterized by a **tortured-depressive** state accompanied by continuous restlessness and fearful agitation. Anxious obsessions generally occur, often with ideas of sin and hypochondriacal fears. At the onset, restlessness and pathological ideation may vary, but if the suffering lasts a long time, this becomes more constant. Complaining then often occurs stereotypically with the same content. In any case the restlessness tends to become tenacious and cannot be overcome. Nor do the patients let themselves be distracted from their complaints. Often it is even impossible to receive answers to simple questions. The inability to influence patients with agitated depression may create a false impression of obstinacy. As the anxiety recedes, the lamenting often becomes oppressive, almost querulous, although an anxious tortured undertone remains recognizable. Agitated depressions may run their course quickly, but tend to be **protracted.** This is particularly true in milder forms in which the oppression outweighs the lamenting; it may extend to several years. Apart from their psychosis patients may show traits of anxious restlessness in their **temperaments.** My investigations are not sufficient, however, to make a clear statement in this regard. I have also not collected enough information about the **physique.** Pyknic constitution seems to be common, even though loss of weight often accompanies the condition.

Hypochondriacal Depression

The listless mood of agitated depression has a somewhat vital character. The second true depression, which I call hypochondriacal depression, has even stronger somatic aspects, but of a quite different nature. In the former, the agitation, so impervious to any mental influence, may affect the body, while in the latter, peculiar misperceptions point directly to bodily organs. The concept of hypochondriasis has an ambiguous meaning; on the one hand, it refers to fears about bodily well-being, on the other, it refers to physical misperceptions for which no organic illness is responsible. Only the latter points to hypochondriacal depression; the former occurs in other depressive states.

The misperceptions of hypochondriacal depression may concern to the whole body, its surface as well at its interior. If there is localization, it is always somewhat indefinite; generally larger areas without specific limits are described as changed. The type of misperception likewise is unspecific. The patient may complain of painful, burning, stinging, drilling, pressing, or tearing sensations, but may also describe the same misperceptions with not just one of these sensations but with all of them. Thus the misperceptions may not be unequivocal at all. Often patients reject all specific descriptions that may be suggested to them, and state that they cannot describe their symptoms clearly as these are not like symptoms of physical illnesses they may have had. They often search out comparisons which reveal that the misperceptions really have a character of their own; their skin felt like rubber, their blood was electric, their nerves were dried up, their intestines were deformed. In speaking of burning, stinging or similar sensations they add that these are different from the usual ones: they are more inside than on the surface. The patients' descriptions make it **obvious that these are misperceptions which a normal person does not have.** Just as in hypochondriacal schizophrenia, sensations may develop in the interior of the body which a healthy person cannot feel. Their origin is of course completely different. In depression, a particular form of mood disorder underlying these sensations must be presumed to be present.

Occasionally misperceptions are felt in the area around the heart. Here the question arises of whether this might not be a sense of oppressive anxiety, as we saw in agitated depression. In hypochondriacal depression the sensations around the heart are never isolated but accompanied by misperceptions elsewhere in the body. When asked if they feel anxious hypochondriacs generally answer in the negative, although they may answer positively when referring to sensations around their heart. Patients with hypochondriacal depression may seem to be anxious and restless when complaining and wailing about their suffering. However, there is never the tortured oppressive restlessness which we see in agitated depression. The tendency to lament varies considerably: when their affect is deeper, patients easily loose hope and complain less. Their restlessness is psychologically much more comprehensible than in agitated depression.

It is also psychologically understandable that hypochondriacal fears may develop around the misperceptions. As stated above, fears about physical well-being have no particular significance as such but are quite common. Patients may not just complain about their symptoms but at the same time express the fear that they have a severe physical illness such as cancer or tuberculosis. Some patients tend to draw such conclusions, others react only to the misperception and complain about it, without thinking about its origin. Additional ideation is almost completely absent. Occasionally certain ideas of inferiority or mild ideas of reference may emerge, but these ideas may remain superficial and are totally outweighed by misperceptions and associated fears. There may be hints of hallucinations in which hypochondriacal fears take on the form of voices, but this presupposes individual predisposition. The hypochondriacal experiences themselves never become delusions; they are always construed as disease phenomena and never as external influence. **This is important for the differentiation from hypochondriacal schizophrenia;** patients with the latter never feel sick, but rather feel molested from outside.

In hypochondriacal depression, aside from the misperceptions, **alienation phenomena** also play a role. In apathetic depression, we will meet a type of alienation in which the patients describe a cooling of emotions, particularly of sympathy with other people. This form of alienation does not develop in hypochondriacal depression. Rather it is directly related to the hypochondriacal complaints. When patients complain, for example, that their whole body is internally changed, it may be inferred that they want to point to the misperceptions. After a more thorough examination it appears that the patients do not feel that their bodies are changed due to new sensations, but, on the contrary, because of lack of feelings. They claim that they are no longer aware of their bodies, their arms, or their legs, that there is no more feeling in them, and that everything seems dead. From the lack of normal emotions and the alienation bodily sensation seems to be changed into too much or too little. The complaints often go further and include sensory phenomena: the patients say that they are no longer touched by events, that their sense of taste, smell, touch, pain, heat or cold, is all gone. Worse still, a patient may claim that he can no longer imagine anything properly, such as the furnishings in his room or the face of the person he just saw. The patient concludes that he is dead, and that he can no longer really die, since his whole body was already dead. He may claim that his personality has become different, no longer the same as it used to be. These are symptoms of **depersonalisation.** Hints of alienation phenomena are found in all hypochondriacal depressions which have reached a certain depth.

Affect, however, does not generally reach the depth it does in melancholia or agitated or self-tortured depressions, although vital layers of the personality are undoubtedly affected, as evidenced by the misperceptions. The patients often appear to be more lachrymose and grieved than deeply depressed. The suicidal risk is considerably less than in the other forms. Despite this, hypochondriacal depression does affect the total sense of well-being. In severe forms, the condition is very painful. Whether the patients complain a lot or a little depends greatly on the rest of their personality, **but inhibition is always absent.**

Case 21: Karl Web, born in 1908, had to leave school for six months at the age of twelve because he was so nervous. He became seriously ill in 1938, expressing many bodily complaints, and was admitted to the Frankfurt Hospital. Here he complained about a tightness around his heart, pressure in his stomach, and weakness and heaviness in his legs. It seemed as if something in his body was pulling him. He had an inner feeling of restlessness, a numb feeling in his arms, and weakness in his legs. At first his limbs felt "extraordinarily light", as if he could easily lift them in the air, and then they felt heavy again. His body was crackling and stretching, as if the skin were about to rip open. He complained a lot, suffered considerably from these misperceptions, and was in a depressed mood. He became better after four weeks, complained less, and was taken home. In the course of the following weeks he became completely well and did not become ill again.

Case 22: Willi Rem, born in 1893, became ill in 1942, claiming that he had accidentally drunk lysol, that his esophagus was burned and that his body contained abscesses. He cried a lot, no longer worked, attempted suicide, and was admitted to the Frankfurt Hospital. He lamented continually. He complained of burning pains in his chest and abdomen, an infection in his throat, and of stinging pains in his teeth. He could stand it no longer; someone should take his life. He claimed cancer was probably developing. After three weeks he was transferred to Institution E., where he continued to complain. He claimed his body was twitching, food was fermenting in his stomach, and he was slowly dying. Sometimes he refused to eat, claiming it increased his pains. He said he had no peace at night. After electric convulsive treatment he seemed at first to improve, but then his old complaints returned: there was irritation in his arms, his fingers, his neck, and his esophagus. He said he had too much mucus in the esophagus, and therefore had to swallow continually. Eight months after his admission to hospital and about a year after the onset of the illness he rapidly became well and was discharged fully recovered.

The two cases presented show the characteristic picture of hypochondriacal depression with depressive mood and the peculiar bodily complaints. In case 21, the depressive mood was not as deep, but the complaints were copious. They were accepted as such, painfully perceived, but not thought about. In case 22 on the other hand, fears always accompanied the complaints. The patient suspected internal disease, lysol burns, cancer, and abnormal mucus secretion as causes of his illness. He also expressed the fear that the bodily phenomena were slowly ruining him. The diagnosis of hypochondriacal depression is not based on these fears, but on the misperceptions out of which the fears develop. The deeper depression of case 22 led to stronger complaints. Temporarily a serious suicidal tendency developed. Phenomena of alienation were hinted at in case 21 when a temporary unusual lightness was felt in the legs. The **alienation** is much clearer in the following two cases.

Case 23: Elise Günt, born in 1889, was depressed for the first time in 1933, with sleeplessness and intestinal complaints. In 1941 she again became ill during her menopause. She became anxious, lachrymose, and complained of intestinal problems. After a few months she was better. In 1942 another depression set in for which she was admitted to the Frankfurt Hospital. She complained that she no longer had any desire to work and suffered from bodily ills. She had the feeling as if something was moving in her body, particularly in between the shoulder blades. There was pressure in her stomach and tension and burning in her head. Her condition became worse and her complaints more intense. She felt a terrible burning in her body; stomach and intestine were pulling, her blood was burning, and her skin was like rubber. She claimed that she had lost her normal feelings, as if all the nerves were dry, nothing more was working inside her, her whole body was dead. She had a feeling of emptiness in her trunk and legs. She could no longer feel pain, heat or cold. If someone put her out naked in the winter, she would not freeze. She claimed she could not die, since everything in her was already dried up and dead. She said that her sense of smell and taste no longer was normal, while sight and hearing were less disturbed. She also said that she could no longer imagine anything, neither her husband nor her acquaintances, absolutely nothing from the outer world. "I have no picture of anything. I can imagine nothing in my head." At times the patient complained considerably more about the alienation phenomena than about the misperceptions. Her very painful condition disappeared after several months; she lost all complaints, was in a good mood, and was discharged home.

Case 24: Mildred Marsch, born in 1914, became depressed in the autumn of 1937. In March 1938 her condition became worse. Everything seemed changed: her thoughts, her body, her environment. Nothing interested her any more; everything seemed to be far away. Her body felt lifeless, as if it no longer belonged to her. Moreover there were many bodily misperceptions: burning sensations in her trunk, itching sensations in her head, feelings of anxiety around her heart, and pains in her back. In August 1938 she was admitted to the Frankfurt Hospital, where all these symptoms were elicited; furthermore when she spoke, it seemed to come from far away, as if it was not her at all. She complained a lot, and often was quite dissatisfied. After two months she was discharged home. The depressed state did not disappear for another year, at which point her **father** was already being treated here. The father became depressed for the first time in 1917 as a prisoner in England; he cried and complained about pains and burning in his chest and back. In 1937 he became ill again, lost all energy and complained about pain in his backs in the areas of his abdomen and his heart. This condition did not improve in Sanatorium H. He was then transferred to the Frankfurt Hospital in April 1939, where he complained a lot and asked for help. He claimed his whole body was burning, head, neck, chest. He felt pierced and burned as if he was stabbed by a thousand knives. Beyond this, everything was so flabby and limp: it seemed as if his body was not being properly nourished by his stomach and intestines. In May he was taken home, but returned in June. He again uttered complaints throughout his whole body: his trunk was burning, pains extended from his back into his head, and he could feel it in his bones as well. He often felt as if his chest were being compressed by his back. He also complained that he no longer had any feeling, that everything was numb and his legs and sinews were all limp. After slight improvement he was transferred to Institution H., where he again complained that his whole body was torn apart. In 1940 his hypochondriacal depressive condition continued.

He complained that he could barely stand the pain. He gradually improved, complained less, and was eventually back to normal. In October 1940 he had recovered and he was discharged.

In case 23 alienation phenomena are particularly salient besides misperceptions; they involve the areas of bodily sensations, the sense organs, and the patient's imaginative capacity. The impression might be gained with this patient that the presence of alienation phenomena represent a more severe form of the disease. In case 24, however, alienation phenomena were present before misperceptions; in any case the patient complained about them while mildly depressed. What is impressive is the father, who was also a patient. In his psychosis, misperceptions dominate the foreground and alienation phenomena are only hinted at. This fits in with the other observation that alienation phenomena are much rarer in men. I am familiar with another family, which does not belong to this investigative series, in which again a woman suffered from hypochondriacal depression with marked alienation phenomena, while her son went through a hypochondriacal depression with no alienation phenomena. Also noteworthy about this case is that the son had anancastic traits in his depression: such traits appear more readily when there is greater self-observation, based on an underlying compulsive constitution. One of my co-workers (Richter 1968) found that hypochondriacal alienation occurred more readily if there was affective irritability and introversion of thinking.

In some cases, besides hypochondriacal and alienation ideas, other fleeting symptoms may appear. One patient repeatedly uttered self-accusations. She maintained that she herself was responsible for her sufferings, that she should have had herself treated differently. It probably was an attempt at explaining the hypochondriacal complaints. In another patient, voices and ideas of reference were reported, but here too there was a clear relationship to hypochondriacal depression, since the voices told the patient that she would lose all her organs but remain alive. Moreover the patient thought that others were avoiding her because she looked so terrible. The voices and ideas of reference concern bodily changes and could be considered as alienation phenomena or misperceptions. If the hypochondriacal symptoms take on the form of voices, there must be an individual predisposition, as we have already found in agitated depression and pure melancholia. The characteristic form of hypochondriacal depression is not affected by these peculiarities, which incidentally are rare. The clear picture is also present in the following case, characterized by a chronic course.

Case 25: Franziska Stil, born in 1878, became ill in 1928 during her menopause: she was melancholic and complained of headaches and stomach pains. After a short stay in Sanatorium K., she was better. In 1937 she was admitted to the Frankfurt Hospital, but appeared to have been ill for some time before. She was full of complaints, expressed many hypochondriacal symptoms, yet after 13 days she was discharged. During the following year, she was in a nursing home. In 1939 she was readmitted to the hospital after a suicide attempt. She was depressed, although at times she was dissatisfied, and she complained incessantly; she had pains all over her body, could not sit or lie down, her arms and back were swollen, her genitals itched, pain came from the top of her head, and she felt stung as if pricked by needles. She was discharged after four months, but returned two months later complaining that she could not stand the pains in her shoulders, arms, and legs. She was discharged again after two months. For a period of time she was in an old age home, but could not tolerate staying there. In 1942 she returned to hospital, complaining of terrible pains in her neck, her head hurt all over, her legs were shivering as if ice was inside, while her arms were burning, her whole body was shaking, and her esophagus was constricted. Her stomach and intestines did not function any more, food lay for days in her stomach. She was very tearful and grumbling. After seven weeks she was transferred to Institution E., where she continued to be complaintive, depressed, and at times faultfinding. Shock treatment did not change her condition. One year after admission she appeared to be somewhat calmer. Her hypochondriacal complaints never disappeared completely and were still present at discharge two and a half years after admission.

Another patient at the time of her admission to the Frankfurt Hospital had been ill for five years. At the time of her discharge she still showed the typical picture of hypochondriacal depression.

In both these cases, the depression lasted for years and perhaps never cleared up completely. The second patient was diagnosed as "hypochondriacal psychopathy"; incorrectly, because until the age of 28 she was not at all hypochondriacal. Only then the picture changed. Apparently such cases are not infrequent, where chronic hypochondriacal depression is misdiagnosed as hypochondriacal psychopathy. If the hypochondria had existed for years, then it is of course difficult to ascertain whether it had always been present – of constitutional nature – or if it developed unexpectedly at a certain point in time. Recently I saw a patient with all the characteristic symptoms: she was moderately depressed and had many bodily misperceptions. She was 53 years old and had had the depression for 20 years. Earlier, up to the age of 33, she was an alert, cheerful person with no hypochondriacal tendency. We have already described patients with a chronic course of agitated depression, but a chronic course seems to be more common in hypochondriacal depression. No matter how long the depression lasts the diagnosis remains the same, with no changes except for some abatement of the symptoms. There is never any progression; the personality remains fully intact. Confusion with hypochondriacal schizophrenia, despite a chronic course, is not possible, partly because the depression is always recognizable behind all the complaints, and partly because the misperceptions are never interpreted as influences coming from outside. Patients with hypochondriacal neuroses are not depressed. This is an important point if confusion between "senso-hypochondriacal neuroses", as I call them (Leonhard 1981b) and hypochondriacal depression is to be avoided.

Hereditary loading with psychoses in patients with hypochondriacal depression is minimal, as we shall see later on. No evidence was found in any of my patients with hypochondriacal depression for the existence of a genotypically bipolar psychosis, even taking temperament into account. A manic element is absent in the subjects and in their relatives. My co-worker E. Schulze (1968) in twelve out of fifteen patients with hypochondriacal depression found a prepsychotic trend towards bodily misperceptions.

Summary

Hypochondriacal depression is characterized by **bodily misperceptions,** the description of which shows them to be very peculiar. Frequently fears about bodily well-being develop, but only misperceptions are essential to the diagnosis. Accompanying **alienation phenomena** are often hinted at and may be marked, affecting bodily sensations, sense perception, and imagination. Alienation due to lack of sympathy, as we will find in apathetic depression, is not present. Other depressive contents are absent, or at most only slightly. Patients generally complain a good deal about their alleged ailments, but without displaying significant excitement. The condition can be extremely painful, but the total personality is less affected by the depression than in cases of melancholia or agitated depression. Consequently the suicidal tendency is low. **Chronic course** is not infrequent. The depressive-hypochondriacal state may never disappear, and nothing essential changes in the picture. As regards **physique** hypochondriacal depressives tend to be predominantly pyknic, but I have not reached a precise conclusion.

Self-Tortured Depression

While in agitated depression the dysphoria is basically without object and in hypochondriacal depression there is a connection with bodily sensation, the dysphoria in self-tortured depression is primarily associated with higher mental ideation. In fact it appears to develop from the mind. At first the depressive mood is not prominent outwardly; the patients may be quiet, merely showing a depressed facial expression. When asked indifferent questions, this behaviour may not change at all. However, **as soon as they begin to talk about their pathological ideas, their depressive affect becomes very noticeable** and they present their convictions with increasing feeling of being tortured. It is difficult to distract a patient in this condition. If distraction does succeed, the affect diminishes. Patients with agitated depression, in contrast, are just as tortured without content as with it; when they keep on complaining about their terrible internal suffering, their behaviour is not different from when they express delusions of sinfulness. In self-tortured depressions, however, the depressive affect becomes more marked as ideas of sin, inferiority, or anxiety are expressed. This close tie to certain conceptual complexes shows that here an emotional level is affected which is clearly on a higher plane than in agitated or in hypochondriacal depression.

Anxiety too plays a major role in self-tortured depression, but only in the form of very specific fears. If distracted the patient's anxiety is no longer recognizable; it does not exist without object. Ideas of having sinned and other ideas of inferiority are more pronounced. Ideas of impoverishment by themselves play no essential role, since external impoverishment as an idea does not contain the same immediate affective tone as personal danger, denigration, or guilt. The idea of impoverishment may more readily move into the foreground in agitated depression since there the affect, in its basic lack of object, tends to seek a content in external perspectives. In self-tortured depression the idea of impoverishment occurs only in very particular circumstances that give it a considerably deeper affective tone: it is not the patient who will suffer impoverishment – that would not be so bad – but rather his close family, his children who need protection, who will suffer. Thus here sympathy with the sufferings of the relatives mixes in and leads to a characteristic symptom of self-tortured depression.

It is just those convictions which normally have a high emotional value which are overvalued in a pathological manner. The emotional tone is often deepened by these convictions through their extreme contents. The patients are quite sure that they have committed not just any sin, but the worst sin that there is. They do not believe themselves to be inferior as some people do, but they are more despicable than everyone else, perhaps more despicable than an animal. They declare that they will not merely suffer and die, but will have to endure the worst tortures of hell, and their family will have to suffer through it all with them. The ideas often appear in this exaggerated form; the patients can never find enough words to describe the terror that is motivating them internally. If contradicted they reply even more intensely that everything is exactly the way they have described. Even if not contradicted they try to convince the doctor of the correctness of their ideas. Due to this continuous stirring up of the deep depressive ideas, the patients appear particularly **self-tortured.** It is as if they must undergo martyrdom again and again with their terrible ideas.

Many ideas carry several emotional components simultaneously. Thus for example, anxiety and ideas of having sinned are simultaneously present when patients declare that terrible punishments await them in hell because of their grave sins; likewise the expression of many patients that their mere presence must be a horror for those around them corresponds to a belief of both inferiority and sinfulness. Other patients' notion that they will cause their whole family to die a terrible death contains anxiety, guilt, and sympathy. It can be seen that such combinations heighten the emotional value of the ideas. Whether a patient is characterized by one or the other type of these ideas depends on individual disposition. In general, ideas of self-denigration outweigh ideas of anxiety.

No other depressive contents are found in self-tortured depression, and hypochondriacal symptoms are absent. Nor do they appear as fears about bodily well-being as found in hypochondriacal and agitated depression. Ideas of alienation likewise are absent in self-tortured depression; neither does the body feel changed, nor does the environment appear distant. Interest in the environment is diminished, since the patients are completely taken up with their ideas. Ideas of reference may be hinted at in the sense that some patients claim that their own despicability is confirmed by the behaviour and speech of people around them. Yet even this form of **ideas of reference,** which might better be called a particular form of ideas of worthlessness, is seldom found in self-tortured depression. **Hallucinations** are similarly rare, but occasionally patients do believe they hear or see tortures that are being prepared for themselves or their families. Probably, again, an individual disposition is a prerequisite for hallucinations and ideas of reference, but they are not of great importance in the total picture.

Patients in presenting their ideas may become very excited; they may moan, cry, even scream out their beliefs. It is difficult to distract them from their ideas, but they do not show the same obstinate tendency as patients with agitated depression. The feeling of being driven is also absent. After having presented their ideas with great intensity they may calm down for a while and apathetically sink into themselves; they may even appear inhibited. However, there is no complete inhibition, since only a small stimulus may cause renewed excitement and replace the apathy. Just as there is no psychomotor inhibition, there is no thought inhibition. Confusion with pure melancholia therefore is not possible. In melancholia the ideas which may be similar are never present so abundantly and with such exaggeration. However there is a resemblance to melancholia or agitated depression. WERNICKE and KLEIST probably were referring to such cases with their concept **of anxiety-melancholia. Suicidal risk** in self-tortured depression is somewhat smaller than in melancholia, but the immediate danger is greater since no inhibition stands in the way of the acting out of the intention.

Case 26: Maria Klin, born in 1888, became ill for the first time in 1908 and was admitted to the psychiatric department of Hospital S. At first she was somewhat excited, then she lay quietly in bed. When questioned she replied that she had sinned against the Holy Ghost and that the Saviour had abandoned her. She became quite desperate about her guilt. At night she occasionally moaned and during the day she often presented her ideas full of woe. Four months later she was discharged somewhat improved. At home she was soon fully recovered and remained well for thirty years. Not until 1939 did she become ill again: she complained of a constant pressure in her chest, wanted to take her life, and again presented religious delusions. She was admitted to the Frankfurt Hospital, stating that she was bad, despicable, godless, and a bunch of devils was inside her and wanted to seduce her to immorality. She said the Devil would fetch her and punish her. She prayed a lot, presented her ideas tearfully and kept on trying to find new words to describe her badness. As long as she was not provoked she remained quiet and kept to herself. After three months she was transferred to Institution W. There at first she was quiet, hardly spoke, and had a frozen depressive facial expression. When questioned she became excited and expressed her ideas woefully She stated that she had performed immoral acts and that her sister must suffer for it. Her sister was probably being beaten and tortured, perhaps in the same institution. "It is terrible, terrible." It was awful that her sister had to suffer for her (the patient's) sins. After she had finished expressing her ideas, she again quietened down. The doctor to whom she presented her ideas usually found her excited while the attendants reported that she was mostly quiet. After six weeks she was transferred to Institution G. She continued the same behaviour for the following months, cried and wailed when presenting her ideas, otherwise she was depressively sunk into herself. Fifteen months after her admission to the Frankfurt Hospital she suddenly recovered, corrected her ideas, and was discharged. Her **mother** had had depressions three times during her life, which were allegedly similar to those of the patient. However she was never hospitalized.

The patient described above showed the characteristic picture of self-tortured depression; she was completely absorbed in her abnormal ideas. Another patient cried out that her children should not be killed. Another patient greatly exaggerated her condition, believing that she was the most depraved creature that had ever been, unworthy of even being looked at. Some patients elicit **ideas of reference.** One patient believed she would undergo capital punishment as extensively published in newspapers. Another patient said that she had killed relatives because of her sins, and added that **voices** told her so, and had announced her death. Ideas of reference and hallucinations only reflect what the patient is otherwise thinking.

Chronic course appears to be rare. I observed only one case in an elderly female patient. Suicides did occur in the families of my cases, but are less frequent than in pure melancholia and agitated depression. I am not able to determine if that was only accidental; suicidal tendency among the subjects was, in any case, great.

Summary

In the center of self-tortured depression we find pathological ideation, including self-accusation, self-denigration, and anxiety directed at the patient himself and, even more, at the family. **The pathological affect appears continually to redevelop in the ideas.** The ideas are often extreme and the patient's claims become hyperbolic: he or she is the worst and most despicable person there had ever been and will be terribly punished. In a self-tortured manner they always return to these ideas, and try to convince others of their correctness, becoming very excited in so doing. When however there is no external stimulus to express the ideas, the patient remains calm and appears to be pensive, probably quietly involved with his ideas. There is no drivenness, as in agitated depression; nor is there inhibition. At any moment excitement can develop from the apathetic quiet manner, and the ideas will be verbalised. In terms of **physique** the leptosome type is not quiet as rare as in other depressive diseases.

Suspicious Depression

We have occasionally found ideas of reference in the depressions already described, but they always appeared to be transitory, had no independent position, and could be understood in terms of depressive ideas which were more prominent. It is a very different case with suspicious depression. Here ideas of reference beside depressed mood dominate the picture. KLEIST differentiated a **"depressive reference psychosis",** while I believed for some time that all these cases belonged to anxiety psychosis.

Just as the bad mood of self-tortured depression seems to renew itself in ideas of inferiority, the suspicious mood develops in the **ideas of reference.**

Occasionally **auditory hallucinations** occur, which however can hardly be separated from ideas of reference to which their contents correspond. Mostly patients take the wrong meaning from actual conversations. But genuine hallucinations do certainly occur.

The ideas of reference always display a **depressive foundation;** the patients make their abnormal observations with depression and anxiety, not with irritability or hostility, as usually found with schizophrenic ideas of reference. It is not always clear where the ideas come from. If a patient observes that other people turn away from him with contempt, his affect may become one of hostility to the environment or, conversely, he may see because of his faults only a justifiable reaction to the presumed criticisms coming from the environment. In this second form, ideas of reference occasionally are intimated in self-tortured depression. They are similar in suspicious depression, but the depressive background is not so clear, since the depression is not so deep. However a patient with suspicious depression is always suffering and never argumentative.

The depressive content of the ideas of reference may vary. In the above example, it reflects a feeling of guilt, which is very common in suspicious depression. The difference from genuine ideas of inferiority or sinfulness is that the pathological content develops in reference to occurrences in the environment from the very beginning. That is the chief characteristic of all the ideas in suspicious depression. In contrast it is not of primary importance whether the depressive content has one or the other form. Alongside the feeling of inferiority and guilt, anxiety plays an important role; threats are deduced from the behaviour of other people. Other phenomena which affectively influence people may also become the content of ideas of reference: misfortune in the family, loss of a job, or impoverishment. These seem to be secondary elaborations which individually are different. The only primary fact is that events in the environment are misinterpreted because of the depressive emotions. The depression is thus connected to intellectual processes. This shows that a higher emotional level must be affected: a level not somatically based but which belongs rather to the intellectual processes.

Correspondingly, this type of depression has a considerably shallower emotional depth than the types discussed above. On the other hand depression may not be relegated to the background, nor may suspiciousness dominate the total picture, and the condition must remain one of depression if the diagnosis is to be correct. Here too the risk of suicide in suspicious depression is not at all low.

On first examination the patients appear depressed. They are either quietly depressed or more anxious, depending on whether they feel themselves despised and dishonored or more threatened. Mild anxious excitement may also appear. States of deep anxiety with severe excitement associated with ideas of reference, however, point to anxiety psychosis. A rapid alternation between relatively freer behaviour and severe anxiety also indicates anxiety psychosis, for suspicious depression, like all pure forms, presents a more even state; there may be a certain fluctuation, but no sudden changes as those typically found in bipolar diseases. Suspicious depression displays excitement of anxious coloration more often at home than in hospital, since the varying and often inappropriate treatment at home may easily precipitate severe reactions.

Case 27: Anna Mül, born in 1909, became ill in 1942. She was anxious, claimed that when people stood together they were talking about her and said she had stolen and was bad. Because she wanted to drown her two children along with herself, she was brought to the Frankfurt Hospital. There she explained that she was unjustifiably being accused, for, although she had once received too much change in a store, she had gone right back and reported the error. Despite this people now treated her poorly in the stores and accused her of shoplifting. Everybody looked at her strangely. There had been something about her in the newspaper, and in hospital she was called a "criminal". She avoided the question of whether she had really heard it: "I don't know, they are always standing together, but I haven't killed anyone." At home she had always claimed that she was going to be taken away. She finally lost control of her anxiety and consequently went into the river with her children. She claimed her husband wanted to get rid of her in order to marry another woman. She was at first very excited, asked for help, said she wanted to go home, but later she was more reserved. After three months the anxious state disappeared and she was discharged. At home she was at first somewhat anxious and suspicious, but then seemed normal. In 1943 she became ill again. Her husband, who in fact did want a divorce, told of scenes of jealously, but above all, her depression came back strongly. Before readmission she went into the river again with her children, but all were saved. In hospital, she confirmed that she wanted to die, that life had no more meaning for her, and that she did not want to leave her children alone. People had stigmatized her only because she had been in the psychiatric hospital. She was always called "idiot". On one occasion she had even said something against Hitler, which everyone knew about now. She was again rather anxious and vexed. After four weeks she was transferred to Institution W. She cried and explained that she could no longer be saved, that she was always afraid of the police, it all must probably be a punishment because she had not gone to church often enough. After three months, that is, in the beginning of 1944, her depression disappeared. She stopped complaining, corrected her depressive ideas, and worked in the institution's housekeeping, since her husband would not take her back. In August 1944 she went through a mild depression which quickly disappeared. In 1945 and early 1946 she again went though depressions, which were not well described in her record, but ideas of reference were always mentioned. Her condition then remained good. She was industrious and had a stable mood; she was discharged in July 1946. Her **father** became ill in 1927, was depressed and believed that he was pressured by his boss in an agricultural business. After this condition had continued in a mild form for months, the depression increased in August. He believed that he could no longer harvest his wheat, and that people wanted to take things away from him. He stayed out all day in the hot sun without a hat, returned confused in the evening, and had to be admitted to Institution L. He was disoriented in time and space, was resistant, clutched people passing by. When he was put on his feet, he let himself fall down. He had to be tube fed. After eight days his condition improved although the disorientation continued. He also misidentified people around him and he was anxious believing that a judge was in the room. Thereafter he rapidly improved and proved to be amnesic about the whole period of acute disease. He still remembered that he had been out in the sun in the field, but he had no memory of the subsequent events. He was discharged six weeks after admission with a diagnosis of amentia due to overexposure. At home he remained healthy and the anxiety present before the amentia was no longer noticeable. In 1939 he was re-examined and found to be normal.

In Case 27, the picture of suspicious depression is very clear. The patient is very depressed, to such a degree that she is ready to drown herself and her children. Her depressive ideas however, all take on the forms of ideas of reference. She has many fears and ideas of sinfulness, but these do not appear as such; rather she is accused of crimes and threatened with punishment by what people and the newspapers say. Her father, too, had suspicious depression. He felt pressured by his boss and developed sunstroke due to his fear that people would harvest his wheat if he did not hurry. The short phase of amentia in the institution, with total amnesia following, was in any case clearly of exogenous origin. The endogenous psychosis which had existed previously appears to have been interrupted by exogenous influences, as is seen after shock therapy and other influences on the brain, for example, strangulation attempts.

The frequency with which ideas of reference occur in other psychoses as well as in suspicious depression often makes a differential diagnosis difficult. This is particularly true when there is no clear recovery. However, a **chronic course** does not negate the diagnosis of suspicious depression, as we have seen in parallel cases with chronic course in other pure forms, and as the following case shows:

Case 28: Hans Ler, born in 1902, became anxious and suspicious in 1936, believing that people considered him inferior. After attempting suicide he was admitted to Hospital H. in June 1936. Here he complained that everything was related to himself; when he saw a clock he had to think: "Your time is up." Among other things, he believed people were accusing him of a crime, he heard voices with similar content, and blamed himself for having lived above his means. In his distressed state he tried to hang himself. He was discharged in December 1936, without showing much improvement. At home he remained anxious and suspicious, continued to refer harmless things to himself, and believed everything was happening because of him. In November 1937 he was admitted to Sanatorium M. Here he complained that everyone wanted him to kill himself. He felt watched and misinterpreted harmless remarks of other patients. In January 1938 he was discharged. The feeling that he was being watched continued. He always believed that people wanted to tell him something from the way they moved and made signs. Despite this, he was able to continue his works as a financial official. In July 1939 he came for an examination at the Frankfurt Hospital. Here he explained that he could sometimes see people making signs about him. He could not rid himself from the feeling that people did not have good intentions about him. He heard unfriendly remarks in the conversations of his colleagues. His mood was still depressed and sometimes he felt like taking his own life.

In this case the picture of suspicious depression is definitely present. Inferiority feelings, self-accusations and fears appear in the form of ideas of reference. The mood is greatly depressed; there are several suicide attempts. The psychosis, however, assumes a chronic course, becoming gradually milder, but not clearing up completely. If the acute reference syndrome is reminiscent of beginning schizophrenia, then it is even more reminiscent if the psychosis does not disappear; however, in schizophrenia beginning with ideas of reference, hostile irritability soon takes over. On the other hand, in schizophrenia one never observes that a depressive mood continues chronically, accompanied solely by ideas of reference with depressive content, i. e. inferiority, sinfulness and fear. The course of suspicious depression is not as strange if one recalls what is found in the other pure depressions. It is a characteristic which is generally more or less shared by the pure forms.

In the families of suspicious depressives there was often a tendency to **mistrust,** indicating a sort of suspicious depression. It is also likely that this tendency represents a type of dilution of the actual psychosis, as we have already found in the other pure forms. **Suicide** occurred in several cases, and in fact appears to be relatively common in suspicious depression.

Summary

In suspicious depression ideas of reference with depressive content are associated with depressed or anxious mood. The patients deduce from the activities of their surroundings that they are considered inferior or sinful or that something bad is awaiting them. Besides ideas of reference, auditory hallucinations occasionally reflect the depressed content. Like other pure depressions, suspicious depression too may assume a chronic course. As long as the combination of depressive mood and ideas of reference with depressive content continues, there is no reason to change the diagnosis. Due to short observation periods confusion with reference syndromes of other origin may occur. Suicide risk appears to be relatively high. Some degree of mistrust, still within normal range, is found among the relatives. In terms of physique, I can only say that my impression has been that the leptosome type was not uncommon.

Apathetic Depression[3]

In hypochondriacal depression we encountered alienation phenomena which related to the perception of the patient's own body and the environment. Now a quite different form of alienation will be described. KLEIST'S **"alienation depression"** may be considered with two meanings. The alienation described here has nothing to do with the hypochondriacal form, rather, there is a sort of cooling of the emotional life. This is less recognizable objectively, but subjectively there is a kind of listlessness. An absence of feeling makes up the content of the repeated complaints of the patients. We have already found this alienation in melancholia, but there it only accompanies other depressive symptoms. In apathetic depression, it dominates the

[3] Several adjectives have appeared in the literature to denote this type of depression: apathetic, anergic, and non-participatory. The translator prefers the first.

picture. The deeper instinctual bodily feelings are less affected by the alienation; the higher and finer ones are more affected, causing the patients to suffer. The patients do not complain that they have lost feelings of hunger, thirst, warmth, cold, desire, pain, and so on, but rather that happiness and sorrow no longer touch them. Apparently these feelings are of a mental nature, that is, emotions which are derived from a higher level of human experience. A very high emotional plane is thus involved. When feelings are more self-centered, more bodily emotions are mixed in and remain relatively untouched; when feelings are more altruistic, i. e. on a mentally higher plane, they are more clearly defined by apathetic depression. Thus **sympathy** in joy and sorrow is peculiarly and severely changed. The patients complain that everything has become so meaningless that they are no longer touched by anything, and particularly that they no longer have any warm feelings for other people, for their family or for their children. They can no longer participate in joy or sorrow. The concept of non-participation expresses this fact. It refers to the lack of mental participation in events in general and in particularly of participation in situations where the normal reaction is sympathy. The term is more precise than the term alienation, which says nothing about how something that used to be close now appears to be alien. In severe cases, the non-participation easily leads to **"depersonalisation",** where the patient claims that he or she is no longer the same person as before, for all feelings have been lost. However, as we saw above, depersonalisation may develop from the hypochondriacal form of alienation as well. Alienation in psychopaths or twilight states should be regarded quite differently.

The cooling of feeling in apathetic depression is at first purely subjective; although the patient claims it is occurring, it cannot be determined objectively. The patients tend to display exaggerated care for their family rather than too little. Perhaps this is an attempt to replace with deeds and words the care and feelings they claim they no longer possess. Taking this into account, a certain degree of non-participation may be observed objectively. In hospital patients are observed to care little about their surroundings and emotionally are apparently hardly touched by them. Family members report that at home the patients were indifferent, had stopped caring, and had to be pressed to pay attention.

It might be presumed that it was a question of inhibition, as in melancholic patients – that however is not the case. When patients speak about their alienation phenomena, they may become quite lively, and complain with irritation and vexation in a way which does not occur in true inhibition. Impoverishment of initiative is thus probably a consequence of impoverishment of participation. This connection is confirmed by remarks of the patients. Combined with a complaint about **lack of feeling,** there is a constant complaint about **lack of will.** The patients do not complain like melancholics that they are no longer competent and can no longer achieve anything, rather they admit right away that they lack will power. They appear to experience this lack as directly as the lack of feeling. Presumably this is basically the same symptom. The higher mental feelings that are disturbed are probably simultaneously the starting point for the disturbance of initiative.

Some patients emphasize the lack of will power more than the lack of feeling. Their external conduct is not different. As long as the patients are not confronted they are rather quiet and lack initiative. However if stimulated and asked how they are doing, they quickly start to complain, and may actually wail severely about their painful condition. They are reminiscent of self-tortured depressives who may also be aroused from their apathetic behaviour to an excited plaintiveness if asked about their problems. In general apathetic depressives are not as excitable as self-tortured depressives.

Weakening of the will may be the cause of the presence of **compulsive brooding** in some patients; other minor symptoms may be noted. Patients may blame themselves for the cooling of their feelings; this may be explained psychologically. Ideas of reference are occasionally present, in which people appear to make accusations. However, these ideas never become prominent. Apathetic depression lacks anxiety, and the mood is rather one of simple depression, which may develop into a feeling of being tormented from within. **Suicidal risk** is often present, even if suicide attempts rarely occur. Generally apathetic depressions are not very deep. This might be related to the patient's claim that it is not just the lack of joy but also of sorrow which is felt with less emotional depth.

Apathetic depression is different from all other pure depressions, since in the others feelings of enjoyment do lose depth while bad feelings do not. This characteristic depends undoubtedly on the special nature of the emotional level which is affected. Just as with initiative the emotions fluctuate with the pro and con of higher thought and increase in depth. If the enjoyment pole disappears the opposite pole of dysphoria also grows weak. I have described this in greater detail in my "Biologische Pychologie" (LEONHARD 1993). In pure melancholia all emotional levels are involved; therefore in that disorder there are similar though less impressive findings as in apathetic depression.

As we shall see later, the number of cases of this depression in my investigative series was considerably less than that of the other pure depressions. Whether that corresponds to objective reality or whether such patients are brought to the hospital less often has not been determined.

Case 29: Nikolaus Traut, born in 1906, felt depressed for the first time in 1937, and was treated by a psychiatrist. In 1940 he became seriously ill and was admitted to the Frankfurt Hospital. He complained that he felt as if there was nothing inside him and he could no longer experience anything. He felt useless since he was no longer interested in anything, he was indifferent about everything and he no longer had any feeling for his relatives. The experience of joy and sorrow was gone. He would rather have had no life than one that was nothing. His condition would not change even if he were to go to the moon, since the "great emptiness" inside him would remain. The patient was quiet, showed little participation in his surroundings, but became livelier when speaking about his inner emptiness. After about five months his depression receded. The desire to work came back and he was discharged in stable condition. His **father** had depressions several times in his life, during which he was pessimistic about the future and felt incompetent. The depressions were transient and he never had to be admitted to hospital.

Case 30: Maria Web, born in 1878, became depressed for the first time in 1927, and was admitted to a sanatorium. There was no medical record of this episode. After ten months the depression cleared up. In 1941 Web went through a short depression. In 1942 she became more seriously ill and was admitted to the Frankfurt hospital. Here she complained about being tired of life and having many worries. Above all she said that she no longer had any real feeling. She did not care if she lived or not and she could no longer feel any joy or sadness. She could not even share any feelings with her son, recently returned from concentration camp. It was as if something had died inside her. She had also lost her will power and could no longer come to any decisions. She blamed herself for it all, particularly about her behaviour to her son, nevertheless there was nothing she could do about it. Most of the time she was quiet, kept to herself, but complained vividly when approached. After three months her condition improved and she was discharged. One year later she was re-examined and appeared to be quiet stable. A **sister,** who had always been anxious, went through a depression every few years, in which she allegedly behaved like the subject. A more precise description was not available; she was never admitted to hospital.

These two cases show the characteristic picture of apathetic depression. Subjective impoverishment of feelings and indecisiveness are prominent. Atypical symptoms are unusual. One patient continued to **brood** about her future, was obsessed with whether she should take her own life and that of her child. In the background of these symptoms the presence of apathetic depression was not to be overlooked; this patient complained that she had no feelings for her husband and her child. Associated with this "raw emotion" she blamed herself. Thus **self-accusations** occur with the subjectively sensed cooling of emotions. **Associated ideas of reference** may appear fleetingly as well.

The following case displays the picture of apathetic depression in a most impressive manner. I present it not only for that reason, but also because of its chronic course.

Case 31: Alfred Puck, born in 1909, visited a psychiatrist in 1931 who found no indication of disease. In 1932 he went to a spa suffering from nervous exhaustion and mental depression. During the following years he worked, but with no enjoyment, and "without inner feeling". He had no mood swings. In 1937 "the light of life was fully extinguished." He was now, as he said, completely emotionless and could not even feel sad. It was as if he was no longer there. He lacked all will power and had to force himself to get up in the morning to go to work. He felt completely indecisive. In 1938 he was admitted to the Frankfurt Hospital, where he presented the above complaints and continuously found new descriptions for his inner emptiness. His soul was really dead and his feeling for other people was gone all together. He spoke in a lively manner when describing his condition, but otherwise sat around without initiative. After two months he was discharged somewhat improved. However his abnormal condition continued. He was still depressed and without energy, but was able to work in his job. In 1940 his condition worsened and he was admitted to the private clinic K. He again complained of lack of feeling, internal stagnation, and lack of relationship to people and things. His complaints became more pronounced and his descriptions very vivid. He used Schiller's verse to describe his internal condition: "The place is burned empty." Afterwards he appeared apathetic and without energy. Three months later he improved and occupied himself spontaneously, and after four months he was discharged. He returned to his work as a business man, but the depression continued with some fluctuations. In 1946 he returned to the Frankfurt Hospital for a short examination because he had bought coffee illegally in order to stimulate his energy. In response to questions he again described his condition thus: "I see beauty, nature, mountains, but I cannot feel it." "Sometimes I feel dead inside and in the morning I don't have the strength to get up." "I have no passions like other people." "I can talk and give orders, but my soul is dead and my body lives on, as if nothing happened." "Not even the horrors of war did I feel like other people." In presenting all that he was lively and talkative. Without stimulation he quickly sank back into himself and behaved apathetically. His mood was always depressed and he often had suicidal ideas without actually making a suicidal attempt. In 1947 he was prosecuted and given a small fine, but his joyless depressive condition remained unchanged.

This patient always used new words and new comparisons to describe his lack of feeling and his lack of will power. His condition varied in depth; in 1940 he was admitted to Institution K. in a particularly bad state, but four months later he was discharged almost recovered. Phasic fluctuations were undoubtedly present, but in general the condition was chronic. The onset was hard to determine, since Puck, like other patients with apathetic depression, regarded his earlier life now appearing in hindsight to be without pleasure as having been joyless all along. What was missing here were the emotions accompanying his earlier experiences. There had been no complete recovery since 1937. The depression in 1947 was not as deep as in 1938 and 1940, but did continue. During the many years it did not change in character; no symptoms were lost and no new ones appeared. Thus there was no question of another condition. A psychopathic condition was ruled out as the depression was much too deep. I would rather assume that some forms of alienation described as psychopathic are in fact apathetic depression with chronic course. The circumstances are surely similar to chronic hypochondriacal depression, easily misunderstood as psychopathic if the earlier history is not carefully examined and the fact that the abnormality began at a certain point in time is overlooked.

It may also happen that alienation phenomena have always existed in a person and could be considered not as a disease but as a psychopathic peculiarity. Even then a combination with apathetic depression is possible. In all pure depressions the characteristic that has also been part of the patient's personality is often a dilution of the disease itself. It is therefore possible that there are people who never develop actual apathetic depression, but who do display the subdepressive constitution of this disease form. Milder phases also occur in patients with this condition. The depression then is not as deep as in the other types of depression.

Of special interest was one of my patients who exceptionally showed both forms of alienation, i. e. hypochondriacal as well as apathetic depression. The family history revealed that the patient's brother had manic-depressive illness. The polymorphic nature of that condition was a factor in the development of the patient's illness in that it affected two levels of emotions.

Summary

Apathetic depression is characterized by depression of mood, accompanied by impoverishment of feelings and will power, more subjectively felt than objectively recognizable. The higher mental emotions, not only joy but sorrow as well, lose their depth. Above all, the ability to sympathize with others is subjectively lost. Patients may complain bitterly, but objectively there is non-participation and loss of initiative. Other depressive symptoms play no role in the clinical picture; however psychologically related self-accusations due to the non-participation do appear often. The depth of the depression is less than in the other pure forms; suicide attempts occur less often. There may be a chronic course without change of the clinical picture. In terms of **physique** I noticed nothing of significance. I have seen apathetic depression in both leptosome and pyknic patients.

Pure Euphorias

Pure euphorias are very rare disorders, as will be seen in the statistical presentation. Because of this rarity, it is not truly possible to prove that they are separate forms, even though the picture is occasionally very impressive. But the concept of pure euphoria is well supported in that just as pure mania corresponds to pure melancholia, there is a pure euphoria for each pure depression. The disorders are not bipolar like manic-depressive illness, but they could be named "antipolar", since each form corresponds to an opposite. Pure mania, too, because of its rarity, should be compared with its opposite, pure melancholia. The symptoms of pure euphorias occasionally resemble those of manic-depressive disease, and, as we will see later, anxiety-happiness psychosis. However in the total picture they are less frequently imitated by polymorphic psychoses than pure depressions.

Unproductive Euphoria

The picture of unproductive euphoria is characterized by a feeling **of profound well-being** which puts the patient into a state of happy contentment. Formerly, in discussing anxiety-happiness psychosis, I described states of "simple joyfulness". They generally represent only short episodes in the course of that disorder, but as such reflect the picture that is typical of the whole course of unproductive euphoria.

The facial expression of the patient displays happiness; it is often a radiant smile. When questioned, the patients confirm that they feel happier than ever. One patient claimed that he had been in a similar mood only once before, which was during the period of his engagement. The feeling of happiness need not have any content; it appears to be of bodily origin, to have a "vital" character. Unproductive euphoria can thus be recognized as the opposite pole of agitated depression, which is also marked by no content, seems to come from inside, and has a "vital" nature. The psychomotor appearance, too, is the opposite: there the patients were seen to be driven by their deep dysphoria, while here they show an even-tempered peacefulness in their satisfied contentment.

Patients with agitated depression look for reasons for the otherwise objectless bad mood; the same phenomena are seen in unproductive euphoria. Here the ideas are expansive: self-elevation, wealth, high position, and also erotic happiness. One patient, a simple soldier, claimed to be a officer, while another proposed to a young woman who hardly knew him, looking forward to the approaching marriage despite her refusal. When discussing such ideas with the patients, it appears that these ideas are not really essential to the overall disorder. The euphoria does not increase or change in any way, nor do the patients place a high value on the acceptance of these ideas. They do maintain their ideas when contradicted, but may also remark, without change in mood, that if the ideas were incorrect, it would make no difference. The ideas are reminiscent of manic delusions of grandeur which seem somewhat playful and which are not taken too seriously by the patients themselves. On the other hand, in exalted euphoria there is a totally different connection of the affect to the ideas, in strong contrast to their unimportance in unproductive euphoria.

Generally only a very few ideas are observed, appearing with the onset of the disorder, loosely maintained subsequently, and only minimally supplemented later. These euphorias always remain **unproductive,** even if the feeling of happiness, objectively observed and subjectively confirmed, goes very deep.

Traits of unproductive euphoria remain unchanged when the content of the ideas is happiness of others, as is much more pronounced in exalted euphoria. In giving to others, man can find happiness to a greater or lesser extent, depending on his personality. Thus it is comprehensible that patients with unproductive euphoria occasionally claim that they want to give others something beautiful, such as peace. The patients are only loosely tied up in these ideas, so that there is no chance of it being confused with exalted euphoria.

Occasionally patients try to act out their ideas; the patient mentioned above proposed to a girl, whose love he expected. In general, however, **motivation is low.** The happy contentment prevents compulsive activity from developing. In hospital, too, the patients are rather quiet; they speak freely with no pressure of speech and they hardly ever ask for their discharge. The patient described below explained that he liked staying in the hospital, for he felt better than ever. Their peaceful behaviour differentiates these patients from those with pure mania, who display pressure of speech and flight of ideas and are always active in one way or another.

Unproductive euphorias generally fade away within a few weeks to a few months but tend to recur periodically. Further observations will show whether there are chronic cases and transitions to normal temperament as observed with pure depressions. A case of unproductive euphoria follows:

Case 32: Karl Gerst, born in 1896, first became ill in 1938. As a policeman he asked some hotel guests for their papers without reason. He was admitted to Institution B. Here he was somewhat self-important, took an interest in other patients, but was otherwise inconspicuous, and was discharged after eight weeks. In 1941 he again began to act strangely. While on patrol duty he began to direct traffic as well, and shook everyone's hand on the street. He was admitted to the Frankfurt Hospital, where he explained that he had been feeling much better and healthier for the past three weeks; that he had never felt as happy as he did then. He had wanted to help people out by directing traffic. He claimed that it was good that he had been brought to the hospital since he could take care of the patients, and was ready to give blood. He urged other patients to work, he liked to enter into conversation without pressure of speech, and always had a radiant and contented facial expression. The euphoria receded after six weeks, and he was discharged in a stable condition.

In Gerst's first illness the euphoric mood was not precisely described, probably because its unproductive character made it less eye-catching. His behaviour in the hotel did show that he was rather expansive. In his second illness he claimed rights not belonging to him, but as could be seen this time these rights were a result of his euphoric mood. In hospital the euphoria was in the foreground; the patient felt better than ever. His activity and his ideas were only playful expressions of his feeling of joy.

Summary

Unproductive euphoria is characterized by a motiveless feeling of happiness appearing to originate in a bodily level of feeling. Only a few ideas associated with the euphoria appear, and they have no great importance for the patient. In keeping with their happy contentment, the patients show little compulsive activity. Occasionally they attempt to put their ideas of happiness into practice but even then are not under pressure.

Hypochondriacal Euphoria

Hypochondriacal euphoria has a remarkable clinical picture which I first observed during two examinations. In the investigative series at that time there were eight such cases with different previous diagnoses (hypochondriasis or hypochondriacal psychopathy); I felt it was important to place these patients into a separate diagnostic category.

The patients' **complaints** are similar to those in hypochondriacal depression; the descriptions are often even more remarkable. Varied and peculiar sensations travel throughout the body which the patients tend to compare with stinging, burning, boring, though they do not actually describe them as such. This becomes clear when the patient is asked for a more exact description and replies that it is as if the brain were cut off, the stomach expanding, or the throat tightened. The claim that internal organs had been displaced is more common in hypochondriacal euphoria than in hypochondriacal depression. These symptoms somewhat resemble schizophrenic sensations, but, as we will see even in patients with a chronic course, the differential diagnosis can easily be made.

The patients have many complaints about these sensations; they are clearly suffering and may even claim that the problems are becoming unendurable, but the basic mood is euphoric. Between the complaints an elevated mood continually reappears which is incompletely covered up by the complaints. Often the patients will complain bitterly while having a cheerful facial expression. At times they become conscious of the contrast and explain that, despite all such complaints, they were able to remain cheerful. The misperceptions show clearly that the same emotional level is affected as in hypochondriacal depression. Apparently both the euphoric and the depressive emotional disorder lead to sensations arising from the somatic realm that are normally unknown. I have treated that point more completely elsewhere (LEONHARD 1970a). Mentally the sensations cannot be subordinated and assimilated, hence they remain a foreign body and are disturbing and bothersome, even if simultaneously the mood is shifted towards the euphoric side.

In connection with these pathological sensations, **alienation phenomena** in the sphere of bodily feelings often develop in hypochondriacal depression. In hypochondriacal euphoria, I found this alienation rarely; altoghether it seems to occur less often here than in depression.

When patients with hypochondriacal euphoria present their complaints they often become very lively and talkative. They emphasize how much they suffer and demand help. When speaking about other topics they remain lively and stimulated, but no longer show pressure of speech. They are difficult to distract, particularly from their complaints. They do not show flight of ideas. They often complain severely about their misperceptions and they may theatrically produce a few tears as if they were superficially depressed. Such behavior may be reactive but passes quickly and the basic euphoric mood reappears. The patients' complaints are often somewhat querulous, but this is more common in chronic cases to be described later. Here I will present a typical case.

Case 33: Betti Rot, born in 1896, first had nervous complaints about her body in 1924, she went to a spa for convalescance and recovered. In 1937 she became ill again and was admitted to Institution S., complaining that her back was throbbing, her head ached, and something was injured in her spine. Her complaints were lively and effervescent. Three months later, in January 1938, she was transferred to the Frankfurt Hospital, where she explained with a smiling face that she was "dead, although alive". She claimed to have a curious feeling in her head, a tearing in the fingers, a throbbing in the back, a pulling in the eyes, and a shrinking in the brain; there must a knot in her head and a cleft in her skull. Both cheerful affect and hypochondriacal complaints continued. The patient called herself a blossoming candidate for death. On one occasion she declared that she no longer had any normal feeling. After ten weeks she was transferred to Institution W. There, too, hypochondriacal complaints were expressed with superficially cheerful affect. After ten months the symptoms improved and she was discharged in stable condition.

In this subject, the contrast between hypochondriacal complaining and the cheerful mood is very impressive. It leads to the patient's joking about herself and calling herself a blossoming candidate for death". Alienation phenomena are hinted at. Her remark that she was "dead, although alive" points to an alienation of the body, confirmed by her later remarks that she no longer had normal feelings.

Only four of my eight cases took this favorable course; the other four were **chronic.** This would have naturally led to doubts about the diagnosis if a chronic course had not been repeatedly found in the pure forms. But even in remembering this observation in the pure forms, it is noteworthy that no less than half the cases took a chronic course. I present one of them.

Case 34: Helene Fis, born in 1874, was healthy until the age of 50. In 1924 she first complained of somatic symptoms. In 1932 she was admitted to the Frankfurt Hospital. In between she had undergone numerous operations because of her symptoms, without anything having been found. She complained of pains in the lungs, the trunk and the neck, palpitations, weakness in the stomach, and itching in the head and neck. She felt as if her stomach was stretched, as if something was rising from her stomach and something was wrong in her throat. The back of her head felt as if she was lying on wires. With all her complaints uttered rather circumstantially and awkwardly she seemed cheerful and in a good mood. She explained that is was a miracle that despite her "pains" she could remain lively and still could laugh. A hernia of the stomach, acquired from an operation, did not bother her at all. She was discharged after five days. In 1936 she returned to the hospital, complaining that she was not receiving enough help for her problems. She sent a similar complaint to the Department of Health. There was no improvement whatsoever. In 1939 she was readmitted to the Frankfurt Hospital, complaining of pains and a "dumb feeling" in her head, pains from her neck to her stomach, pressure in her trunk, and so on; she demanded an operation. Although she moaned about her "terrible pains", she was in a good mood, spoke in a lively manner and was very animated and cheerful. After three months she died of bronchopneumonia at the age of 65 without her mental condition having changed at all.

Until the age of 50, Fis was healthy and could take care of her household. The sudden onset at a certain point of time contradicts, as we also saw in chronic hypochondriacal depression, the opinion that she might be a hypochondriacal psychopath. In both hypochondriacal euphoria and depression, the unchanging state with no recognizable progress during the whole course speaks against a diagnosis of schizophrenia. Furthermore, the chronically euphoric mood with no affective dampening would in no way pass for schizophrenia. The sensations themselves often have the grotesque form reminiscent of schizophrenic symptoms, but are never perceived as external influences. Hypochondriacal euphorics doubt as little as hypochondriacal depressives that the sensations are signs of disease processes. In chronic cases the picture corresponds entirely to that of hypochondriacal euphorics with phasic course. No other interpretation is possible. The tendency towards a chronic course, to a certain extent present in all pure forms, appears to be particularly great in hypochondriacal euphoria. Such patients come to the hospital relatively often because rather querulously they demand help for their complaints. This is only natural in view of the bothersome symptoms. The condition more likely belongs to hypochondriacal euphoria than to depression, because in the former the patients tend to look for fault in others more than patients do in the latter. Even when argumentative, hypochondriacal euphorics never think of blaming their surroundings for their sensations, as schizophrenics do; they only ask for help. If the chronic cases are included, hypochondriacal euphoria is not all that rare. I had to examine such cases several times; their previous physicians did not know how to diagnose these people who complained so much, but who appeared cheerful and satisfied.

Summary

In hypochondriacal euphoria **misperceptions** of various kinds combine with **elevated mood.** The patients complain in a lively manner about their problems and seem to suffer, but their behaviour, their facial expression and their statements reflect a euphoric mood. When asked they will confirm that they are in a good mood and hardly affected by their symptoms. More often than other pure forms hypochondriacal euphoria tends towards a **chronic course.** The complaints then become somewhat querulous, but the euphoric mood remains.

Exalted Euphoria

In unproductive and hypochondriacal euphoria the elevated mood seemed to be based in the body, with no connection to higher mental contents. This is not the case in exalted euphoria. As in the condition of the opposite pole, i. e. self-tortured depression, in which the mood is enhanced through the depressive ideas, in exalted euphoria the mood is enhanced through the euphoric ideas; just as extreme as the ideas of inferiority are in self-tortured depression are the ideas of grandeur in exalted euphoria. This shows that an emotional level is affected in which mental contents and not bodily processes are based. The ideas of exalted euphoria are very reminiscent of those which we will meet in connection with anxiety-happiness psychosis. In both conditions patients occasionally claim that their beliefs come from a higher power. The enthusiastic ideas have an entirely different character from the more playful ideas of unproductive euphoria. The inspiration psychosis of KLEIST, which corresponds to the expansive autopsychosis with autochthonous ideas of WERNICKE, may resemble both exalted euphoria and anxiety-happiness psychosis.

Patients with exalted euphoria also have an elevated mood but not a steady peaceful feeling of happiness as in unproductive euphoria. Instead the feeling increases greatly when the ideas are discussed. The patients become rather enthusiastic, even ecstatic, and express themselves with deep inner emotion. When the examiner turns away from the patients or leads the conversation towards a neutral theme, the elevated mood sinks and the patients appear considerably less enthusiastic than patients with unproductive euphoria. During neutral discussions the euphoria is less recognizable than the ceremonial posture and affected manner of speech, which characterize a patient during his enthusiasm. This kind of euphoria has a rather viscous character. The notion of cheerfulness, which despite its inprecision can be used to describe the enthusiasm of manic and unproductive euphorics, no longer fits in with the elevation of the mood of exalted euphoria. Flippancy, an aspect of the term "cheerfulness", is absent in these patients. Instead they are ecstatically filled with joy. As regards content the ideas of exalted euphorics are nearly always simultaneously egoistic and altruistic; the former in that the patients feel powerful, wealthy, and erotically superior, the latter in that they wish to use their power for the good of others. They repeatedly claim to be very influential and may call themselves leaders of humanity who will bring happiness not only to themselves but to others as well – perhaps bliss in this world or the next, as well as eternal peace or eternal good health. They often trace their idea back to God, who has blessed them and has given them these ideas. In less religious periods they may themselves feel very competent and called upon to perform great things. In milder form they are content with appearing as bearer of happiness in their immediate surroundings, announcing that they would save their relatives from misery or cure the other patients in the hospital. A purely egoistic form of ideas, that is, a pure megalomania, does not appear in these patients – a purely altruistic form is more likely. They may expressly declare that they are not doing this for themselves, that they only want to help others. It appears from the pathological affect that altruistic motives have moved more into the foreground than egoistic motives. At any rate ideas of bringing happiness are in first place and delusions of grandeur in second place. This also applies to the ideas expressed by patients with anxiety-happiness psychosis as well as inspiration psychosis, as emphasized by KLEIST.

The claim of the patients that their ideas come from God is based partially on **hallucinatory experiences** seemingly appearing at the peak of the affect. They are primarily visual and are reminiscent of dream experiences. God or a saint appears to the patients and his mere presence, or words that accompany the vision, lets them know that they have great duties to perform. Although the patients claim these experiences to be reality, they are actually pseudohallucinations rather than genuine ones. They are never sensed as ego-alien, but only reflect what is felt by the patients from inside. The patients talk about it without any astonishment, but with the same enthusiasm with which they talk about other ideas.

There are no other symptoms in exalted euphoria; psychomotor excitement, as seen in anxiety-happiness psychosis, is absent. Thought is only impaired in that it is hard to have the patients pay attention to anything else besides their ideas. The following case illustrates the clinical picture of exalted euphoria.

Case 35: Heinrich Horch, born in 1907, became ill in 1940. At night he heard the voice of God: "You are an unbelieving wicked man, but I will reveal myself especially to you." He was admitted to the Frankfurt Hospital, declaring that he had to write a book about the Bible, that he would give others a great experience. Later he declared himself to be a universal man who had unusual powers, including great physical strength. He requested that blood taken from him be equally distributed among the patients so that they might become better. He wanted to help all people. After three weeks he recovered and was discharged. His **father** had always been a very religious sectarian. At the age of 60 this religiosity became pathological. His behaviour was very peculiar, he wanted to convert everyone, and explained that it was a gift of God. Later he calmed down.

This patient described the characteristic picture of exalted euphoria. The dramatic posture and ceremonious speech show that he was completely filled with his ideas. The feeling of happiness about his own power and greatness is always mixed with the feeling of happiness about being able to do good to others. Pseudohallucinatory experiences occurred at the onset. Of great interest is the inheritance from his father who clearly went through a phase of exalted euphoria. Beyond that the father all his life seemed to have had the psychosis in diluted form in his excessive religiosity. This condition represented only a pathological exaggeration of his normal personality. We made similar observations in the pure depressions. Another patient who during the course of twenty years had four episodes of exalted euphoria declared: "The joy about that which we will all experience is indescribably large. – I need your help, the help of the whole world. – Germany will become greater and more powerful, I shall take over the reigns of government. – I will be under the protection of the heavenly father. – The free choice of people will decide if I will follow Hindenburg or not; I shall willingly follow the voice of the people. – I know that I have a great duty to perform." On one occasion when he was somewhat less ecstatic he called himself "Master" promising to build the largest organ in the world.

The tendency of pure depressions occasionally to assume a chronic course was not observed in my early cases of exalted euphoria. But later in my investigative series in Berlin, I met a patient who did have a chronic course.

Summary

In exalted euphoria the feeling of happiness is tied to ideas whose content is simultaneously **self-elevation** and the **happiness of others.** The patients lose themselves in these ideas, often with a high degree of ecstatic emotion, but when they are distracted from them, their affect diminishes; the euphoria need not be present continually. However when speaking about neutral topics the patients maintain a ceremonious manner, reflecting the particular nature of the affect. Not uncommonly they claim that the ideas were imparted to them by a higher power, sometimes by way of **pseudohallucinatory experiences.**

Confabulatory Euphoria

In suspicious depression we presented a pure form in which depressive ideation arose from circumstances pointing to intellectual activity. We find a similar phenomenon in confabulatory euphoria, described by KLEIST as expansive confabulosis. Reasoning seems to play a consistent role, filling the pathological affect with content. However there is less connectiveness and evidence of greater production of independent imagination. This is surely an expression of the different kinds of mental activity in the two affected directions; depressive affect leads more to pondering about facts, while euphoric affect leads to free fantasy.

The euphoria is not so deep in confabulatory as in exalted euphoria. This also parallels what we found in the depressions where in the suspicious form abnormal affect was considerably less pronounced than in the self-tortured form. The euphoria of the confabulatory form is rather mild, it does not flow as freely as in mania. Nor can the confabulations be called playful, since they are permeated by the affect.

The confabulations are full of accounts of happy experiences. This is quite obvious from the accompanying self-elevation. The patients relate, for example, how they held discussions for this or that important personality, what great power, honour and wealth they themselves possess, and they talk about the people who had praised them. Patients love to create sensations, by reporting travels full of adventures, dangerous sea voyages, fights with robbers and wild animals. Encounters with God may also occur, including advice imparted by God. Patients may become quite enthusiastic and in this way resemble exalted euphoria. However if they are allowed to continue talking, the difference becomes clear: while patients with exalted euphoria maintain their ideas and anchor their whole affect in them, the confabulators quickly move on and have many other fantastic tales to add. These stories seem to be based on errors of memory, as expressed by the term confabulation, but they are probably more often merely fantastic thoughts, lacking the character of experience. This is particularly true when the patients claim some high position. Such ideas do not always originate from errors of memory. Presumably in the pathological and exaggerated fantasy activity genuine confabulations of an experimental character are mixed with fantastic notions derived from these patients' thoughts. Illusions may also be present, but these are difficult to differentiate from the confabulatory perceptions.

The patients enjoy telling their stories, becoming lively and talkative. They easily digress from questions posed to them and return to their confabulations to which they are affectively closer. Their lively nature is reminiscent of mania, but they do not resemble schizophrenics. Confabulatory paraphrenics do display a relatively natural manner, as we shall see later, but not the liveliness of confabulatory euphorics. In schizophrenics confabulations are rather rigid, more like memory errors repeatedly presented in the same way, and are only slowly replaced by new ones. In contrast the confabulations of euphorics are more flowing, less rigid, permeated with transitory notions without the character of experience. Generally the right diagnosis may be made based on the immediate clinical picture without having to observe the progress of the condition. Furthermore, dilutions of confabulatory euphoria occur, appearing only as eccentricities and corresponding to the subdepressive and subeuphoric temperaments, as has been discussed with reference to the pure forms. The **father** of a confabulatory euphoric, not part of our investigative series, had such a confabulatory nature. He busied himself with calculations to determine the onset of the Kingdom of Christ, described the natural events which would precede it, prophesied anarchy and poverty, and classified nations according to their religious importance. This was observed on the side, while his son's disorder was examined. The father could continue to work and did not appear disabled. He was not at all schizoid, but quite open and approachable. An example of confabulatory euphoria follows.

Case 36: Hans Fers, born in 1903, first became ill in 1917, telling wild stories: he owned a villa, he could ride a horse, a fat girl took him into the forest, he was a doctor, a medical officer in the army by profession. He was admitted to hospital, where he calmed down very quickly so that he was discharged after four weeks. In almost every year that followed he was reported to have "spoken wildly". In 1926 he was readmitted, talking about wild animals which he had seen: "I do not know if it was a dream or reality." He saw signs like Joan of Arc, heard inner voices "for the reconstruction of the German Reich in golden times". He claimed to have visions, day and night, seeing them clearly. He was going to be king of the world republic. "I place the crown myself on my noble head. My coronation will take place here in Frankfurt by the aristocracy of France and Germany." The soldiers would be trained "as toys" and not for war. He would marry the most beautiful woman. He had ridden a horse through flames to reconcile Germany and France. "The people's monument in Leipzig is buried now." He called an attendant "Richard Lionheart with the bitten cheek". He was very loquacious and erratic. After three months his condition abated. He became reasonable, declared his ideas to have been nonsense, and he was discharged. In 1929 he relapsed, intending to apply to the Foreign Legion, and was readmitted. He declared: "I cannot help it that a treasure came to me in the enchanted castle"; "I must carry the axis of the earth"; "meteors are exploding the earth." Voices told him that Prussia had become friendly with the Jews, he was set up in America, and King Fuad wanted to complete peace negotiations in Berlin. After three weeks, he was transferred to Institution E., where he continued to confabulate, claiming: "I rode through the fire in spirit as a German Joan of Arc." After three months he again calmed down, but remained in the institution for a year since he apparently had minor relapses (the medical record is scanty). Then he recovered and was discharged. In 1932 he married. In 1934 he was again readmitted, declaring: "I am Kaiser Friedrich the Third, King of the World Republic"; "in the Masurian Lakes the dead are resurrected swimming to the surface and wake up alive": "I will lay the keel for a new ship as a memorial for the Turkish war of the Persians against the Arabians"; "air ships will be built on the Bodensee"; "there is a garden in the desert, the train travels through for the children of the desert." After seven weeks he calmed down and was discharged. In 1939 he was rehospitalized, having declared that the movement of the earth had changed everything, that horses lying on the battlefield of Verdun came to life again, that God had told him he had to conquer Africa and the Balkans. Four months later he was well again and went home. In 1941 he was drafted into the Army, but was released when he relapsed again and had to be readmitted to hospital. He claimed that he had to make plans for a funicular to France; in the Foreign Legion he had been on guard in Africa; German long distance artillery had shot off a corner of the sun and diminished its shining. "During the Frankfurt Fair red and green balls came down along with bolts of lightning which lit up the dark clouds." After three months he became reasonable, corrected his ideas, trying to make them more comprehensible by saying that he read a lot, and while sick he mixed up what he had read with reality. He related his experience with the fiery horse to horses with wings in a Frankfurt theatre. When sick he always claimed to have a feeling of happiness, often also a feeling of being graced by God. He was discharged, returned to work and seemed normal. In 1944 he returned to hospital, declaring that a piece of the moon was missing, that the earth might develop holes and tracks. "Rockets go into the earth and combine with the fiery mass, then the earth's crust will slowly dissolve"; "Time has diminished and moves around worse." After three weeks he again improved and was discharged. He returned eight months later, again having begun to confabulate at home. In hospital he soon improved, expressed himself freely and sensibly, and was discharged after three weeks. In 1945 he was again readmitted, claiming that he had been in Rome, then had gone to Africa, and from there to America. The horse he had used he later exhibited in Frankfurt out of gratitude. He showed marked pressure of speech and continued to talk when no one was listening. After six weeks he was well again. As his condition improved, he seemed mildly disgruntled; he did not like to be reminded of his stories, and was visibly upset that he had spoken such nonsense. Yet while sick he always had a happily elevated to mildly enthused facial expression. In 1946 once again he was readmitted, speaking of an alleged trip to Africa and Asia. After two weeks he was transferred to Institution E. Discharged from there he was readmitted repeatedly, always with the same confabulatory symptoms, which lasted for a few weeks, after which his condition returned to normal.

The psychosis of subject Fers is very impressive, since many phases repeatedly present the clinical picture of confabulatory euphoria. In the earlier phases Fers was often somewhat ecstatic, rather like a patient with exalted euphoria, if the many confabulations had not been present. Later he confabulated expansively without a discernible ecstatic mood. Eventually the fantastic confabulations solely dominated the picture. That was probably an expression of the patient's aging; he was not as effusive in middle age as he had been in his youth. In every phase the characteristic picture of confabulatory euphoria was present. The long observation period, revealing recovery from every phase, rules out confusion with a schizophrenic disorder. Elsewhere (LEONHARD 1981a) I have described a patient with a similarly impressive clinical picture; his mother too had confabulatory euphoria.

Summary

Confabulatory euphoria is characterized by **fantastic stories and elevated mood.** Sometimes the stories are based on errors of memory, at other times they are entirely imaginary. The stories are of a joyous nature, often including self-praise or merely following a fantastic sensational pattern. Misperceptions can play a role but it is usually hard to determine whether something was truly misperceived or only confabulated.

Indifferent Euphoria[4]

The inclusion of indifferent euphoria rests more on theory than on practical experience. If, as we have seen up to now, every form of pure euphoria corresponds to a similar form of pure depression, then apathetic depression certainly must have its opposite pole. However it seems to be very rare. Apathetic depression itself is uncommon, rarer than the other forms of depressions. If indifferent euphoria is rare at the same rate as the other euphorias, then it will hardly ever be observed. In my first investigative series I had only one patient.

After theoretical considerations, which I have discussed elsewhere (LEONHARD 1970a), I believe that this disorder at the opposite pole of apathetic depression likewise is associated with weakening of emotion and initiative. The two forms differ in their basic mood, but not in their content, just as hypochondriacal depression and hypochondriacal euphoria differ in affect but not in content. The following case corresponds to this theoretical expectation.

Case 37: Hermine Os, born in 1905, became ill in 1941, losing all her energy. She could no longer find the energy for anything, neither for work, nor writing, nor reading. She perceived herself to be quiete lame. She was admitted to the Frankfurt Hospital, complaining above all that she was so indifferent, felt everything was so hazy, she could not participate in any activity with her children. Yet while complaining she was not depressed, rather she smiled contentedly. After four weeks she was transferred to Sanatorium H. There likewise she complained of lack of energy while appearing cheerful. After three months she was better and could be discharged. At home she felt well for a few months until she relapsed and had to be readmitted to the Frankfurt Hospital, one year after her previous admission, presenting the same picture as before. She complained that she could not complete anything, had no energy, and no longer felt right. She was poorly motivated, preferred to lie in bed, but in conversations smiled contentedly. After eight weeks her condition improved, she developed initiative, and could be discharged.

From this clinical picture I felt the correct diagnosis was indifferent euphoria, since the complaint of impoverishment of emotions and willpower were combined with euphoric mood. Non-participation was objectively more prominent than in depression, most likely because euphoric mood is more readily associated with it than depressive mood. The diagnosis of hebephrenia is ruled out not only by the late onset but the phasic course and natural manner of the patient, who remained somehow warm despite the subjective emotional blunting; her condition was more reminiscent of a quiet hypomaniac. Hebephrenics seldom notice their own emotional impoverishment.

Case 38: In my investigative series in Berlin I also found a case of indifferent euphoria. Marie Sch., born in 1937, a pharmacy assistant, ordinarily had an equable temperament.

---

4 The German adjective "teilnahmsarm" literally translated into English would be "participation poor"; the translator prefers "indifferent" (compare footnote on "apathetic depression").

She was not outgoing but always had girlfriends. In 1962 she came to see me because, as she said, she had been feeling different for about six months, and that the change had occurred "like a sudden shock". She complained that her inner feelings were not there anymore. It did not matter whether she was saying something funny or something sad. She could perceive everything she heard or touched and could decide if something was beautiful, but everything seemed strange to her and she lacked any true feelings. When she said something, it was as if in her inner thoughts the things of which she spoke seemed quite remote. She also felt that her face remained emotionless when she told stories with feeling and that her hands were expressionless when she gestured. From the conduct of other people she would recognize their liveliness while she herself felt only her own indifference. The others had never made any comments, probably not having noticed any change; and even her parents did not notice the abnormal state. While the patient made this report, she seemed lively, fresh, and cheerful. I advised her to be admitted to the hospital for further observation, but she did not accept this advice, since she did not really feel sick. When I inquired into her condition a year later she wrote back that she felt better. In the meantime she had married. She turned down my invitation to talk to her again.

In this patient too, the lack of participation could be recognized. When she claimed that her facial expressions and gestures were no longer natural, she was clearly referring to lack of feelings ordinarily accompanying expressive movements. Yet at the same time her posture was lively and her mood cheerful. Despite her impressive description of her condition she did not feel seriously ill. During this single interview her apparent lack of drive might have been covered by her cheerful nature. Perhaps indifferent euphoria is observed so rarely because the patients in general feel no need to see a doctor. The above patient appeared in no way abnormal to the people in her surroundings.

Summary

If I have outlined indifferent euphoria correctly, it is characterized by subjectively felt and objectively recognizable impoverishment of emotions and willpower combined with euphoric mood.

The Cycloid Psychoses

Psychotic syndromes of cycloid psychoses have been described by both WERNICKE (1900) and KLEIST (1928). My concept of these disorders is somewhat different. For WERNICKE anxiety-happiness psychosis appears in two separate forms: anxiety psychosis on the one hand, and "expansive autopsychosis with autochthonous ideas" on the other. WERNICKE developed the concept of "motility psychosis", but he did not separate periodic catatonia from it. Later KLEIST and FÜNFGELD (1936) described the differential diagnosis from schizophrenic disorders. Wernicke denoted the two poles of the disorder as "cyclic motility psychosis". Confusion psychosis appeared to him as either "periodic maniacal autopsychosis" or "agitated confusion", inhibited confusion psychosis as "intrapsychic akinesia". In the latter KLEIST spoke of "confused stupor". The concept "confusion" comes from MEYNERT. KLEIST included anxiety psychosis and happiness psychosis which he considered as "inspiration psychosis" in his "paranoid marginal psychoses", while he grouped motility psychosis and confusion psychosis with "cycloid marginal psychoses".

The cycloid psychoses, of which three are to be distinguished, are characterized by complete recovery from each phase. If this does not hold the diagnosis is usually wrong. A chronic course corresponding to chronic mania is very rare. It may be observed that patients with cycloid psychoses rarely lose some of their vigor after repeated phases and hospitalizations. This also happens in manic-depressive illness, and can be seen as a reaction to this disorder. Recovery from cycloid psychoses was confirmed by thorough follow-up examinations carried out by Frau VON TROSTORFF and myself (LEONHARD and VON TROSTORFF 1964). I emphasize the follow-ups because the cases reported below were all observed until the end of each phase; some of them could be followed over a long period of time. PERRIS (1974) has paid much attention to the cycloid psychoses in recent times and has confirmed their separate status.

Nevertheless, in spite of remissions occurring in cycloid psychoses it must be emphasized that there is a certain relationship between them and some forms of schizophrenia – the **unsystematic schizophrenias.** Parallel to the three cycloid psychoses the three unsystematic schizophrenias manifest certain symptomatic similarities. Perhaps here and there the disease process involves similar brain functions, but the genetic differences clearly call for separate categorization. This I will consider in detail. The cycloid psychoses are not at all similar to other forms of schizophrenia; in systematic and combined systematic schizophrenia the clinical picture is completely different.

Anxiety-Happiness Psychosis

Happiness psychoses are considerably rarer than **anxiety psychoses.** This parallels manic-depressive illness in which depressive phases are more common than mania. However with closer observation it seems that if actual ecstatic phases are absent anxiety may be interrupted by a state of happiness lasting for a short time, perhaps an hour or even less, during which ideas of being summoned or of salvation are expressed. These transitory states are very useful in confirming the bipolarity of the psychosis. However they are not necessary for the diagnosis; anxiety-happiness psychosis can be recognized from the picture of the anxiety phase alone.

Fears such as being tortured or killed, of losing relatives are characteristic. They are rarely present in pure form, but rather appear in association with **paranoid symptoms.** The anxiety is characterized by distrust and ideas of reference. It develops not only from within, but patients also see it constantly reconfirmed by events in their environment. For example they believe that the police is following them, that a man standing before them is threatening them, that their houses are surrounded, that patients in the hospital are avoiding them, or that employees are talking about their pending arrest. Accompanying these **ideas of reference,** less precise **reinterpretations** occasionally appear in which events in the environment seem to be threatening without definite foundation. If this latter form of idea construction with accompanying perplexity begins to predominate, then the differential diagnosis leans toward confusion psychosis, where we will again meet ideas of significance. However despite the perplexed traits, the basic mood of anxiety psychosis remains primarily anxious. There are transitions from ideas of reference to **illusionary** and **hallucinatory experiences.** If patients claim, for example, that they deduced from the gestures of people in conversation that these people were discussing the patients' imminent execution, then we are dealing with ideas of reference; if they claim to have understood what had been said, then it is a case of illusions; if they claim to have heard discussions of their execution though no words were spoken at all, then we must assume the presence of hallucinations. All these phenomena occur in anxiety psychosis and are often difficult to sort out. Sometimes patients refer to olfactory experiences where it is difficult to differentiate whether odors are being misinterpreted or if they are being hallucinated. Similarly patients may complain that the food is poisoned. The absence of precise symptoms, the fluctuation between ideas of reference and hallucinations is characteristic for anxiety psychosis.

Hypochondriacal experiences are often found: itching, feeling of heat, something like shaking of the body, and the like. Sometimes patients claim that they are being bodily **influenced.** However they seldom confirm that these phenomena affecting them have external origin; rather they describe sensations similar to those in hypochondriacal depression. If patients speak more definitely of being influenced from outside, that they are perhaps hypnotized, a possible diagnosis is affective paraphrenia, as we will see later.

However, simple misperceptions are hardly less frequent in anxiety psychosis than ideas of reference. They may even dominate the picture. It is the nature of polymorphism in bipolar psychoses that first one symptom will predominate and then another. Furthermore, varied degrees of inferiority, self-accusation and occasionally alienation phenomena may appear.

Paranoid symptoms are considered to be prognostically unfavorable by many psychiatrists; such may even be considered to be characteristic of early schizophrenia. However the more pronounced the underlying anxiety is the less justified is this negative prognosis, as we will see more clearly in relation to affective paraphrenia. The lively, often extraordinarily severe anxiety differentiates the condition from suspicious depression, as mentioned before.

The direct expression of anxiety varies considerably. Patients may wail, moan, scream, or plead in a high degree of excitement and repel every attempt to approach them; this may far exceed the excitement of agitated depression. They may however be rigid and motionless, revealing their anxiety only in their facial expression. The rigidity may suddenly turn into excitement. To what extent these are psychologically different reactions to anxiety on one hand, or psychomotor symptoms on the other, is not always determinable. Certainly both exist.

Just as we found some symptoms of manic-depressive illness overlapping other polymorphic psychoses, in anxiety-happiness psychosis we find the same phenomenon. Psychomotor symptoms overlap those of motility psychosis. In anxiety psychosis akinesia may appear as a rule for only a short period of time, even though a high degree of anxiety does not psychologically explain such rigidity. Perplexed forms of stupor are also found, bordering on the symptomatology of confusion psychosis. Symptoms which resemble the excited pole of motility psychosis or confusion psychosis may appear in anxiety psychosis. Situations then arise comparable to the mixed states of manic-depressive illness. At times it is hard to determine readily which form of polymorphic psychosis is present. However with the lessening of these phenomena the accompanying symptoms generally recede as well, and the anxiety psychosis with its characteristic picture becomes clear.

Milder forms of the disease may resemble melancholia, if depression accompanies anxiety. Ideas of inferiority and sinfulness may be present. The polymorphism of anxiety-happiness psychosis may also resemble that of manic-depressive illness. Of the pure forms, agitated depression is the most likely to be imitated, as discussed in relation to that disorder. Only rarely do paranoid symptoms recede enough for agitated depression to be confirmed. When less anxiety is present and ideas of reference predominate, suspicious depression may be simulated.

Happiness psychosis, the other pole, may resemble exalted euphoria. As in the latter similar ideas appear with the feeling of happiness of an ecstatic character. Patients feel elevated, often immeasurably, even to a divine level; they do not want merely to feel overjoyed by themselves, they also want to make others happy. They may express ideas of being called, of great fortune, or of salvation. The calling to a higher duty often comes from God, and the idea itself is perceived as having been **inspired by God.** If the affective waves are less strong, patients may assume the role of helpers, without reference to a higher duty – helpers against diseases for example. In a broad sense, all such ideas may be summarized as **ideas of happiness,** meaning both the patient's own happiness as well as that of others. Sometimes women do not act like this themselves, but want to be effective through their child, like Mary through Jesus, or through a man, usually a prominent one, whose marriage proposal they await with the feeling of happiness, marked by erotic undertones. Even unreligious persons may express religious ideas. On the other hand patients commonly feel called upon to undertake social or political tasks, wanting to bring justice or eternal peace. The picture of happiness psychosis, like that of exalted euphoria, is complemented by **pseudohallucinatory experiences,** particularly those that include meetings with God or a saint. During these appearances, which are visual, auditory, or both, patients receive their alleged mission. Often they reinterpret actual observations, so that ecstatic ideas of reference develop, replacing their anxious experiences. Thus a harmless remark may be perceived as a calling.

At the height of the disorder, as may also occur in the anxiety phase, slight or more definite incoherence of the thought process may appear, resembling symptoms of excited confusion psychosis. Hyperkinetic traits as seen in motility psychosis may also appear. Furthermore rigidity of postures is not infrequently found in happiness psychosis. Should this seem to look like a lofty gesture, for example, that of blessing, this can be understood psychologically as a consequence of the ecstatic affect. It may also happen that the patients remain nearly motionless for long periods, not saying a word, behaviour that does not seem to make any sense. It could then be suspected that, aside from the ecstatic affect still recognizable in the facial expression, a stuporous state has developed as an independent syndrome comparable to either akinetic motility psychosis or inhibited confusion psychosis. Perplexity with elevated mood could point toward one of the euphorias, while rigid posture may indicate a form of akinesia. In milder form this cycloid psychosis may imitate manic-depressive illness. Both as post-phasic fluctuations and as interim phases lasting somewhat longer, pictures resembling mania or hypomania may be found, characterized by elevated mood, overactivity, and flight of ideas. If specific emotional levels are affected certain forms of pure euphoria other than the exalted form may be imitated, such as unproductive euphoria or confabulatory euphoria.

The affect of anxiety-happiness psychosis has another peculiar characteristic besides its tendency to become excessive. It may mount suddenly and then abate just as quickly. Both anxiety and ecstasy have this **fluctuating character.** One hour the patients may run through the room with the greatest anxiety , while in the next they may calmly reply to questions and perhaps try to prove their calmness with a smile. Again, in a state of highest ecstasy they may assume a lofty posture, preaching to other patients, then suddenly becoming calm and able to participate in the activities on the ward. In exalted euphoria and self-tortured depression we saw affect rise when the patients' ideas were discussed with them. The same phenomenon is found to a lesser degree in anxiety-happiness psychosis. Mood fluctuations are precipitated less through the patient's ideas than other stimuli. Unusual occurrences on the ward, such as taking a blood sample or giving an injection, may cause anxiety attacks; the visit of a clergyman, or another patient's complaining, may produce waves of ecstatic emotions. Basically however, the affect of anxiety-happiness psychosis is labile from within.

Not infrequently this leads to the peculiar phenomenon in which the most severe ideas of anxiety are expressed without inner excitement; the same applies to ecstatic ideas. When the patients smile while expressing their ideas of anxiety, they are generally trying to hide their anxiety; but even when that is known, the contrast of the smile to the content of the ideas is so striking that smiling ought to be impossible. In contrast to this patients may express ecstatic ideas with a mildly displeased face. Here too the unhappy expression probably has the purpose of disguising how important the ideas really are for the patient. But this tendency to disguise shows that the affect is no longer associated with the ideas. The explanation for the remarkable phenomenon is that the patients arrive at both their ecstatic and their anxious ideas while in the extremes of their affect, and, after the usually rapid abatement of the affect, do not correct the ideas but only become rather unsure. Pronounced ecstasy lasts considerably less long than anxiety, so that the contrast with the ideas is more common. A similar fluctuation of the affect is not found in the pure forms. Lability of mood is a common feature of manic-depressive disease, but not to such an extent as in anxiety-happiness psychosis.

The mood fluctuation does not only have to occur between one pole and the center; it can occur between the two poles, even between the greatest extremes of anxiety and ecstasy. Patients who have just expressed the most severe fears may very rapidly assume ecstatic postures, declare that it is all behind them and that now they and the rest of humanity are going to be saved. This type of affect change has an influence on the idea construction in a particular manner. As we have just seen, convictions that arise during a pathological affect disappear only slowly after the affect itself has disappeared. On the other hand since anxiety ideas cannot conform to an ecstatic affect nor ecstatic ideas to an anxious affect, there may be a transition from one to the other during which the ideas are **double-faced.** For example, the patients believe they are approaching a great destiny, full of severe suffering but leading to a great reward. Depending on the mood at that moment, one or the other aspect moves into the foreground. For this type of ideation, religion with its doctrines of punishment and salvation offers a very suitable content. Anxious and ecstatic symptoms are particularly apparent when patients believe that they must die a "sacrificial death" for others. Religious coloration is also often found in anxiety psychoses when there is no recognizable affective swing toward the ecstatic pole. Presumably the affective swing is hinted at and produces the religious content of these ideas. Neither in manic-depressive illness nor in other pure forms of depression do ideas of anxiety and inferiority take on the character of sin and divine punishment as often as in anxiety psychosis.

Despite the polymorphism of the psychosis it should be emphasized that in general the clinical picture is dominated either by anxiety with paranoid and hypochondriacal features or by feelings and ideas of happiness.

Case 39: Helene Hem, born in 1884, first became ill in 1931 at the beginning of her menopause. She became depressed, developed feelings of anxiety, and complained of inner unrest and obsessive brooding. After six weeks of treatment at Spa K., she improved and was discharged. Nine months later she had to be readmitted. This time she complained of lack of will-power, feelings of anxiety, inner unrest, and prickling in the eyes. Her facial expression was rigid. Two months later she was discharged and remained well except for depressive periods that developed every fall and lasted until spring. In 1940 she became ill again, following influenza; this time she was excited and had to be admitted to the Frankfurt Hospital. Here she declared that she was not sick: "I am happier than the happiest human being." She claimed that Mrs. K., who had been following her and had made her nervous had been taken to hell by the devil. The elevation of her mood took on a markedly ceremonious form. She spoke only in monotones, as if she was praying a litany, and sang religious songs with eyes closed. When asked if she was blessed she nodded beatifically. Among other things she claimed: "The most holy one is coming soon. The mother of God has appeared to me. She wore a garment of heavenly blue, stood in heaven and waved to me"; "The day of the last judgement will be tomorrow. You will be astonished at what will happen. Stars will fall from heaven. The moon will no longer shine. The sun will be darkened and the dead will arise. I am happier than ever." She urged the other patients to kneel and await the most holy one. The ecstasy remained at this pitch for seven days and then gradually receded. However a ceremonious elevated mood remained longer. Finally this mood too disappeared. The patient became rational, corrected her ideas and explained that she could not have acted otherwise. She had seen the mother of God and had received the knowledge that she must save mankind. Eight weeks after admission she seemed a bit depressed but was discharged. At home the depression gradually increased; six months later she had to be readmitted after having expressed suicidal ideas. She was very anxious and suspicious, believing that people were talking about her and wanted to do something to her. She also complained of pains in the chest, prickling in the head, as if fire were inside, and prickling throughout the whole body. She cried and moaned a lot, then stood around with no energy. After five weeks she improved somewhat, and was discharged. At home she was well on the whole, although at times somewhat depressed. In 1944 she again became ill, was anxious and believed that people were following her. As she wanted to take her life she was readmitted to hospital. She complained that she could no longer work and that her thoughts were not right. Her whole body felt like fire or prickling needles; she claimed that anxiety was torturing her, but she did not know why. At times she complained vividly, at other times she was almost motionless. After four weeks she was transferred to Institution W. Here at first she was anxious, complaining and hypochondriacal, then she rapidly improved. After three months she was considered to have recovered completely. She had complete insight. At her own request she remained in the institution as a nursing assistant. In March 1946 she was temporarily depressed after she had influenza, but then remained quite normal and she was discharged. Her sister became ill in 1925, three months post-partum. She was anxiously excited, could no longer work, complained of inner unrest, and wanted to kill herself and her child. After this condition had lasted three months, she was admitted to Institution M., where she seemed more affectively labile than depressed. She complained that she had changed internally, felt an antipathy toward her child, and felt as if something were wobbling inside her head. Her deep depression seemed to have disappeared. Her mood normalized quickly and discharge followed.

The happiness psychosis of this patient is very impressive. The ecstatic affect is easily recognized from the ceremonious posture and exalted religious ideas. At the height of the illness visual hallucinations appeared in which the patient saw the Mother of God. Likewise her anxiety psychosis, with ideas of reference and hypochondriacal symptoms, is typical, with its contrast to the happiness phase. Another patient's condition was characterized by rapidly alternating affect. She begged not to be shot dead, but right away said that she was not afraid to die, and even wanted to sacrifice herself in order to rid the world of hatred. Another female patient called herself a sinner, but at the same time offered herself to die as a sacrifice for the sake of redeeming the world and bringing peace to all mankind. A male patient expressed the belief that he would be guilty of the downfall of the world, at the same time hearing a divine voice assigning important tasks to him.

If the ecstatic pole is missing, investigation of the family nevertheless may confirm its genetic presence. In the following case the illness of her brother shows that the patient who had only anxiety phases was in fact afflicted with anxiety-happiness psychosis.

Case 40: Katharine Klin, born in 1905, first became ill in 1936. She no longer wanted to work, believed that people were watching her and were thinking that she had had an abortion. She believed that she had a bad influence on others, declaring that she should be shot. She was admitted to the Frankfurt Hospital because of her increasing anxious excitement. She cried copiously that she wanted to die, the doctor was recording everything incorrectly, the other patients were laughing about her, people thought she was a criminal, and that she was going to be electrified. Alternating with excitement she at times showed stuporous behaviour. After four months she calmed down and was discharged. In 1942 she again became ill, expressing suicidal thoughts, and was readmitted to the hospital. This time she was despairing, explaining that she was a great sinner; for a long time she would not answer questions, but stared into space. Then she would run around in anxious agitation. After three weeks she was transferred to Institution G., where her condition alternated between excitement and calmness continued. Yet she always remained anxious, making suicide gestures while she was either in an excited or an outwardly inhibited phase. During a remission she explained that everything had seemed changed and that she had thought that everything referred to herself. At that time from her facial expression it could be deduced that she had a wrong impression of the environment. After nine months she improved, her mood normalized, and she was discharged. Her **brother** became ill in 1934 after appendectomy and was admitted to the Frankfurt Hospital. He explained that the world had changed and was standing still. Simultaneously he had an elevated mood and he misjudged his environment. He thought a fellow patient was Hitler. Soon his mood became more anxious; he declared that the chimneys had stopped smoking and that it was probably his fault. Then an ecstatic phase followed during which he claimed that rays came from his hands, he had divine power, and he would establish a new religion. Periodically he developed incoherent pressure of speech. He sang, whistled, and made gymnastic movements. He spoke of angelic figures and of a spiritual face that he had seen. Although he appeared outwardly euphoric and laughed, he repeatedly spoke of a sin that he had committed. After eight weeks he calmed down and was discharged. However he returned to hospital after a week, explaining that people had greeted him peculiarly, that a bridge had appeared golden and seemed to belong to a different part of the world. He also claimed that a log on which he sat seemed to be pulsating from below, he felt as if he was swaying. His condition varied between high and low; occasionally he was afraid of being thrown into a pit. After three weeks he again appeared normal and was discharged. Two months later he was admitted to Institution G. in a state of ecstasy; the world appeared to him to be full of miracles. Four months later he was improved and was discharged. At home he remained well for several years until 1940 when he relapsed again. He became very active sexually with his wife, and was overly excited. He said his neck was burning. He was readmitted to Institution G., explaining that there had been terrible noises everywhere, the ticking of the clocks and the crumpling of the pillow had been so peculiar, and there had been a sweetish taste in his mouth. He believed that the Copernican system was no longer correct. He felt rays in his body emanating from surrounding objects. In talking like this he was still anxious, but soon became euphoric and somewhat pathetic. He then stated that he wanted to remain a German. After three months he was improved and explained that he had had curious notions and had heard voices; that it had been a relapse of his earlier mental disease. He was discharged in an equable mood with very good insight.

In the cases presented there were no symptoms which might have led to any other diagnosis than anxiety-happiness psychosis. But overlap with the other two cycloid psychoses occurs quite frequently. I could have presented cases where it could not be determined which of the cycloid psychoses was to be diagnosed. One such an example is two sisters, one of whom had anxiety-happiness psychosis, whereas the other had confusional psychosis. It was not possible to determine which of the two conditions was genetically present in reality. This circumstance should not lead to the conclusion that genetically there was no difference. Furthermore it should not make it easier to just diagnose a cycloid psychosis. In every case the effort should be made to make a more accurate diagnosis. For if in spite of such overlaps the characteristic symptoms can be ascertained the certainty of the correct diagnosis is established. Under no circumstances should the diagnosis of a cycloid psychosis be made if the clinical picture is not clear, rather it should always contain the typical symptoms. Even more so, it should not be permitted to diagnose a cycloid psychosis merely because the illness has a benign course. The course merely confirms the previous diagnosis. Apart from overlaps within cycloid psychoses some others should be taken into consideration. Overlap with **manic-depressive illness** has already been discussed in detail. Differential diagnosis from **affective paraphrenia** will be dealt with later.

Summary

One pole of anxiety-happiness psychosis is characterized by anxiety, accompanied by distrust, ideas of reference, hypochondriacal ideas, ideas of inferiority, and not infrequently sensory illusions as well as occasionally feelings of being influenced. **The other pole** is characterized by ecstatic mood and feelings of happiness, at times accompanied by ideas of reference and sensory illusions. The basic symptoms need not always be in the foreground. There may be pictures with more or less clear traits of manic-depressive illness, confusion psychosis, or motility psychosis. This is an expression of the polymorphism of anxiety-happiness psychosis. The anxious phases are much more common than the ecstatic, so that cases of only anxiety psychosis are frequent, while those of only happiness psychosis are rare. The **duration** of the various conditions is similar to that of manic-depressive illness, whereas a sudden change between anxious and ecstatic mood occurs often in the same phase. **Prepsychotic** traits are often found, hinting at a more severe disorder later, that is, a tendency towards becoming greatly excited, the so-called exuberant temperament (LEONHARD 1970b).

Excited-Inhibited Confusion Psychosis

The **basic disorder** of confusion psychosis lies in the thought process, which becomes **incoherent** during excitement and does not progress during inhibition. The incoherence here is different from that in confused mania; it can no longer be regarded as an abundance of flight of ideas. Flight of ideas do occur, but are not more frequent at lower levels of the disorder; on the contrary, they are rarer. Confused patients with lesser degrees of excitement characteristically keep on speaking of things not relevant to the theme, for instance, of experiences at home or in the hospital not at all related to the immediate topic. The listener does not understand why the patients have chosen a certain topic, but often there is no lack of logic. Thus the doctor coming to see the patient expects the patient to talk about his health, about events in the ward or about a visit he has received; instead the patient talks about experiences he had years ago somewhere else. Why this happens is unclear and cannot be determined by questions. Every question brings on a new story, the connection to the momentary situation remaining open. I call this type of thought disorder **incoherence of thematic choice.** Here confusion psychosis does not imply distractibility by momentary events in the environment.

If the thinking disorder is less marked – mild cases occur frequently – the theme is not entirely abandoned, rather through partial comprehension of its content it determines the theme that follows. For example, if the patient is asked for the difference between a tree and a bush, instead of a logical answer the patient may talk about the berrybushes in the garden at home. If the patient is asked for the difference between to give a present and to borrow something, the patient may talk about presents given on the occasion of celebrations. My term for this is **digressive choice of theme.** The distinction from digressive flight of ideas is that each thought is held on to for a while, whereas true flight of ideas continually produces new notions.

Excited thinking leads directly to verbal excitement; thus **incoherent pressure of speech** is an essential symptom of excited confusion psychosis. On the other hand, when with higher degrees of inhibition thought stands still, verbalizations also stop so that **mutism** arises. It simultaneously hides **thought inhibition** which cannot be immediately determined in a mute person. It can be deduced however from the fact that the patients are only stuporous in the sense that complicated movements requiring thought, are absent. Automatic movements and simple movements in response to commands are preserved. When requested the patients will dress and undress themselves. They feed themselves and go to the bathroom on their own. All movements are somewhat slowed down, most likely because the drive behind the thought is impaired. There is impoverishment of facial and other expressive movements to the extent expected from the emptiness of thought, but there is no rigidity. All this helps in the differentiation of the stupor from **akinesia,** which will be discussed under motility psychosis.

Due to inadequacy of their thoughts patients in the inhibited phase are unable to judge correctly events in their surroundings, thus becoming **perplexed.** The few verbal expressions they offer generally indicate that they do not know what is going on or what it all means. The **perplexity** can often be recognized from the searching, questioning anxious facial expression.

Both excitement and inhibition of thought lead to abnormal ideas. In **excitement** patients easily display misidentification of persons which however is generally transitory and often playful. This is not the same as the peculiar declarations of some schizophrenics that people in their surroundings are acquaintances. Transitory ideas of reference and auditory hallucinations are also common. On the other hand, during **stupor** at first nothing can be discovered about pathological ideation, although the patients may later report their experiences from hindsight. In milder forms when patients are not completely mute but only taciturn, pathological ideation may be observed. The patients no longer understand the connections in the events around them, everything appears peculiar. They look for the significance of the events which they no longer understand. This perplexed questioning for meaning characterizes **ideas of significance.** Incomprehensible events often arouse a fear that something is being planned against them. Thus ideas of significance may turn into **ideas of reference.** This is particularly true when anxiety as an independent symptom accompanies perplexity. In this case ideas of reference appear even with only mild inhibition. Misdiagnosis of schizophrenia should not be made. Ideas of reference and of significance are part of the schizophrenic symptomatology only if more severe inhibition or anxiety is absent. Patients with excited-inhibited confusion psychosis may claim that something is being planned but is hidden from them, everything is disguised, that it is all only a rehearsal, magic is being practiced, the food tasted strange, so many people were coming and going, the cars were going so fast, etc. **Hallucinations** often occur, primarily auditory, but visual and somatopsychic forms may also appear.

Here too some unusual pictures are found due to the polymorphism of the psychosis. If the intensity of the disorder is low, **manic-depressive illness** may be imitated; the patient may appear to be quite manic. If there is **confused mania** it may not be possible to decide whether the phase belongs to manic-depressive disease or to confusion psychosis. If the patients remain approachable, respond to everything, and observe their surroundings despite the confusion, then mania is likely, since confused patients often lose the capacity of taking in their surroundings sensibly. Elevated affect is also evidence of mania. Misidentification of people nearly always points to confusion psychosis. Similarly it may be impossible to differentiate inhibited confusion from **stuporous depression** seen in cases of manic-depressive illness. While depression may point to one and perplexity to the other, the emotional state is often hard to judge in the presence of stupor; severely depressive coloration is not rare in confusion psychosis.

Commonly the affect is more anxious than just depressed, in this way overlap of confusion psychosis with **anxiety psychosis** occurs. The formation of ideas in confusion psychosis takes on an anxious coloration here. Ideas of reference and of significance take on a threatening character. Significant events in the surroundings appear to the patients to be preparations for his arrest or mistreatment. Other symptoms, similar to those in anxiety psychosis, such as ideas of inferiority or hypochondriacal ideas may develop. If the anxiety is not combined with inhibited but with excited confusion psychosis, a type of mixed condition occurs. These anxious confusion psychoses are very common, much more common than confused anxiety psychoses, i. e. cases which are essentially anxiety psychoses but which display traits of confusion psychosis. However the anxiety is generally only transitory and is readily relieved by elevated as well as indifferent mood.

Occasionally a mild **ecstatic mood** develops in inhibited confusion psychosis, recognizable from some remarks made and the joyous facial expression of the patient. In excited confusion psychoses, ecstatic emotional upheavals are much more common. Since conversely patients with happiness psychoses, as we have seen, may become incoherent differential diagnosis at that time may be impossible. In happiness psychoses incoherence is generally only of short duration, and in confusion psychoses ecstasy is less pronounced, thus observation of the course of the condition generally results in a clear picture.

Even so, differential diagnosis is complicated by **rapid affective fluctuation** (which we found in anxiety-happiness psychosis) commonly occurring in confusion psychoses. It is however less a change between anxiety and ecstasy and more a simple lability of the mood between tearfulness and joyfulness. In this form change of affect is particularly significant for confusion psychosis.

Frequently symptoms of confusion psychosis overlap with those of **motility psychosis.** Akinetic symptoms may accompany stupor and hyperkinetic symptoms may accompany incoherence. Traits of motility psychosis may in fact temporarily dominate the picture. Patients may at first present the characteristic picture of excited confusion psychosis, and shortly thereafter the picture of hyperkinetic motility psychosis. Differential diagnosis is then impossible and may even remain in doubt when the whole phase is observed. Occasionally a mixture of opposite poles occurs. In incoherent excitement with pressure of speech, rigidity of posture and facial expression may reveal an accompanying akinesia; in perplexed stupor without change of the perplexity or lack of initiative, purposeless movements may signal the presence of hyperkinesia. Their considerable uniformity reveals the underlying thought inhibition.

The frequent overlap of the symptomatic pictures of confusion psychosis and motility psychosis might arouse the suspicion that they might be two forms of the same disorder. Nevertheless marked differences arise from the statistics. Observation of the relatives shows that disorders within the family are generally of the same type and that there is no indiscriminate fluctuation between the picture of confusion psychosis and that of motility psychosis. In our investigation of twins, as we shall see later, there was a strikingly opposite trend between these two psychoses. – The following are descriptions of two cases of confusion psychosis with other cases in the family.

Case 41: Margarete Hil, born in 1913, became ill in 1941. She was excited, spoke a lot, broke a window pane, and was admitted to the Frankfurt Hospital. Here she was cheerful, loquacious, and displayed flight of ideas. She misidentified people around her and confabulated that she had met the doctor three years earlier in a wine bar. Then her mood became more irritable, alternating between cheerfulness, irritability, and tearfulness. Her thought process became quite incoherent; for example, she said: "I will also pull your feet – yes, a fallen woman – I love freedom – yesterday Kurt stood before the door, he had his heart in the right place – no, we don't need to die..." She gave the doctor a false name and claimed he had had an affair with her previously. She claimed that other people had several names and that a certain nurse was her biological sister. Frequently she scolded in a confused manner. After four months, the excitement receded and she was discharged. She calmed down at home and remained well.

A **sister who is also a subject,** Katharina Mark, born in 1909, became ill in 1949, wandering aimlessly through the city and bothering passersby with her confused speech; she was admitted to the Frankfurt Hospital, where she resisted everything, struck out around herself, and claimed her husband was standing in the garden; her mood changed continuously between irritability and friendliness. She did not reply to questions appropriately, but showed confused pressure of speech: "Deutschland, Deutschland über alles – the Kaiser has a guilty conscience – where is my husband, where is the Kaiser, where is the pastor – who has already seen the good Lord – blessed are they who do not see but yet believe – I want a ring even if it is only made of tin – then I'll take my clothes and go – give me water to wash my hands..." She moaned and cried but soon thereafter her mood became elevated, partially amused and partially ecstatic. On one occasion she declared "I am Hitler", and saluted rigidly. She rapidly improved under shock treatment with cardiazol, and was discharged six weeks after admission.

The two sisters had the typical picture of excited confusion psychosis. One of them appeared to have gone through a brief ecstatic phase, in that she claimed to be Hitler; otherwise there were only mild disturbances of mood, compatible with the picture of confusion psychosis. Another patient considered herself to be Hitler's daughter; her ecstatic phase was of short duration as part of her inhibited confusion psychosis. A description of her case follows.

Case 42: Helene Heus, born in 1914, became ill in 1940, showing extreme anxiety, hearing herself being spoken about and cursed. She was admitted to the Frankfurt Hospital where she was stuporous. She rarely answered questions, in a manner revealing a perplexed anxious mood. She remained in this stuporous state, and instead of answering looked around quite perplexed. Under cardiazol shock treatment she gradually improved. Four months after admission she had recovered and could be discharged. At home she returned to her occupation. After ten months she relapsed and was readmitted to the Frankfurt Hospital. She was again stuporous, spoke rarely, and displayed a perplexed facial expression without anxiety. Rather, she seemed euphoric, her few remarks hinted at a jovial mood. She described herself as Hitler's daughter and explained that this would be made known throughout all Germany. Under cardiazol shock treatment the symptoms receded. At first she was still shy and embarrassed, but later she felt well and her behaviour was unremarkable. She was discharged barely three months after admission.

In the first phase of the patient's disorder, apart from inhibition there was anxiety, in the second phase ecstasy. As most of the time she was mute it was not possible to ascertain her pathological ideas. After her recovery no report of her mental state was available. – The following patient illustrates a bipolar course.

Case 43: Elisabeth Schrib, born in 1989, first became ill in 1932, and was admitted to the Frankfurt Hospital. She was excited, developed pressure of speech, and recited in an affected tone of voice. The mood varied, erotic tendencies appeared, and a hyperkinetic restlessness developed with capering and gesticulating as well as a uniform beating motion and repetition of individual words. She then became nearly mute, appeared anxious, and complained about masks that she saw everywhere; even the nurses were only masks. Excitement followed, this time mostly verbal, with flight of ideas and incoherence. After five months she calmed down and could be discharged. In 1939 she became ill again, going into an "attack of frenzy", necessitating readmission to the Frankfurt Hospital. She cried, screamed, jumped out of bed looking frightened, and developed incoherent pressure of speech. She then became inhibited and perplexed, declaring she wanted to have nothing to do with electrical things. She continued to alternate between incoherent-excited and perplexed-inhibited states. After ten weeks she calmed down. She was only slightly labile in her mood at that point and could be discharged. However she returned six days later, very much inhibited, stating that everything at home had been so strange, so empty. The advertisements on the street had seemed so funny. Her husband had been different, the lenses in his glasses had turned around all of a sudden. "How strange, when I touched him, he was so empty – the people walk so strangely; every now and then there are mix-ups." To her it was a complete mystery how it all had happened. She also had the feeling that she herself and other persons were being manipulated. After four months this perplexed-inhibited state disappeared; she was friendly and open, displayed no abnormality, and was discharged. A **brother** became ill in 1929. He was afraid of being poisoned, and was admitted to the Frankfurt Hospital. Here he was depressed, moaned that he had made mistakes, that his brain had failed, and that he had suddenly seen spirits before his eyes. His friend was persecuting him and his sisters were not healthy as could be seen by the color of their faces. He said people thought he was a homosexual and they described him as a Jew. Food smelled strange, he was distrustful, holding himself back. Then suddenly he recovered with insight into his condition and ten weeks after admission he was discharged against advice. At home he improved further and was well for many years. In 1939 he again became ill and committed suicide before he could be brought back for observation.

This female patient first showed hyperkinetic symptoms; later there were no symptoms incompatible with excited-inhibited confusion psychosis. The last phase is very well described making it possible to recognize the presence of inhibited confusion psychosis with impressive ideas of significance. The patient spoke with partial insight at the time when inhibition was minimal. Her declaration that at times she felt guided fits in with the syndrome of perplexed significance. – Her brother showed confusion with inhibition and anxiety.

In the chapter on anxiety-happiness psychosis I mentioned two sisters of whom one had that disorder, while the other had confusional psychosis. In the following case, too, differential diagnosis would have been impossible if both sisters had not had lengthy religious affiliations. One was a nun, the other was a member of a religious congregation. The religious attitude which they brought with them as they became psychotic partially explained their ecstatic religious ideas to the extent that the symptoms of confusion received a greater weight as an expression of their disorder.

Case 44: Matilde Wig, teacher and nun, was born in 1901. She first became ill in 1936, was excited, claimed to be graced by God, and declared that the Holy Father had given her a mission. She stated that she was clairvoyant, could prophesy, had received the stigmata and her body was changed; since she had angina the marrow was flowing out of her bones. She was admitted to Institution R., where she was very cheerful and excited, spoke without interruption with flight of ideas in a confused manner. At times she shouted, danced, and believed that she had received the wounds of Christ. After six weeks her condition improved rapidly, she became calm, friendly and composed, and was discharged. She returned to teaching, but after three months she relapsed. She again spoke of divine grace, made mysterious allusions, always seemed to be hearing something, and significantly placed her finger on her mouth. She then became very excited, and had to be readmitted seven months after her discharge. At first anxious she was afraid both of the other patients and of dying. She resisted everything, was very excited, screamed confusedly for hours at a time, admitted hearing voices, and spoke of queer bodily sensations. Gradually she became calm, but remained anxious and seemed somewhat confused. She was discharged six months after admission, but at home soon became excited again and had to be readmitted. She spoke of divine signs, heard the voice of the prioress, explaining that it all came from the lower part of her abdomen. Pressure of speech, flight of ideas, incoherence and confused excitement lasted for six months with fluctuations, then the patient calmed down. She remained well for three years. In 1941 she again became ill, was excited, spoke a lot, displayed elevated mood, and was brought to the Frankfurt Hospital. There she ran aimlessly through the ward, did not respond to any questions, and spoke in total confusion. Her mood was generally cheerful, intermittently also anxious. On one occasion in a state of perplexed excitement she screamed "I am full of lice and bugs." Her condition fluctuated for three months, after which she settled down and could be discharged.

Her **sister** Aloisia Wig, born in 1911, is also a subject. She became ill in 1941; she prayed in church with conspicuous pathos, explained that she had offered herself to God as atonement and would die. She was admitted to the Frankfurt Hospital where she was greatly excited from the beginning, got out of bed, hit the nurses, was distracted by all impressions from her surroundings and developed a senseless pressure of speech which included many religious notions; for example: she was in the middle of the cathedral, she must sacrifice herself for the world; she was waiting for him who would come and to whom she would become engaged to be married. Her mood was consistently elevated. Under cardiazol shock treatment the excitement lessened. Pressure of speech continued but gradually became less incoherent and more digressive. She then became mildly inhibited, somewhat perplexed, producing a connection between the name of her doctor and the name of her religious congregation. Finally, three months after admission the psychosis remitted, and she was discharged.

Religious coloration of the clinical picture may of course be the product of anxiety-happiness psychosis, but if the religious attitude already dominated the healthy personality it cannot seriously be regarded as a symptom of the illness. Thus both sisters were given the same diagnosis of confusion psychosis.

In the history of one of the two sisters and in that of some other patients mention is made of **flight of ideas.** Partially these may overlap with a manic state, but more frequently incoherence of the thought process overshadows flight of ideas.

Summary

Confusion psychosis is accompanied in the excited phase by incoherence, and in the inhibited phase by inhibition of the thought process. At lower degrees of excitement only "incoherence of thematic choice" appears. Excitement of thought is connected with pressure of speech, and thought inhibition with verbal impoverishment, even mutism. In the excited phase symptoms include misidentification of persons and often ideas of reference and hallucinations, particularly auditory. In the inhibited phase, perplexexity may be accompanied by ideas of significance and of reference, less often by hallucinations. Frequently the picture is not pure; traits belonging to motility psychosis, anxiety-happiness psychosis, and manic-depressive illness may be present. The course varies: sometimes only one pole of the disorder is observed, sometimes the poles follow one another. Not infrequently there is rapid alternation between the two poles. Frau von Trostorff (1966) found a frequent prepsychotic characteristic in the form of digressive or, on the contrary, retarded thinking within the range of the temperament. We did not undertake more intensive investigations into any other prepsychotic characteristics. In terms of **physique** the pyknic form seems to predominate.

Hyperkinetic-Akinetic Motility Psychosis

The hyperkinesia of motility psychosis represents a psychomotor form of excitement and does not depend on disorder of thought or feeling, as do the excitements of confusion and anxiety-happiness psychosis. Therefore the movement disorders seen in this psychosis do not depend on higher mental activity, but are more or less reflexive, i. e. movements of expression and reaction. The former appear in connection with certain mental states, and the latter because of their pathological exaggeration, called **short-circuit movements** by KLEIST, are immediate reactions to sensory impressions. At times the former predominate, at other times the latter, at still other times both occur equally. Correspondingly the patients gesticulate, i. e. make winking, threatening, beckoning, resisting, forbidding, or encouraging movements, etc. Some of their head and shoulder movements express other mental contents. They may stomp their feet as if they were quite impatient. Their facial expressions reflect joy, sorrow, eroticism, anger, worry, disappointment and much more. With their reactive movements they show hyperkinesia by grabbing their bodies, their hair, or their clothes, shaking their beds, ripping out mattrasses, climbing on chair or tables, rattling the doors, banging on the walls, clutching other patients, and so on. Expressive and reactive movements combine in various gymnastic and dancing displays.

If hyperkinesia is moderate, movements of expression and reaction remain quite natural. With increasing excitement, the movements lose some of their naturalness, and become exaggerated though not greatly distorted. Only in severe hyperkinesia does "grimacing" and a similar disorganized restlessness of the rest of the body occur. Under such circumstances the prognosis in terms of life is often unfavorable since the picture may become one of "fatal catatonia" (STAUDER) or "dangerous hyperkinesia" (NEELE); however the prognosis in general remains favorable. Disorganized expressive and reactive movements point toward a worse prognosis when they appear with only moderate restlessness. This will be discussed under periodic catatonia.

Speech is little affected in pure hyperkinesia. Verbal expressions require preparatory thought and are not increased by the psychomotor excitement. On the other hand unarticulated screaming, often of an expressive nature (for example, howling as an expression of anger) is common. Speech may even be inhibited by movements of expression and reaction, so that the patients appear dumb. WERNICKE described this as **"silent hyperkinesia"**. On the other hand occasionally in motility psychosis short disconnected sentences may be emitted. These may be verbal expressions arising during hyperkinesia without preparatory thought as unarticulated sounds. In such cases overlap with confusion psychosis does not have to be considered.

The opposite pole of hyperkinesia, i. e. **akinesia,** represents a pure psychomotor form of inhibition, recognizable in both cases by the fact that involuntary movements are affected. In perplexed stupor only those activities cease which require preparatory thought; reactive movements are maintained, and expressive movements are only slightly diminished, because impoverishment of thought occurs. In contrast, in akinesia reactive movements disappear and expressive movements become rigid. The patients do not respond to the simplest demands which ordinarily are followed automatically. They no longer get up, no longer get dressed, and stop going to the toilet. They do not ward off stimulation of the skin unless it is strong and deep reflexes come into play. Posture is stiff, the torso, the head and the arms are expressionless and the face is rigid. In incomplete akinesia voluntary movements are still carried out, but akinesia is recognizable from the stiffness of the posture and facial expression and in the slowed reaction time. Thus even in mild cases the differential diagnosis from stupor of confusion psychosis may generally be made easily. Activities based on the patient's own initiative, which disappear in stupor, are also active in akinesia, since impulses perhaps still arising centrally may no longer lead to realization due to the psychomotor disturbance. Occasionally in akinesia persistent postures and holding against objects are present, on the other hand, a general muscular flexibility may be found. Probably this depends on the rest of the condition, particularly the affective element, whether reflexive movements occur or not.

The characteristic picture of hyperkinetic-akinetic motility psychosis is often complicated by traits of other polymorphic psychoses. **Confusion symptoms** are particularly common. Hyperkinesia may be combined with incoherent pressure of speech if the kind of excitement seen in confusion psychosis is present. Misidentification of persons, ideas of reference and hallucinations occur as well. On the other hand akinesia may be associated with perplexity, and as later questioning may reveal, perplexed idea construction. In "mixed" cases thought inhibition may lead to hyperkinesia; as with confusion psychosis movements become more uniform. Akinesia may be combined with excitement of thinking which occasionally breaks through if the akinesia is of moderate degree – incoherent pressure of speech then appears with stiff posture and rigid facial expression.

Not infrequently symptoms of **anxiety-happiness psychosis** accompany motility psychosis. Hyperkinesia may be accompanied by elevated mood and feelings of happiness or by anxiety and anxious ideas of reference. Anxiety and less frequently ecstasy may accompany akinesia. Both are hard to recognize from the rigid facial expression, so sometimes only later questioning provides hints about the affective fluctuations.

Finally there are also pictures reminiscent of mania and melancholia. If cheerful affect accompanies moderate hyperkinesia, differential diagnosis with mania may be difficult. Such conditions develop particularly in the early stages or recession of motility psychoses. In general however the distinction between manic compulsive overactivity and the considerably more primitive hyperkinesia remains clear. On the other hand akinesia may display transition to stuporous depression. – Differential diagnosis from periodic catatonia will be considered in more detail later.

Several cases will now be described; first a patient with hyperkinesia, then a patient who in the course of decades had more phases of akinesia than hyperkinesia.

Case 45: Maria Streub, born in 1914, became ill in 1938. She was in an elevated mood, sang a lot, and showed pressure of speech easily losing the thread. She was admitted to the Frankfurt Hospital where she displayed severe hyperkinesia. She sang, danced, threw herself on the floor, made pleading and other pathetic gestures; she hit out around her, stamped her feet, made faces, blinked with her eyes, hit the doctor on the head and then threw herself into his arms. With vivid movements she exhibited herself at first as excessively cheerful, then impertinent and erotic, and finally angry. With cardiazol shock treatment she improved rapidly; after a short relapse following termination of treatment she became calm and remained calm for the next three months. She maintained a lively, cheery temperament which she had had prior to her illness.

Case 46: Käthe Dint, born in 1897, first became ill in 1916. At times she spoke a lot and at other times not at all. She was admitted to Institution M. displaying flight of ideas, haughtiness and impertinence. She was diagnosed as a psychopath, and was discharged after two months. At home her condition continued to vary. In 1919 she was readmitted to Institution M. where she was again lively and talkative, intermittently mildly in a bad mood. In 1920 a similar condition necessitated her return to the institution. In 1923 she became more severely ill. At home she was related to have been excited, in Institution M. she was at first severely inhibited. She lay rigidly in bed, kept he eyes shut, resisted passive movements, was incontinent of urine, and had to be tube-fed. After a few days she suddenly became excited, jumped around, climbed up on the window, and spoke no sensible word, but screamed in great distress. Then a state of rigidity with catalepsy followed. This alternating behaviour took place several times. Five months after admission she was considered improved and she was discharged. At home she continued to fluctuate in a mild form. In 1925 she was readmitted to Institution M. and once again was rigid, motionless, and did not speak. After a few weeks she became excited, ran about naked, and washed herself for hours at a time. Then she calmed down, explaining that she had misrecognized her surroundings and that during her inhibition she had experienced many terrible as well as beautiful things; on one occasion she had sworn five oaths, on another had heard voices of angels. Three months after admission she recovered and was discharged. At home she remained unremarkable for a year, but in 1926 she was admitted again. At times she was mute and akinetic, at others more normal. She again recounted her experiences during the akinesia; on one occasion she had had the feeling that she was the universe and wanted to turn off the air in order to become a human being again. After six months of fluctuations between inhibition, excitement and reasonable behaviour she was discharged, having made a rapid recovery. She was normal for a year, industrious and helpful. In 1927 she was again excited; she wanted to jump out of a window to put an end to her life of sin, and had to be returned to Institution M. At times she was stuporous, at other times cheerful. For most of 1928 she was in a "catatonic stupor", although intermittently friendly and open. In 1929 she went through a long period of akinesia, after which she slowly recovered. In 1930 she was "mentally well and cheerful". Mild depressive phases occurred, but she was discharged in good condition. She worked well at home and had many interests. After six months she again became depressed and complained of bodily sensations. In the Institution she was akinetic, hid under her blanket, and was mute. She clenched her teeth and kept her eyes closed. She then became excited, began hitting out around herself, and jumped about. Three months after readmission she had improved again and was quite unremarkable. She explained that during this last phase she had thought that the world was coming to an end. She was readmitted during the same year, 1931, lying down stiff, motionless, and had to be tube-fed. This condition lasted a few weeks, then mild fluctuations began. In 1932 she was discharged. At home she at first behaved normally, but was returned to the Institution after a few weeks because she had stopped speaking. Soon after she was talking again and explained that she had thought she was in a different world, either a darker one or a lighter one. She then alternated between mild states of inhibition and excitement. In 1933 she was discharged. She remained well for several years; she seemed normal except for periods of moodiness. In 1936 she again spent two months in the Institution in a "catatonic stupor". She then remained well at home for eight months. In 1937 she returned to the Institution with stupor, rigid posture, and fixed facial expression. The akinesia lasted for several months, again necessitating tube-feeding. The patient then became alert and active and was discharged. She was readmitted in 1939, excited and running around; she then became stuporous. In 1940 she alternated between being stuporous and being overactive, lively, and approachable. Although she still talked a lot in 1941, she was discharged. Five months later she was admitted to the Frankfurt Hospital where she was mute, but strongly resisted having her blood tested. She then became deeply akinetic, lay rigidly, kept her head lifted off the bed, and had to be tube-fed which she permitted without resisting. After a few weeks the akinesia began to recede, but soon returned. Three months after admission she became well. She was vivacious, talkative, had a lively facial expression, joked and wrote natural, emotional warm letters to her family. No personality change could be found.

This patient is of particular interest because in the course of more than two decades she had many periods of severe akinesia which were understandably interpreted as catatonic; but 25 years after her first psychosis she displayed no defect at all. On the contrary, after the disappearance of the akinesia she experienced in the Frankfurt Hospital in 1941 she was remarkably lively, sympathetic, and in good contact, i. e. of light hypomanic temperament.

I will now present a patient with overlapping symptoms of motility with confusion psychosis.

Case 47: Paula Tom, born in 1921, first became ill in 1938 and was admitted to the Frankfurt Hospital. She displayed pressure of speech, flight of ideas, and incoherence. Her mood was cheerful, but easily shifted to tearfulness. Although her motor system was active there was no real hyperkinesia. After ten weeks she improved and was discharged. In 1941 she again became ill and was readmitted to the Frankfurt Hospital. This time there was severe hyperkinesia, she hit out, turned her bed over, gesticulated, grimaced, rolled about, and spoke only disconnected single words and sentences. Under shock treatment she became quieter and similar to her condition at her first admission. She developed pressure of speech, flight of ideas, incoherence, and was cheerful in alternation with tearfulness. A relapse of hyperkinesia occurred, but lasted only a few days. Finally she calmed down and was discharged six months after admission. A **sister** was in the Frankfurt Hospital in 1927 at the age of 17. She displayed hyperkinesia with screaming, kicking and rolling about, and many gestures and facial expressions. After a few days she became quieter and was taken home by her family. At home she remained well.

In this case the picture in the first phase resembled that of confusion psychosis. In the second phase the pure picture of hyperkinetic motility psychosis became evident. The hyperkinesia was severe. Even more severe was the sister's condition. It is not uncommon for it to last only a few days. Motility psychoses are often of short duration, but may recur frequently. WERNICKE described a form of motility psychosis recurring at menstruation.

Two cases follow showing overlap with anxiety-happiness psychosis. In the first case there was anxiety with mistrust, in the second the illness began with an ecstatic phase.

Case 48: Elisabeth Dor, born in 1908, became ill in 1942. She was anxious, believing somebody wanted to poison her, and she was admitted to the Frankfurt Hospital. There she was perplexed, looked about anxiously, and asked, "What am I accused of?" She distrusted her husband believing that he had incited other people against her, that he had been unfaithful, and that he wanted to get rid of her. She became increasingly restless, whined, jumped out of bed, ran around, wrung her hands, gesticulated continuously, held on to other patients, and climbed up on the window sill. After this excited condition had lasted three months suddenly she became akinetic. She was quite motionless, kept her eyes closed, had to be fed during which she barely opened her mouth; when she was brought to a sitting position she let her head fall forward. The phase continued four weeks, then excitement returned, which however stayed within moderate limits and gradually abated. She was well six months after admission, happy, lively, industrious, and could be discharged.

Case 49: Maria Trös, born in 1923, became ill in 1941. She had an elevated mood, seemed ecstatic, claiming that she had heard the voice of the Mother of God. She wanted to die for Germany. She was admitted to the Frankfurt Hospital where she laughed and cried, continued to express ideas of happiness, but from the start was predominately hyperkinetic. She whistled, sang, did gymnastics, threw her feet in the air, gesticulated with her hands, hit out around her, and made many mimic movements. This condition disappeared after three weeks, leaving her calm, cool and collected, and she was discharged five weeks after admission. A **sister** had a "similar disease" during adolescence. For a while she had been "crazy", confused and excited. She recovered without having to be hospitalized.

In case 48, apart from the accompanying affective symptoms, the course of the disorder is especially impressive in that severe hyperkinesia suddenly turned into extreme akinesia.

In all cycloid psychoses, if the disorder becomes very severe, **clouding of consciousness** may occur. In the following patient there is a hint of this in that visual hallucinations appear during an acute phase of akinetic motility psychosis. Similar to this are the **dreamlike symptoms** described in detail in case 46.

Case 50: Friedrich Hoff, born in 1901, became ill in 1940. He was quiet, hardly spoke, did not react to questions, wanted to cry but could not; he was admitted to the Frankfurt Hospital. There he moved very slowly, did not speak, and became gradually more severely inhibited until he was completely rigid, resistant, with an expressionless face. The akinesia lasted for four weeks, then receded rapidly. He explained then that during this period of time he had always had anxiety and had been afraid that something had happened to his family. He had also seen many pictures. On one occasion a whole trial took place in which he was condemned for the reason that he was no longer of any use to the world. He then came back to normal in mood and psychomotor function, and was discharged.

The akinesia in this patient was of very brief duration. Apparently both hyperkinesia and akinesia if severe tend to have a short course. Furthermore patients with "malignant hyperkinesia" recover rapidly if death from exhaustion is prevented.

Summary

The **excited pole** of motility psychosis is characterized by restlessness composed mainly of expressive and reactive movements. Likewise in the **inhibited pole,** reactive and expressive movements are affected; voluntary movements only in so far as they contain the psychomotor elements. In milder cases the disorder is recognized in the rigid posture and facial expression despite the continued presence of voluntary movements. In this way mild cases of akinesia can be differentiated from the stupor of confusion psychosis. Incoherent pressure of speech often accompanies hyperkinesia. Incoherence of the kind where disconnected words are blurted out with interruptions is, in fact, typical of motility psychosis. However the presence of pressure of speech points to an element of confusion psychosis. A circular fluctuation of the psychosis between the two poles is quite common; **akinesia** is rarer than **hyperkinesia.** On the other hand the duration of akinesia is often longer and may last for months whereas hyperkinesia lasts less long, usually only a few weeks. In terms of **physique,** without more precise investigations it can only be said that the leptosomic element is considerably more prevalent here than in confusion psychosis. **Prepsychotic** signs frequently manifest a "temperament of motility" with abundance of expressive movements and at times a dance-like nature (VON TROSTORFF, 1966).

The Unsystematic Schizophrenias

Systematic and unsystematic schizophrenias have essentially nothing to do with each other. The common name is justifiable only in terms of tradition, because since KRAEPELIN and BLEULER all endogenous psychoses leading to defect have been grouped as schizophrenias. The deep parallels of the unsystematic schizophrenias are much closer to the cycloid psychoses than to the systematic schizophrenias. This relationship is above all emphasized by the fact that each of the cycloid forms corresponds to one of an unsystematic schizophrenia. There are relationships between anxiety-happiness psychosis and affective paraphrenia, between motility psychosis and periodic catatonia, and between confusion psychosis and cataphasia. The differential diagnosis is often difficult. On the other hand there rarely is a problem in differentiating between a systematic and an unsystematic schizophrenia. Not only are the symptomatic pictures completely different, but their course as well is quite different. The systematic forms display an insidious onset with a gradually progressive course, while the unsystematic forms may go into remission or may even be clearly periodic. Periodic catatonia may produce as many attacks as manic-depressive illness. Bipolarity, too, is characteristic of the unsystematic schizophrenias.

Affective Paraphrenia[5]

Among the paraphrenias described by KRAEPELIN there is **paraphrenia systematica,** characterized at first by ideas of reference which gradually progress into a more severe picture with hallucinations, errors of memory and delusions of grandeur. Affective paraphrenia has a clear relationship to this clinical picture although the latter cannot be called "systematic" in the sense that I use the word. In my opinion "systematic" implies involvement of the brain, whereas for KRAEPELIN there is systematization of delusions in his concept of systematic paraphrenia. If only milder cases are considered, affective paraphrenia also corresponds to the progressive reference psychosis of KLEIST. However, not only the delusional element should be emphasized, but the affective disturbance is of special significance as well. This disorder establishes the relationship with anxiety-happiness psychosis and is often the nucleus for the diagnosis in chronic states. Pathological affect, which SPECHT

---

[5] May also be translated as "affect-laden paraphrenia"

(1901) described in paranoia, is decisive for affective paraphrenia, while it hardly plays a role in other forms of paranoid schizophrenia.

At the onset the affective disturbance is often very obvious. The clinical picture may be dominated by **anxiety** associated with false ideas similar to what we have seen in anxiety psychosis. Ideas of reference and hallucinations accompany the anxiety. In many cases however from the outset the thinking disorder may not be entirely derived from the anxiety. The ideas of reference may be illogical, perhaps even absurd, and thus thereby hint at the beginning of defect. The hallucinations may at the beginning be not only threatening or humiliating, they may also contain simple scoldings or indifferent messages. In cases with hypochondriacal symptoms, in place of straight fear or straight **misperception,** hypochondriacal hallucinations related to the outside world may come to the fore. Patients may show a doubting attitude similar to that which may exist in anxiety psychosis, but the prognosis is worse when the patients are convinced that external influences are operating on their bodies. The greater the anxiety, the more it can be blamed for any abnormality of ideas. If however the abnormality is great despite relatively little anxiety, then affective paraphrenia should be suspected rather than anxiety psychosis. The paraphrenia may also start with **ecstatic state,** leading to confusion with happiness psychosis. Patients' hallucinations of divine appearance or other forms of calling are not grounds for an unfavorable prognosis. Happiness psychosis may also be associated with ecstatic ideas of reference. On the other hand, evidence for beginning affective paraphrenia would include hallucinations losing their ecstatic quality and ideas of reference assuming illogical forms.

The affective element of affective paraphrenia changes in the course of the disorder. Anxiety gradually becomes **irritability;** ideas of reference lose their anxious content and develop into hostile reinterpretation of events in the patients' surroundings. This may be called an **irritated reference syndrome;** frequently but not necessarily other abnormal ideas and hallucinations may be associated with it. There are also cases where right from the beginning there is more irritability than anxiety. If the affect is at first ecstatic, it may remain so, although its depth will decrease while elaboration of ideas increases. The expansive character of ideas (to be discussed below) may still be connected to the ecstatic coloration, although it is sometimes found later as a defect phenomenon in anxious or irritable forms. As in anxiety-happiness psychosis both affective poles may alternate, and ecstatic ideas may be mixed with those of the irritated reference syndrome.

With the further course of the disorder the picture is no longer dependent on the trend of the original affect. Often there is a predominance of an unhappy irritable or a joyous elevated mood throughout the whole course, but the pathological ideation becomes fantastic in the later stages. Hallucinations may develop massively in all areas; the patient may be disturbed by voices and bodily sensations; visual phenomena may appear; olfactory and gustatory hallucinations are not rare. Frequently misidentifications occur. Persons in the surroundings are given names and positions which are incomprehensible to the normal observer since there are generally no objective relationships between the real persons and the identity given them by the patients. The new identities named are often highly placed and well known personalities. Furthermore in later stages ideas of grandeur are often found. These are sometimes accompanied by errors of memory. All this points out that the affective disorder is accompanied by a profound change which above all impairs the ability to think logically. Such a fantastic picture appears to resemble what we will later meet in the systematic form of fantastic schizophrenia. Yet the differentiation remains possible.

KRAEPELIN claimed that in cases he called paraphrenic the total personality remained more intact than in other schizophrenias. That may be so in mild but is unlikely in severe cases. Patients may become very dull, losing interest in everything and reacting with indifference to their institutionalization. The personality is no longer preserved. It is true that patients with affective paraphrenia, later than most other schizophrenics or even never, may eventually incur severe damage to their personality. Something else is more important: not whether the affective life underlying the personality is good or bad, but the relationship of the affect to the thought disorder. Just as ideas at the onset are either anxious or ecstatic, they always maintain an affective background. No matter how dull the patients have become, their ideas are always presented with affect. If this is not recognized right away, one need only contradict the patient to observe an immediate increase of the affect. In these circumstances they may speak of their persecution with severe irritation, or of their self-elevation with enthusiasm. Thus it may be recognized that the **pathological affect** still determines the direction of the disease, although the fantastic nature of the ideas may prove that the defect affects the thought process. This predominance of the affect in the ideas justifies my speaking of **affective paraphrenia.** Since emotional reactions to normal stimuli are decreased, but remain strong in connection with the pathological ideas, the impression is gained of a rigid, one-track affectivity no longer capable of modulation.

These patients clearly differ from patients with systematic fantastic schizophrenia in the manner with which they present their ideas. The latter always present their ideas without any profound affective participation; they speak of ideas of persecution in the same tone of voice in which they speak of their ideas of grandeur. When contradicted they may perhaps become superficially upset, but no profound agitation develops. It is safe to joke about their fantastic idea construction, while with affective paraphrenics it is easy to precipitate the same reaction as in paranoiacs with whom, as we will see, affective paraphrenics have a relationship.

With modern **psychopharmacological treatment** the excitability of patients with affective paraphrenia is diminished. They are then at times willing to accept contradictions, still retaining however an inner tension that is easily recognizable. Sometimes they deny statements that they previously had made when they were more delusional. This too points to the persistence of inner affect. Since the latter has decreased as a result of medication it no longer is expressed without inhibition, but it is still sufficiently present to prevent the ideas from being brought into doubt during discussions with the patient, making it impossible to have a satisfactory conversation. It must be admitted that generally speaking with modern drug therapy the diagnosis of affective paraphrenia is more difficult than that of other endogenous psychoses. The medication attenuates the affect, thereby pushing back the most important symptoms for the correct differential diagnosis.

Apart from affectivity, the differentiation of fantastic ideas expressed by patients with affective paraphrenia from those expressed by patients with systematic fantastic paraphrenia is variable. Rarely are the paranoid delusions in affective paraphrenia quite as obvious as those found in the systematic type. Hallucinations may be present, but they are not very pronounced, or they affect only one sense, such as hearing, while somatic delusions are not observed. Delusions of one kind or another may be less well developed compared with hallucinations. Alternatively one symptom may be in the foreground while other traits may only be intimated. These characteristics of the hallucinatory paranoid picture may help to differentiate affective paraphrenia from the systematic forms, but the decisive factor remains the affective element.

Severe cases of defect are much more common in women than in men. Of several psychoses in one family sometimes the male patient has a simple paranoid while the female has a fantastic syndrome. I cannot determine if this might be explained psychologically on the basis of a man's stronger logical tendency offering more resistance to the disease, or if the disease stops progressing earlier. Affective paraphrenia is altogether more frequent in women. Overall the fantastic symptomatic form develops in about one third of the cases of affective paraphrenia. In another third the pathological process comes to a halt in the transitional phase between the reference syndrome and the fantastic picture. As variable as single symptoms may occur in the fantastic stage, one or another fantastic trait may give an indication in this direction without the disorder having passed beyond the transitional phase. In the final third, affective paraphrenia stops progressing while still presenting a simple paranoid picture. In that case there may be a chronic irritated reference syndrome or, if the basic mood is more elevated, a chronic megalomania.

If the pathological process stops still earlier delusions of persecution or of grandeur may be found, i. e. the picture of **paranoia** as described by KRAEPELIN. Since the affective disorder of paraphrenia may tend toward both poles, delusions may also readily develop in both directions, thus delusions of persecution and grandeur may appear simultaneously. Characteristically delusions of grandeur in KRAEPELIN'S paranoia often have an ecstatic coloration; such patients often claim to be prophets or healers, such as we find in affective paraphrenia. Therefore I believe that KRAEPELIN'S cases of paranoia belong to affective paraphrenia. While other forms of schizophrenia produce no similar pictures, in affective paraphrenia are found all degrees of severe schizophrenic forms to the mildest cases, which if considered by themselves would hardly appear to be schizophrenic. This hypothesis was demonstrated remarkably clearly in the **Stot family,** as described by me with its 18 cases of affective paraphrenia (LEONHARD 1950). This approach does not however touch the psychopathic forms of paranoia, which appear to be derived from a psychopathic constitution and include most cases of **querulant delusional disorder.**

The dual nature of the delusions which we have already observed in paranoia is also characteristic of the later stages of affective paraphrenia. With the progression of the disorder to the fantastic stage, simultaneous delusions of grandeur and persecution are nearly always found. In these later stages some patients, in conversation with them about their ideas, may become more easily irritated, others more enraptured.

The variability of the symptoms in the course of the development of affective paraphrenia shows the **unsystematic character** of the disorder. This is confirmed by the not infrequent admixture of traits of related psychoses. In the paranoid anxiety syndrome symptoms of confusion are not only common in the acute phase, but also occur not infrequently in the end stages. Such paraphrenia with confusion is reminiscent of cataphasia. Catatonic traits too may appear in the framework of affective paraphrenia. Furthermore patients with pressure of speech and flight of ideas are not rare, showing a manic element. Again this occurs both in acute and end stages. On the other hand melancholic symptoms occur only in the acute stages; they are absent in the end stages. In relation to the systematic schizophrenias there is not only a similarity with the fantastic form, but a confabulatory picture may occasionally be seen as well.

The course of affective paraphrenia is partly chronic, partly remitting. I shall illustrate the picture with cases of psychotic relatives. First a case with milder defect.

Case 50a: Luise Kin, born in 1888, first became ill in 1922. She believed that she had committed a crime and would be arrested by the police and beheaded. She was admitted to the Frankfurt Hospital where she stated that people had treated her strangely and had shouted insults at her, that she had been written about in the newspapers, and that she had probably been watched by a man when masturbating. Besides being anxious she was sullen and resistant. She was sent to a care home but returned to hospital in 1923. She still awaited a sentence of death for having acted immorally and claimed that people had been watching her and had greeted her strangely. This time she was more clearly irritable. She remained in hospital throughout 1924, working in an office where she was industrious but lacking initiative. In 1925 she was occasionally irritable, referred harmless remarks made by others to herself and wanted to give herself up to the court. At that point she was transferred to Institution H. There she continued to express her ideas of reference, the police and the court were her enemies – for this she had "very many secret proofs". She became excited when she spoke of her ideas, but otherwise was friendly and somewhat dull. In 1926 she was a bit morose, but was able to work in an office. For a while she received home care, but in 1928 she had to be readmitted because she scolded a lot. Soon thereafter she was placed in a different home. However in 1930, being generally dissatisfied as well as impertinent and haughty, she was again returned to the institution. During the following years she was friendly most of the time, but became offended and upset at the slightest provocation. Then it appeared that she still maintained her old ideas of reference; the house was being prepared for court proceedings involving the masturbation and her death sentence was to be pronounced soon. In 1938 she said it was terrible of the police to keep her in an institution for years instead of having immediately questioned her about the charge. Furthermore various daily activities gave rise to paranoid reinterpretation; she was being given bad food and accused of staying in bed out of laziness. She claimed, "every day more impudence". When I examined her with SCHWAB in 1939 she had to be questioned very cautiously since she easily became angry. She spoke of alleged persecutions and of "much irony, mocking, and contempt", as well as "impudence and meanness". She cited as an example that the nurses scratched their heads shamelessly and told her many things "around the back of the flower". She still insisted that court proceedings and a death sentence were being arranged for her.

A **brother,** Rudolf Kin, born in 1879, became ill in 1912 and was admitted to the Frankfurt Hospital. He was anxious, trembled, claiming that the Jesuits were persecuting him and that a voice had prophesied misfortune for him. Then a voice told him that he was the saviour of mankind. At times he cried out for help, at others he lay stiff in bed. Having calmed down he was discharged, but was readmitted three months later after he had destroyed a room at home. He spoke quietly and with affectation; then he was transferred to Institution H. In 1913 he assumed military postures, claiming that he was being given wrong clothing; he was very long-winded. In 1914 he was very irritable, flailed his arms, jerked his shoulders when excited, and had a way of expressing himself with many errors; for example, he said, "I am not for the consumption of property...", and asked for his "under-rights". At times he would scream "stupid fools" excitedly to the wall. Arguing with voices, aggressiveness, and peculiar verbal utterances were repeatedly mentioned. Once he remarked, "I lay a mentally complete claim on my thoughts. What I say is education. I am totally oriented." In 1915 head-shaking was observed, a defence perhaps against the sensory hallucinations. In 1916 he accompanied his head-shaking with pushing motions of his arms. He said, "my thoughts are high German, but the thoughts concerning my discharge are English, French and low German." He claimed that his nephew had been there and had left him with an English, French and German thought. He scolded a lot and enraged, screamed at the doctor. He died of heart failure.

A **sister**, Therese Ju, born in 1885, became affected in 1932, talking of an allegedly wealthy uncle in America. Gradually her ideas became more fantastic. She claimed that Hitler was her husband's step-brother, that her sons were important men, and that a cousin was the "main guardian of Germany". She claimed her husband had founded the Labor Service for young girls to receive a moral upbringing. At first she managed to stay at home, but in 1941 she became more excited, talking to herself "in ecstasy", and was consequently admitted to the Frankfurt Hospital. There she claimed that her husband had been with Hitler when they were both customs officials on the Dutch border and that she had known Hitler as a child. A Doctor K., member of an organized band, had performed operations on her mother. "My mother was shown her own womb so that she would have more children." She spoke with a great deal of affect. When she talked of the great deeds her relatives had performed with Hitler she became enraptured and quiet ecstatic. On the ward her haughty manner was conspicuous. After five weeks she was transferred to Institution W. where her fantastic and paranoid symptoms continued.

In this patient the affective paraphrenia progressed only to the level of a simple paranoid picture. The delusions were purely persecutory, once the psychosis had been ushered in by anxiety. In an irritable stage she continuously reinterpreted events in her surroundings and referred them to herself indicating the presence of the **irritable reference syndrome** mentioned above. It may be noticed that there were hints of systematization. She maintained with anxiety that she would be prosecuted for masturbation and she continued to claim with increasingly intense excitement that the prosecution should not have been delayed. Hospitalization always brought her back to this complaint. Her sister presented a fantastic confabulatory picture without hallucinations. Her **brother** had cataphasic traits, also seen in the other unsystematic schizophrenias, but the fantastic traits are still clear. What despite their differing pictures unites all these psychoses is the **affect** associated with the delusions. The patient and her brother became extremely irritated when they presented their complaints, while the sister in expressing her ideas became frankly ecstatic.

The next case in a somewhat different form showed a mild picture of affective paraphrenia with a more definite systematization than the previous case.

Case 51: Antonie Daf, born in 1874, had always been sensitive and suspicious. In 1907 she believed that people had been sent to spy on her. In a popular theatre which she often attended the director always watched her during the performance. All kinds of other curious perceptions strengthened her belief that the people from the theatre wanted to do something with her. Other members of the audience bothered her by crowding around her seat. The director's wife watched her at certain points in the play. She began to hate his wife and wrote many letters to the director, finally accusing his wife of unfaithfulness with a Jewish lawyer, with whom she allegedly saw the wife in a photograph. She was admitted to Psychiatric Hospital M. in 1912. Here she said the theatre director S. had greeted her conspicuously, pretending to be writing while in his box seat in order to give her signals to write to him. When he performed and when he took his bows he always looked at her. From this she took it that he was becoming increasingly interested in her. On one occasion the director saw her and became so excited that he had to wipe the sweat from his brow. Meanwhile his wife watched her closely. He was not always at the theatre, but when he was, the theatre was twice as beautiful. His wife tried to hold him back and tried to block her out of his field of vision. For this reason he looked at her with the saddest of eyes. Then he began to put advertisements in the newspaper which referred to her. She recognized that these were meant for her because they were so nicely composed. In order to keep on seeing him she went to the theatre so often that her income from her activities as masseuse and embroiderer became insufficient. She said that her profits decreased because she was being exposed to harassments. Men were coming to her with no money, laughing contemptuously or taking off their clothes in order to annoy her. At first she did not know why she had these difficulties. However when people in the theatre during the intermission watched her and knocked against her she knew it was all because of the director. She therefore wrote to him repeatedly asking him to put a stop to it all. A man followed her for a while, and after she wrote to the director about him, the man disappeared. From that she deduced that the man had been working for the director. Although he did not reply to her letter she thought he had invited her via the newspaper advertisement to go and see him, so she went to his office. There he turned away from her as if she were poison. Because she continued to write to him she was ordered by the police to stop. Since she could not abide by this order she was taken to an official doctor who told her that the director thought she was crazy. She continued however to deduce from the newspaper advertisements that he was trying to tell her that he loved her. He even hinted that he wanted a divorce. Since he could not get a divorce because of his child, she felt obliged to tell him he was not the child's father. She had once seen the child in the theatre and it had resembled lawyer K. whose housekeeper was her girlfriend. Later she had seen a photograph of the lawyer with the director's wife. Finally the police took her to the hospital. There she noticed a peculiar smell. At home, too, she had frequently noticed curious odors. After two weeks she was transferred to Institution E. Here she protested against her admission and wrote to the theatre director asking him to free her. She was discharged after four months. In 1926 she was admitted to Hospital S. because of abdominal pains. She said that the director S. had loved her for many years and wanted to marry her, but was prohibited from doing so by his relatives. She was moved to the psychiatric ward. She continued to relate her ideas. After her discharge in 1912 she stated that the Catholic masses were no longer being conducted properly. She then wrote to a Cardinal who put things back in order. Her influence had led to a German being named Cardinal after the previous one had died. In 1917 she was told that S. was in Weimar, but that was a lie since she had always seen him playing a role in the theatre. "All his spies tell him when she goes to the theatre and then he acts for her because he loves her." He often came to her as the mailman or as a pedlar. At first she did not recognize him, but she did later. He tried to meet her in other disguises, too. As before he continued to communicate with her through newspaper advertisements. Finally she felt it was all lasting too long, and she wrote an energetic letter; but she received only unfriendly advertisements as a response. At that point she tried to sue for damages for the lost years and went to all the offices up to the Minister of Justice, but was always turned away. Meanwhile the director had decided to marry her and was only waiting for her to receive recognition so that she would be of his social level. The Pope had been so pleased with her ecclesiastical suggestion that he was prepared to give her recognition. However Cardinal B. hid the letter. She had known since 1918 that even God wanted her to receive her rights. In this connection she remarked that right from the beginning, about 1907, she had felt influenced by telepathy. In 1923 she was nearly poisoned. The family in whose home she had rented a room had put something in her food making her ugly and shapeless.

Daf spoke in a lively manner with adequate affect and a smattering of humor. If the improbability of her story was pointed out, she became angry and said, "Now you are being stupid." She would then stop talking. She refused to go home and accept a job, rather she would wait until Mr. S. at last would marry her. She was transferred to Institution E. There she wanted to hold the institution responsible for a disorder of her wrist. In theatrical performances she always recognized director S. in one role or another. She also saw him in people who passed by the window and heard his voice on the radio. She liked to stir up trouble and defended the interests of other patients. About 1935 director S. was replaced in her fantasies by the movie actor Hans Albers, whom she wanted to marry. On one occasion she saw "ultra rays" being directed against her nerves. On another she saw a cross in the church. Every Sunday she wrote to "her Hans Albers", otherwise she was industrious. When B. SCHULZ and I examined her in 1939 she maintained all her ideas (SCHULZ and LEONHARD 1940). Apparently she still meant director S. when she referred to Hans Albers. She had been told that the director had died, but she knew that he really was Hans Albers. Although she was 65 she still wanted to marry him because "it was true love". Whenever she was unhappy he made announcements on the radio to let her know that he was still there. Her manner of relating her story was lively, talkative, and open. She was somewhat digressive, without actually displaying flight of ideas. Her gestures and facial expressions were natural. Her mood was good and she was happy in her certainty that director S. would soon be coming to liberate her. There was no evidence of flattening of affect and her whole manner was a bit hypomanic.

Several relatives were mentally ill. I could only find out details about her **mother's brother,** Mathias Halb, born in 1840. He was admitted to Institution M. in 1872, having already been ill for six years. "The condition began with complete mental confusion, at times leading to frenzy, so that his wife had to get out of his way." In the institution he was agitated, broke a window with his bed pan, and tried to escape. Otherwise his expression was empty and indifferent. When questioned he gave many wrong answers. He often had to be secluded because of his uncontrolled agitation. This state of affairs continued during the following years. In 1875 the medical record states: "foolishness, sits with a stupid half-smiling expression all day, quite calm, apathetic, and non-participatory. He responds to questions with silence or talks nonsense quietly." In the years that followed there are repeated reports of his ripping up his clothes and bedding. In 1879 he was taken home in the last stages of pulmonary tuberculosis.

From the clinical picture of patient Daf given in detail here one can certainly speak of a true **paranoia** in KRAEPELIN'S sense of the word. In any case I could find nothing directly contradicting this interpretation. Her replacing director S. with Hans Albers in her later years is explicable psychologically. When she learned that S. had died, she had to find a replacement. More remarkable is that she had earlier believed to have seen S. in various disguises. This did not take the form of true misidentification because it was always later that she realized that the visitor might have been S. Of course her ideas were at times quite fantastic; for example, her expectation of personal recognition by the Pope, also her belief that the Cardinal was against such recognition and had therefore intercepted her letter. Similar ideas are found in many cases of paranoia. Nor do I see any special significance in the occasional reference to olfactory sensations. Since Daf, as she claimed, had always had a sensitive sense of smell, hers were not true hallucinations, but rather reinterpretations of actually perceived odors. Otherwise there were no hallucinations.

In the course of her disease over several decades Daf displayed **systematized delusions** with preserved affect, preserved activity, and preserved intelligence, i. e. a preserved total personality. Thus the criteria which KRAEPELIN gives for his definition of paranoia were met. Systematized delusions involving director S. were present from the onset to the end. When the patient began to express plans for religious reform, a new delusion seemed to be developing. It is also possible that she had a mild ecstatic fluctuation leading in this direction. These ideas were however quickly incorporated into the old delusional system. As a reward for her suggestions she expected recognition from the church which would elevate her to the social level of director S. to permit their marriage to take place. In this way a connection was established. On the whole the delusions were more expansive than persecutory, since they were dominated by the belief that she was loved by S., who stood far above her socially. The ideas of persecution however were not a secondary reaction to her lack of success. Rather the self-references and reinterpretations show that delusions of persecution had an independent significance in the pathological process.

The inclusion of this clinical picture in affective paraphrenia, which otherwise often presents extremely fantastic pictures, is due to the fact that all gradations from the mildest to the more severe exist. These gradations are often found even within one family. The **Stot family,** mentioned above, presented many pictures corresponding to or at least remarkably similar to KRAEPELIN'S paranoia, but also including cases of chronic irritated reference psychosis as well as clearly fantastic forms. It is unclear whether the catatonia in the family of patient Daf is nosologically related to her own psychosis. Since several of Daf's relatives were psychotic, it is possible that periodic catatonia which has a heavy hereditary loading was present in the family as a second component. Of the 18 schizophrenics in the Stot family, I myself found three to be catatonic. I will here only speak of catatonic traits, since in patients' fluctuations between excitement and inhibition depressive, anxious, manic-like, or ecstatic phases may readily be recognized.

If cases of pure paranoia within the framework of affective paraphrenia are rare, it is because paranoia is generally a rare condition. Women are hardly ever observed to have the pure form. In affective paraphrenia the affective element tends to be more severe in women than in men, and progresses more frequently to fantastic forms.

In cases of affective paraphrenia with systematized delusions acute phases do not enter the clinical picture. Deeper affective fluctuations in one direction or the other destroy the interconnections because they always produce new ideas. Daf may have had an ecstatic spell with the appearance of her religious ideas but it was not major. Her erotic delusions were accompanied by a corresponding feeling of happiness, but without fluctuation.

In contrast to the mild paraphrenia of the above patient I shall now present a case with pronounced **fantastic symptoms.**

Case 52: Berta Dieh, born in 1872, wrote an insulting card in 1900 to a woman who had allegedly ridiculed her. Similar complaints were repeated indicating that she had ideas of reference. More severe symptoms appeared in 1929, at which time the patient believed her family was going to get rid of her. In 1930 she wrote to the police asking if it was true that "Reichspräsident von Hindenburg had registered me as his wife in the documents of both the church and the state, and, on which day and date that had happened?" As a consequence she was admitted to the Frankfurt Hospital. There she demanded immediate discharge and appealed to Hindenburg's son. She accused her husband of being involved with his niece and other women. She said that Hindenburg had made her his bride years before. Hindenburg's grandchildren had committed incest with her relatives. She had met the Queen of Holland years before. After four weeks she was transferred to Institution H. There she had somatic symptoms and claimed that she was developing cancer of the mouth. She was irritated, hit the nurse, gave the doctor false names, and claimed Hindenburg had been arrested for questioning. In 1931 she scolded at lot. In 1932 she claimed that the institution was a brothel, that the women were being sexually exploited, and that there was poison in the food. She threatened the doctor with execution. In 1933 she said that Hindenburg wanted to get rid of her and had promised the doctors 10,000 marks if they killed her. In 1934 she claimed that a Frankfurt dentist had given her a "fungus" disease. In 1935 the "observation post" kept telling her things – for example, that she would be murdered. In 1936 she had terrible pains due to poison in the food. In 1937 she believed her husband was in the institution. She had olfactory sensations, and argued about her discharge. In 1938 she complained constantly about the "poisonous substances". In 1939 she grumbled a lot in general. SCHWAB and I examined her in 1940. She spoke of an alleged meeting with Kaiser Wilhelm II, and misidentified people in her surroundings, believing she was surrounded by princes who were allegedly locked up in the institution. She spoke of voices from above and many sensations, some quite fantastic. She claimed the swellings of her glands were going into her brain. She became very excited, emphatically demanded her discharge and was insulted when nobody listened to her; she broke down in tears when notes were dictated about the interview, and refused to give further information if she was to be so misrepresented and insulted.

A **sister,** Friederike Ma, born in 1871, was admitted to Institution A. in 1905, after having harassed a man with protestations of love. At first she was cheerful and said she felt as well as a fish in water. She carried on erotic discussions, on one occasion she cried, and she also reported ideas of reference. Customers in her store had bothered her, always changing their minds but never buying anything. Her mood continued to vary, being generally cheerful and occasionally depressive. She signed her name using her alleged lover's last name and the title "millionaire's wife". She continued to scold a lot and to claim that she was being harassed. In 1906, she maintained her love delusion and called the doctors "sons of whores". In 1907 she wrote to the Kaiser, asking him to save her. She believed that she had been poisoned and hypnotized. In 1908 she was hostile towards her surroundings. In 1909 she called herself "Princess Luise Friederike of Pfalz-Zweibrücken beim Rhein und Meisenheim". She had a lot of bodily complaints. In 1910 in a letter she asked about her "treasure", meaning her alleged former bridegroom. In 1912 she was cantankerous and scolded a lot. She claimed that money had been stolen from her. In 1913 she demanded to go to her alleged husband and in a letter addressed the wife of the Kaiser as "Dear Sister-in-Law". In 1914 she wrote to the Kaiser calling him her brother. During the following years she again took offense and easily became annoyed. In 1920 she was friendly and called herself the sister of the Kaiser. In 1921 she was again more irritated and claimed that the doctor was conspiring against her. She defended herself against alleged sexual harassment with obscene words. She called herself the "Princess of Prussia" and was very haughty. She was taken home in this condition in the year 1922. Later she entered an institution that kept no medical record and where she later died.

This patient (case 52) had at first a reference syndrome, lasting for years, before a fantastic picture developed characteristic of affective paraphrenia. The fantastic symptoms here may be distinguished from those of fantastic paraphrenia of the systematic group. Missing are absurd ideas which are physically impossible. What is more important is the way affect is expressed. When her ideas were not taken seriously, the patient was extraordinarily sensitive, burst into tears and refused to give any further information. The affect however related only to her world of ideas. Otherwise she was quite dull, adjusted indifferently to her institutionalization, and was not interested in anything apart from the world of her delusions. Her sister at first offered a paranoid picture with delusions of persecution and of love. Later the fantastic element was manifested by the immensity of the ideas of grandeur.

In the cases described anxious and ecstatic symptoms were often found. In the following case they are very prominent; in addition there are elements resembling cataphasia.

Case 53: Josef Ho, born in 1892, became ill in 1933 and was admitted to the psychiatric ward of Hospital S. He was anxious and expressed many ideas of reference. He claimed that he had been spied upon, his colleagues had met and talked about him, and that something was not right in his relationship to his wife, a situation he had deduced from the hand gesture of a colleague. He stated that the railroad police had probably spread rumours about him since a train guard had once pointed at him and said, "He is going to lose his job." He would probably be forced to leave his job and be punished. He had become so upset about it all that he had wanted to take his life; he had made two suicide attempts. He claimed that he was being maligned on the ward and had been accused of a great breach of his professional integrity. During the following days he became more irritable and attacked patients by whom he felt injured. He heard voices night and day announcing his discharge. He then developed severe anxiety during which he rammed his head against a window and cut himself. Ten weeks after admission he was transferred to Institution E. There he was very anxious and heard voices saying that the Supreme Court had sentenced him to death. This condition continued for three months at the end of which the patient became somewhat calmer, but continued to hallucinate. He was discharged against medical advice. Six months later he was admitted to Psychiatric Hospital M. Again he was depressed, expressed ideas of reference, and heard accusatory voices. He was transferred to Institution E. For another six months he went through fluctuating degrees of anxiety and then became ecstatic. He thought he was a "God-Man", and declared himself to be Christ. He was ready to give his life as an apostle and martyr. He also heard the voice of God. This condition lasted a few months. He was then irritated, frequently scolded his voices and uttered "scatter-brained speech". In 1935 he was irritated, had ideas of reference, and became aggressive. In 1936 he was still irritable, resistant, and suspected poison in his food. In 1937 he was plagued by scolding voices. Again and again his irritated states were interrupted by periods of good and friendly behaviour. In 1938 the voices accused him of performing abortions. In that connection he frequently became excited; then he was again friendly. In 1939 he was irritable and hallucinated. When B. SCHULZ and I examined him in 1939, he was rather lofty, displaying elevated self-consciousness, believing himself to be blessed by God and hearing God's voice. He also admitted hearing other voices. His speech was often unclear and he was hard to fixate on specific questions. Test questions revealed severe thought disorder: (Box/basket?), "Box, it depends if it's four-cornered, if it's wooden; it's all six times six is thirty-six, then you have 150 to 300 places." (Question repeated), "It's a question, it's something new; one could also say what's the difference between a table and a chair." (Table/chair?), "The other is four-cornered, other tables are round, just as with chair, different chairs and benches, making different thoughts, I can't say what it's related to." (Stairs/ladder?), "When I go upstairs I think of a stepladder or of the Mount of Olives, a ladder for example for prayers on the Mount of Olives." (No rose without thorns?), "Yes, the rose, I already know the roses, different roses are needed for the step altar, the high altar, roses without thorns." A **brother,** Alois Ho, born in 1882, was admitted to the psychiatric ward of Hospital S. in 1936. He had been suffering for some time from feelings of inferiority. He had anxious fears and at times did not dare to eat. In the evening he stood around for a long time and rubbed his collar-stud against his arm. He had also made two suicide attempts. He explained the rubbing by saying that he thought there was sweat on the collar-stud. Everything had to be in good order. Furthermore he had become irritable and overly sensitive. In the office a superior had treated him badly. He was anxious but had no severe psychotic symptoms and was discharged eight days later with the diagnosis of "abnormal reaction".

The subject's brother probably had a mild paraphrenic psychosis. His anxiety and irritability corresponded to the symptoms of the subject. The latter had affective paraphrenia characterized by fluctuations of anxiety and ecstasy, irritability, and hallucinations of a paranoid type. Besides there was a thought disorder reminiscent of **cataphasia.** We have already found schizophasic admixtures in the paraphrenic picture of Rudolf Kin, brother of **subject 50a.** There are overlaps in all three forms of unsystematic schizophrenia. Occasionally it is impossible to arrive at a clearcut differential diagnosis. The disorder in patient Ho, above, with the affective fluctuations had a remitting course.

Summary

Affective paraphrenia has a remitting or insidious course, with early expression of **ideas of reference.** Moreover affect fluctuations of **anxiety** or **ecstasy** are often seen during the onset of the disorder and not infrequently during its later course. Moods are always associated with pathological idea construction; anxiety with self-references and hallucinations, ecstasy with false perceptions, particularly ideas of happiness. At the onset it is hard to differentiate between affective paraphrenia and benign **anxiety-happiness psychosis,** but generally it soon becomes recognizable that the delusions and hallucinations are no longer derived from anxiety and ecstasy, but become illogical. Somatic delusions which patients with anxiety-happiness psychosis usually relate to their abnormal mental state in affective paraphrenia have a hallucinatory character with patients feeling influenced from outside. Affective fluctuations are often associated with irritability which grows out of anxiety. This is an irritated reference syndrome, which contains fewer anxious but more hostile reinterpretations of the environment. At this stage affective paraphrenia may be arrested. Similarly mild ecstatic mood may persist and produce chronic delusions of grandeur. Finally, and more commonly, affect may be disturbed in both directions, with delusions of persecution and of grandeur existing simultaneously. Systematization of the delusions may occur at times to a high degree, producing the picture of KRAEPELIN'S paranoia.

Often affective paraphrenia does not stop at this stage, but progresses further. The illogical component, present as a tendency from the start, becomes more pronounced so that eventually **fantastic delusions** of grandeur, errors of memory, misidentification, absurd ideas, and hallucinations in all spheres are present. These traits are rarely as evenly developed as in the systematic forms of fantastic schizophrenia. Some may be absent, while others may dominate the foreground. The key to the diagnosis is in the **affect.** While patients with systematic fantastic paraphrenia appear to have no emotional ties to their inner world of delusions and consequently talk about it without affective display, in patients with affective paraphrenia the delusions remain strongly anchored in the affect. Patients may speak of their ideas with deep irritation or with pride and enthusiasm. Similarly the paranoid affect is maintained in the formation of absurd ideas. Apart from their delusions the patients may become quite dull.

Cataphasia (Schizophasia)

The clinical picture of cataphasia derives from schizophasia, described at first by KRAEPELIN and then by KLEIST. KRAEPELIN pointed out severe confusion of verbal expressions yet comprehensible activities, while KLEIST emphasized the presence of neologisms and mix-ups of words, seldom mentioned in KRAEPELIN'S descriptions. My own approach was at first closer to KRAEPELIN'S, since he described a very circumscribed clinical picture seen fairly frequently. But surely his description is too narrow. I came to this realisation through family investigations which displayed considerable variability in the picture and made me pay more attention to the verbal lapses discussed by KLEIST. The final description of the clinical picture was achieved by my recognizing the **inhibited form** of the disorder (LEONHARD 1961). My colleague OTREMBA (1963) was able to confirm the presence of both forms in a number of impressive cases.

Since the term schizophasia refers essentially to the excited form, I chose the term **cataphasia** in order to point out the disintegration of verbal functions in both the excited and the inhibited states. This trend towards both poles of the disorder relates cataphasia to the two other **unsystematic schizophrenias,** since both affective paraphrenia and periodic catatonia have a the tendency to spread toward two poles. The parallel is complete in that the end state of cataphasia may reveal the predominance of the one or the other pole, or a defect syndrome may appear with an absence of symptoms of either pole.

I do not assume, as KRAEPELIN and KLEIST did, that in cataphasia speech is disturbed independently from thought; instead a general thought disorder can usually be detected. It is not particularly significant that the patients behave sensibly in daily life, since usual activities demand little thought. Rather it is a question of affectivity and activity which may in fact remain quite well preserved. I see it is the thought disorder which strongly affects speech as the central symptom of cataphasia. Thereby a relationship to **confusion psychosis** is revealed. We found affective paraphrenia to be related to anxiety-happiness psychosis, and periodic catatonia to motility psychosis. Similarly, on the basis of the thought disorder, cataphasia may be compared with confusion psychosis.

At the start there may be a problem of making the correct differential diagnosis between the two disorders. Confused excitement and phases of perplexed inhibition are also found in cataphasia. One can however generally recognize early the profound disorder of thought which goes beyond incoherence into **logical derailment.** On the other hand, in the inhibited phase there is a **numbing of reactivity** going beyond perplexity.

In the excited form of cataphasia the picture is dominated **by confused pressure of speech,** especially so in severe cases. The patients' speech is vivacious – they turn toward the investigator as if they have something interesting to say, but one cannot understand a single word. It is as if they were speaking a foreign language. New words and concepts constantly appear where the interrelationship and the connection with what was previously discussed are incomprehensible. The grammatical order is regularly destroyed – sentences are left incomplete, declensions and conjugations are incorrect, a noun appears unexpectedly in the genitive and not the nominative, and a verb is in the infinitive when it should be in the third person. Often it is not clear which words belong in one sentence, i. e. a flowing speech is heard without knowing where one sentence begins and another ends. If a transcript of a patient's speech is read, commas and periods may be inserted almost arbitrarily. Such forms of cataphasia may be seen in severe end-states or in excited phases. The latter are comparable to the excited phases of confusion psychosis though here the logical errors are much deeper.

In several cases I noticed that there was great **uniformity of verbal expression** during the acute phases. The same words appeared in varied but generally senseless connections. At first I considered the possibility of a combination with periodic catatonia in which iteration is very common; however it appears that the uniformity, which at times even resembles verbigeration, represents a separate symptom of cataphasia. Here perhaps stimulation and inhibition of thought are somehow mixed together. The patients I saw displayed no other catatonic traits.

When confusion is a symptom of an acute phase the patients are not capable of any sensible activity. On the other hand in the defect phase they behave in an orderly way despite their confused speech and carry out activities assigned to them in a reasonable manner. But this contrast should not be overemphasized since the usual tasks given to the patients are simple ones. Their action can sometimes be quite illogical. One of the patients whom I presented at a case conference did not at all behave appropriately: at one point he suddenly lay down on the floor of the auditorium. On another occasion in an examining room he took off all his clothes although he should have realized from the situation that he was not going to have a physical examination. The relatively appropriate activity should be looked at in parallel to another phenomenon essential to cataphasia. The patients can speak much more sensibly about concrete topics than they do in free speech. When speaking about the day's activities they often do so quite correctly. Apparently thought is still capable of sufficient orderliness when it is concretely stimulated by the immediate experiences of day to day life. **This is a characteristic trait of cataphasic thought disorder.** The same patient who has just given some sensible replies may suddenly begin to speak in complete confusion.

There are many transitional stages between this most severe form of excited cataphasia and the mild, beginning changes. Depending on the degree of the disorder one can carry on shorter or longer discussions before illogical phrases appear. Sometimes in simple conversation the disorder is only apparent in isolated unusual ways of expression. In such cases the phenomenon may be revealed by giving the patient **intelligence questions,** asking him, for example, to explain differences or to interpret proverbs. The logical order breaks down much more easily during such tasks than it does in simple questions. The answers often show that the intellectual performance actually began correctly, that is, from the answers given it can be seen that the solution hovered internally, but at the same time the thought ran off the track in the wrong direction. At times the distinct impression is gained that thought becomes misdirected only during verbal formulation and not in the thought process itself.

The **"psychic-experimental test"**, not to be confused with an intelligence test, should not be omitted when an endogenous psychosis is suspected, since it is important in all forms to determine whether and what kind of formal thought disorder is present. I have a regular scheme that I use in every case. There are six differentiation questions with concrete concepts, as follows: 1. box/basket, 2. stairs/ladder, 3. brook/pond, 4. mountain/mountain range, 5. tree/bush, 6. child/dwarf. There are three differentiation questions with abstract concepts: 1. error/lie, 2. stingy/thrifty, 3. lend/give. There are six composition questions with sentence construction from three concepts: 1. child/meadow/flowers, 2. field/hunter/rabbit, 3. sun/curtain/room, 4. rider/horse/ditch, 5. car/curve/tree, 6. peasant/harvest/ rain. In addition there are seven proverbs. It makes no difference if the patient has never heard of the proverbs, because a healthy mind can still make reasonable attempts to understand a proverb or simply explain that the meaning is unclear. The same is true if a patient cannot answer a question due to lack of intelligence. In formal thought disorder, illogical answers will be given.

I present some of the answers of two patients in whom the experimental test readily demonstrated a severe disturbance with mixing up of words, displacement of words and neologisms:

(Box/basket), "Both have different qualities in their 'finishedness'." (Stairs/ladder), "They must be placed on the plans." (Brook/pond), "The brook has an off-shoot." (Tree/bush), "The start of the tree and the bush is the same." The other patient: "Bush is a 'shrub-like' growth, a growth closed in itself." (Child/dwarf), "A dwarf is a stable factor, a child is a yielding factor." (Thrifty/stingy), "A thrifty person is 'going peacefully', and stingy is a 'false-looking' manner." The previous patient said: "The stingy person is tricky, the thrifty man is sufficient." To the proverb, "The apple does not fall far from the tree", one replied: "A black-out of the apple, so that it does not get into other people without buying or obtaining." The other replied: "For man inheritance comes from father and mother, and does not fall far from it." A third patient's answers with short formulations are particularly impressive in that despite the verbal peculiarities the correct ideation can be recognized: (Box/basket), "Box is in a rectangular position." (Stairs/ladder), "Stairs are an open way." (Brook/pond), "Pond is a stand-still." (Mountain/mountain range), "Mountain is a one-time climb." (Error/lie), "Error is material-covering." Here the patient is probably thinking that contrary to a lie in an error something is covered up, i. e. unknown.

If the disorder assumes the form of **thought inhibition,** verbalisations dry up more and more so that now thought derailments can no longer be discerned. In an extremely inhibited form of cataphasia the patients stop speaking altogether. In that case the diagnosis and differential diagnosis become quite difficult; especially differentiating between cataphasia and the perplexed stupor of confusion psychosis.

As is perplexed stupor, **mutism** is in the foreground, while simple reactive movements continue. The facial expression of the patient does not, however, display only the perplexity of inhibited confusion psychosis, but a simultaneously recognizable emptiness or **numbness of expression** as well. The patients follow events in their surroundings not with inner anxiety or internal questioning, but usually do not follow them at all, remaining dull and withdrawing into themselves. I have observed some patients who turned their face to me but whose look were not meaningful and had a somewhat compulsive quality. When addressed, instead of replying, they **stared at the questioner.** When I stepped aside they followed me with their gaze. As I looked at them they did not change their position but looked at me in return. Several of them contracted their brows as if they were concentrating on trying to remember something. Some maintained this facial expression indefinitely. Probably this is the result of thought inhibition giving rise to a more forceful mental effort. Yet the emptiness and dullness of expression is maintained with uniformity and without modulation. A true thought process is apparently no longer present.

In spite of the dullness of these patients' behaviour they are not catatonic. Psychomotor function in a narrow sense is unchanged, though the patients move slowly. Just as the mutism of inhibited confusion psychosis does not correspond to the akinesia of motility psychosis the mutism of cataphasia does not correspond to catatonic stupor.

The fact that in cataphasia there is not only inhibition but also a logical disturbance of the thought process may be demonstrated when stuporous patients do say a few words, which may be completely incomprehensible. I was particular impressed by a female patient whom I was able to observe for many years. She at times showed pressure of speech of which nothing could be understood, whereas at other times she was stuporous and showed the dull expression typical of inhibited cataphasia. Her mutism was interrupted quite unexpectedly by quite confused utterances. It was evident that the disturbance of thought continued during the stupor.

In describing excited and inhibited phases of cataphasia separately I substantiate the fact that these pictures occur frequently. When neither pressure nor poverty of speech is present this in-between state while the disturbance of thought and speech persists is to be considered as a **defect state.**

During casual conversations with patients their **affect** at first appears to be well preserved. They are usually friendly, speaking frankly with persons in their surroundings. But this apparently open attitude does not reveal the underlying lack of deeper emotions. Sometimes the existence of cataphasia may be suspected from the flatness of the way the patients express their satisfaction, before it is noticed that there is a thought and speech disorder, which does not have to appear in the beginning of the conversation. This equanimity clearly helps to differentiate cataphasic from affective paraphrenic patients, the latter revealing a certain affective tension even in the absence of irritation. A cheerful basic mood is not seen in all cases, in fact the patient may be in a bad mood; even then the flatness of affect is present. A friendly-satisfied attitude is seen only in the excited form of cataphasia.

Activities of patients with the excited form are generally performed well. Patients not only may agree to carry out tasks assigned to them but also spontaneously look for things to do. But the more the picture shifts towards the inhibited form the more poverty of speech replaces loquacity, the more their activities as well as their affectivity diminish.

In the inhibited phase of cataphasia **ideas of reference** are characteristic, probably because of a similar basis as in inhibited confusion psychosis of which cataphasia is a "malicious relative". In cases where ideas of reference are prominent symptoms for a long time the diagnosis is not necessarily affective paraphrenia. But such ideas occur less often in cataphasia, probably because the flat affect inhibits the formation of ideas.

In the excited form of cataphasia **confabulations** may appear, mostly with an expansive tinge. Patients may declare that they have traveled around the world, that they have experienced many sensational events, that they are of noble origin, that as children they had been exchanged, etc. One patient claimed to have flown an atomic airplane in World War I, to have driven the Americans from the Mediterranean, and then to have planted cotton in Africa. Another patient boasted like Jesus to have walked for half an hour on Lake Galilee. There is no need to consider an overlap with the delusions of patients with affective paraphrenia. These delusions are found only in lively excited cataphasic patients who do not show tension. Probably while the patients are in an excited and confused phase the delusions arise with the emergence of fantasies which are sustained by a cheerful or even ecstatic affect.

The characteristic pictures of cataphasia may be colored by traits of other forms of unsystematic schizophrenia or of cycloid psychoses. Genuine **catatonic symptoms** occur in the form of parakinetic restless movements or in the form of akinetic states with rigidity. During the onset **affective fluctuations,** like those in anxiety-happiness psychosis or in affective paraphrenia, i. e. anxiety and ecstasy with corresponding ideation, are fairly common.

The **course** in cataphasia is similar to that of affective paraphrenia. Slowly progressive as well as remitting cases are not infrequent; some good remissions are observed reaching clinical recovery, enabling the patients to go back to work.

For the sake of illustration I present a case of a patient with excited cataphasia, who displayed extreme ideational as well as verbal and above all grammatical disturbances. The word "dictator" in the first sentence was occasioned by my having asked the patient to take dictation:

„I do not always want to play the Dictator. I was released from the Hoechst factory for the third time and I returned a menu which was not refunded. I see a lot of money and try to get it. Because of income from high work he was good. He had two tickets. A West German demands a hundred million marks, owner of a brick factory. We have to pay Master Luzius, we often lose. That would be a separation. The income, the foreign income loses high, ask Wagner's Printing Company, 10 pennies small and maybe 30 years. Two subscribers are having a hard time. I have men here, racially pure, Wolf at the exhibit, loses the household money. The test of race and woman's hand would be a lordly animal. I have been in the house for a year and have been vaccinated right and left, and whoever isn't a cannibal, is over 30 years old. But that is often different, when he is fed poorly, other people are dissatisfied. I wouldn't know now what is going on. I mean, at four o' clock there are potatoes instead of bread. I said, when I put it somewhere and we need it, then I know where it is. I can't serve the other one this way, he thinks; his name is Arnold (a fellow patient who worked washing dishes). Sometimes I could just say nothing, now that it's been six weeks and an escape attempt, it must be three weeks ago, it's about time. It is only, that he has a place to stay here. The people have been duped in the Hoechst factory, he has bad beds or high machines. Whoever travels a lot, keeps getting dumber. That's the way it is. Lots of people don't want a ticket on the height. I don't bother with the price of gold at its height. Four billion gold marks, as minister of war, it's not enough for two or three. If it has to be bought, then one must the height. Yes, I don't know, what it here, I don't know at all, with whom I am here. The swastika government has Red Cross goods. He was a beggar, he need not risk the leadership. I mean he also has his note book, his writing. He was arrested in the sixteenth precinct, they also took Napoleon's shield, Blum's bottled beer company and brewery filling. Small and big too against high men. People have bad animals on the street. It is often, one is called, it has to do with Arnold. Luther permitted, he is allowed to sleep in the sleeping room. If the eyes were closed against higher people. If you need three high machines for rest, then it is occupied. I always said, if two high machines are, then I can have my opinion. No, it had to be there. I took a bath right away and didn't walk around naked. They often want animals for bath animals. There are often towels. The animal coexists. That is the saying, he get's for free." (Whom do you mean?), "The dog is sick and is being operated on. It is difficult to warm up the dog's food."

Three brothers elicited various symptoms compatible with cataphasia: severe confusion, stupor, in acute phases anxiety with ideas of reference, ecstasy with ideas of happiness. There were intimations of confabulation.

Case 54: Johannes Vog, born in 1889, became ill in 1924 and was admitted to the Frankfurt Hospital very anxious and perplexed. He felt that everything was changed, that people had strange eyes, were talking about him and stared at him, and that the newspapers carried stories about him. He also felt influenced and hypnotized. At times he was very anxious, moaned and wrung his hands. He was discharged after eight months, but returned soon after, still perplexed and anxious. He was now hearing many voices: "Two voices are arguing over my head." He complained endlessly and on one occasion declared he felt like the middle of the world. He was often confused in his utterances. For example he declared, "Oh doctor, I have only one thing in the middle, that would be myself and my family." He gave inappropriate answers to test questions (The one who digs a pit for others ...?[6]). "That is something, I would like to say, double powers were involved." He was transferred to Institution H. where he remained perplexed and talked in a confused manner. In 1926 he left without permission, and was returned to the Frankfurt Hospital. Everything still seemed strange to him. He spoke of hints and allusions, had ideas of reference and heard voices. Transferred to the Institution, he once again left without permission and was readmitted to the Frankfurt Hospital, where he was even more confused. He felt influences on his body, heard voices and declared, "On my command the world stood still. I stopped the rain." In the institution records of the following years, there were repeated descriptions of his completely confused speech, with gesticulations, and cheerful mood. For example, in 1935 on one occasion he said, "No, I understand that everything will be all right as soon as I am outside, that right away the family will be the concept

[6] German proverb: "The one who digs a pit for others falls into it himself".

of all concepts. In steadfast loyalty, it is said. I was told that it was a trick of the friend of the theatre, of Caruso." Despite this confusion, he busied himself steadily in institutional tasks. When MEYER and I reexamined him in 1939, he approached us with lively facial expression and gestures, and spoke as if he had something important to say. In fact we could understand nothing. (Why are you here?) "The nurse demanded it, they could no longer help themselves, because of the error of the new world order all the animals were under a spell. We want to praise ourselves with new months." (What does that mean?) "Of private affairs, marriage of reason, family, our stories become right." (What?). "Thought and being, to be or not to be ... Walter Hauch, they played tricks on him! Alphabetically one says 25, 172. It is not the same with D. They tried it with D, they should have had to disband. The earth, the wheat, the spring. The pilots always play tricks on one another ..." (Basket/box?) "Round and four-cornered, there are also long, full laundry baskets, they are the jokes of bottle basket champagne." (Tree/bush?) "The tree is more practical than the bush." (Child/dwarf?) "That is a dumb comparison, the child, the ones who are astonished, that is the intelligence of thought and being." (Necessity is the mother of invention?[7]) "Those are unordered things, that was, I believe, horse raising and cows and pigs, all of them were not in order."

A **brother,** Jakob Vog, born in 1878, became ill in 1912 and was admitted to Institution K., where he expressed ideas of grandeur, called himself captain and an aristocratic land owner. He was generally cheerful but at times irritated. The ideas of grandeur continued during the following years as did the bad moods. The patient occupied himself industriously in institutional tasks. Occasionally there were mistaken activities – on one occasion he left the institution and tried to buy cows and a horse. There are also statements in his record about "eccentric" speech. After 1918 there were no fluctuations of mood, but the ideas of grandeur continued. In 1922 he was reported to have presented his ideas in a totally confused form. In 1923 he wanted to buy a dog from a peasant. In 1924 it was not possible to converse with him comprehensibly. He hardly ever responded correctly to simple questions, and sometimes he presented very bizarre topics. In the records of the following years he was reported occasionally to have become irritated and moody. Generally however he was "quietly satisfied" although quite confused. In 1930 he was described as "mildly manic", he "gabbled much nonsense and occasionally displayed ideas of grandeur". There are many references to voices he said he heard. On one occasion, from his confused expressions it seemed that he thought the horses on the farm belonged to him. He did not change during the following years, remaining industrious, but continued to be so confused that all communication was incomprehensible.

A **second brother,** Heinrich Vog, born in 1893, became ill in 1921. He claimed that an angel had appeared to him, and that the sinful world would repent; God would lead Germany to victory. He was admitted to Institution K. where he claimed that there were four dragons at home that he had to kill. He was held to be manic with states of confusion and excitement. A depressive state with inhibition followed the mania and deepened into stupor. After a few weeks there was another period of excitement, with religious visions and confusion. He was also reported to have heard voices from the ceiling. The confusion increased so much that no more communication was possible; soon he became stuporous again. He continued to alternate, the phases lasting from a few weeks to months; occasionally the fluctuations occurred within one day. At the height of stupor, he lay motionless in bed, said nothing, and starred rigidly into space. Later he sometimes recounted ecstatic experiences which he had had during the stupor; for instance that God had appeared to him. The ecstatic component is rarely mentioned in his records after 1925.

[7] Nearest English equivalent to the German: "Not bricht Eisen."

Excitement was always accompanied by severe confusion, and in 1926 he was described as "mildly manic". The mood was no longer ecstatic, but friendly and cheerful. On one occasion while stuporous he smiled to himself. Beginning in 1928 catatonic excitement was mentioned though severe confusion was still emphasized. In that year ideas of salvation were once again elicited in the confused speech. From 1931 on he several times became aggressive while excited. During stupor negativistic symptoms appeared. Around that time he produced confused speech even during his quiet periods. The excited and inhibited periods were no longer separable, both became less severe, and the patient became increasingly dull. In 1937 he grimaced during stupor and picked at his body. His confused speech took on the form of talking to himself. During the following years he was always dull, unapproachable, off-putting; he mumbled occasionally to himself and for short periods was often loud and disruptive – more in the form of impulsive excitement than of longer lasting phases. Stuporous states were still recognizable, by now they had a clearly akinetic character – for example, the patient lay in bed with his head lifted off the pillow and grimaced. In the records he was always called catatonic.

Here the subject was an excited cataphasic. This was essentially true for his brother Jakob as well. The latter expressed ideas of grandeur at the onset. Because of these ideas and his confused loquacity he was sometimes described as "mildly manic". Signs of anxiety-happiness psychosis are also recognizable in his brother Heinrich, who was ecstatic at first. Several times he too was described as manic. During the further course this patient varied much more than his brothers. In his excited phases he was confused and his inhibitions became stuporous. While his condition could be assumed to be a case of cataphasia with a periodic course, there were apparently genuine catatonic mixtures at the end, indicated by the patient lying in bed with his head uplifted, grimacing. Apparently there was an overlap of two unsystematic schizophrenias in which catatonic features were added to those of inhibited cataphasia.

Confabulations not clearly seen in these three brothers did appear definitely in the following case.

Case 55: Adolf Roth, born in 1880, probably had early symptoms with ideas of reference. He first became seriously ill in 1930 and saw God in a cloud. He said, "Since that time, I was omniscient. I knew how volcanoes arose, and I could predict accidents, floods, and natural catastrophes. Since I have become omniscient through the breath of God, through revelation I sent out warning letters throughout the world, to warn the people." He spoke enthusiastically of his calling and assumed the posture of a preacher. There were frequent errors of thought and speech. (Island?) "A mass that can stand in rivers and in stagnant ponds." (Radio?) "A discovery or an invention connected with the forces of nature." (Binet picture, "Window Promenade"?) "He must want to make a joke, to scare the ladies." He explained, "Man has to live in old fashion again, without railways, as in the times of Christ, but that cannot happen due to the discovery of electricity." He was transferred to Institution H., where he continued to speak confusedly with neologisms. The elevated mood continued at first, but after a few months a mildly depressive state with inhibition set in, which soon passed. In 1936 he changed and was partially in a good mood and partially sullen. Although he corrected his ideas his thought disorder continued. (Lies have short legs?[8]) "Lies are kinds of things, which never produce anything com-

[8] Translated literally from the German: "Lügen haben kurze Beine."

plete and assume that the truth can bring forth something complete, like lies with short legs. That is not half and not whole, through truth there are no short legs. But through truth there can be short legs, too. I can't think that way, it will all come to me later." He left without premission and was readmitted to the Frankfurt Hospital. He spoke of nightly visions and voices, and was again transferred to Institution H. He left without permission again and alternated between the hospital and the institution. At times he was depressed and expressed ideas of reference, and at other times he was ecstatic. In a questionnaire in 1937 he described his qualities as "Special toy aeroplane and zeppelin producer, independent." In a letter he wrote that as an aeroplane producer he wanted to be a skilled man and not a quack. Furthermore he declared, "I am now thinking of being discharged and doing other than aeroplane work." He again spent much time with inventions, constructed models, wrote confused notes and expressed ideas of grandeur when he was calmer. In 1939 he left the institution without permission and never returned.

A **father's brother**, Franz Roth, born in 1849, was an alcoholic who went through two phases of alcoholic delirium, and with epileptic seizures during his alcoholism. After withdrawal treatment he was hypomanic, for example, at age 78 he still went dancing. With interruptions he was in Institution W. from 1906 to 1931. From time to time his excitement increased as in mild mania. At those times he was unusually busy, talked a lot, laughed, joked, and performed practical jokes. Later he was somewhat more calm and reasonable. He died in the institution of apoplexy.

It may be assumed that the father's brother also had a predisposition toward cataphasia that was so clear in the subject himself. The manic fluctuations were certainly not independent of it. As I have stated above, cycloid psychoses gradually deviate, in a certain sense, from manic-depressive psychosis, i. e. signifying a more profound disorder. In this way it may be agreed that a patient with unsystematic schizophrenia running an unusually mild course may resemble a patient with manic episodes. The two brothers of case 54 likewise showed manic symptoms during the milder cataphasic phases.

The subject himself (case 55) cannot be described as having had either an excited or an inhibited form. The essential symptom of cataphasia in this case was the thinking disorder, even in the absence of clear-cut excitement or inhibition. A tendency to confabulation was also present.

Summary

Cataphasia appears in two forms, the **excited** and the **inhibited.** The former is characterized by confused pressure of speech, in which verbalizations become incomprehensible in severe cases. The patients' behaviour however remains generally sensible and their activities are well preserved, but their affect is flattened. The **inhibited phase** in some cases is characterized by mutism, in milder cases by poverty of speech. The **thought disorder** of cataphasia, its central symptom, is reminiscent of that seen in confusion psychosis, but in the excited phase it goes well beyond simple incoherence, and in the inhibited form, well beyond simple thought inhibition. In excited cataphasia severe logical blunders are found, as exemplified by the **patients' speech** with wrong choice of words, neologisms, and lack of grammatical correctness. In the inhibited form errors in logic also

occur but the principal symptom is thought inhibition which as in confusion psychosis may lead to perplexity. In the patients' facial expression an internal dullness can be recognized. Furthermore instead of answering patients tend to stare fixedly at the examiner. In excited cataphasia **confabulations** occur frequently, in the inhibited form, **ideas of reference.** Besides the excited and inhibited forms there are patients who display severe thought and speech disorder but who are neither talkative nor taciturn, i. e. they display only the central defect syndrome of the disorder.
As in affective paraphrenia, cataphasia often contains admixtures of other unsystematic schizophrenias or of cycloid psychoses; **anxious** and **ecstatic mood fluctuations** occur quite frequently. The **course** of cataphasia can be insidiously progressive. More often remissions occur, sometimes with a periodic course.

Periodic Catatonia

I have described periodic catatonia as an independent disorder, with considerable hereditary loading (LEONHARD 1941). In common with the polymorphic phasic psychoses, **bipolarity** is even more marked than in affective paraphrenia and cataphasia. Its phases are characterized by psychomotor excitement and inhibition which often alternate. **Hyperkinesia** and **akinesia** are seen as in motility psychosis; pictures more reminiscent of perplexed stupor or excited confusion are also found. Anxious and ecstatic states are less frequent, but they do occur. There may be a mixture of symptoms which we have so far not seen; the cycloid psychoses at any time and in each type showed either inhibition or excitement, never both together. No patient had rigid akinesia of his posture and hyperkinetic restlessness of his extremities at the same time. However, similar symptoms are found occurring together rather frequently in periodic catatonia. Such a scattering of symptoms rules out the notion of "mixed states", on the contrary it is clearly seen that the disorder goes beyond the limits previously observed. It is no longer a matter of excitement or inhibition of one psychic area, rather it is a qualitative change leading to simultaneous occurrences of excitement and inhibition. Hence in periodic catatonia there is a destructive process which despite other similarities is absent in motility psychosis. It must be said however that during its first outbursts periodic catatonia may be so similar to motility psychosis that it can easily be misdiagnosed.

If in the absence of pure hyperkinesia or pure akinesia increased motor restlessness is accompanied simultaneously by inhibition of the psychomotor system, **uniformity** generally develops. Patients with akinetic traits, i. e. rigidity of posture and movement generally show this uniformity if there is simultaneous motor restlessness. **Stereotypes** and **iterations** thus become important symptoms of periodic catatonia.

The combination of hyperkinetic and akinetic symptoms may be seen in some patients who lie stiff in bed but iteratively strike with one arm, others stand in the room with rigid expression and posture while making rocking movements with their body or showing only facial grimacing. The picture shifts further towards akinesia when the rigid posture is only occasionally broken by certain movements – for example, pattering along. Hyperkinesia may recede even further when motor restlessness produces only certain postures which are not relieved. These **posture stereotypies** are also very characteristic for periodic catatonia.

Another typical feature is when akinesia is accompanied by hyperkinesia in the form of **impulsive actions,** often called instinctive actions. Akinetic patients may suddenly jump out of bed, run through the room, knock objects over, become aggressive, or suddenly scream and right away return to a rigid posture. Furthermore akinesia may be accompanied by an "instinctive" resistance in which every external stimulus produces an opposite reaction. This **negativism** in periodic catatonia is a particularly characteristic phenomenon accompanying the akinesia. In motility psychoses opposite tensions appear as a defence against anxiety; facial expression shows the anxiety, here it is not a question of negativism.

In the symptoms just described akinesia is in the foreground while hyperkinetic phenomena are only superimposed. Hyperkinesia may predominate however, in which case akinetic traits only determine how the movements are carried out. The movements become stiff and staccato, they are not quick or graceful nor do they flow into one another. This type of movement is **parakinetic.** The admixture of an akinetic element is proven by marked rigidity of body parts not immediately involved in movements. If the extremities are moving, the face at that moment may be stiff. If the face shows movements, the arms may be held stiff. If both face and arms are moving, the trunk may be rigid.

At the same time the movements lose their expressive character. As I have discussed elsewhere (LEONHARD 1976b) facial movements and gestures normally have a very fixed form. Every small deviation destroys them. Likewise their expressive character is lost if the internal course of the movements is changed. Expressive movements may play a role in the hyperkinesia of periodic catatonia, in so far as expressive and reactive movements are concerned, the latter occurring without higher conscious motivation; but because of the disturbed course of their expression their content is no longer recognizable. This disorder has less effect on reactive movements: if patients handle their beds, climb on chairs, or push at doors, though the movements are rigid, the reactive character of the movements can still be recognized. On the other hand facial movements become contentless and appear as grimaces; movements of the arms arising from gestures look like directionless poking in the air. Primitive hitting, pushing or shoving movements no longer expressive nor reactive may enter the picture. Facial or bodily tics may appear, giving the movements a parakinetic character.

The combination of hyperkinetic and akinetic symptoms was illustrated in remarkable fashion by two patients. During the examination there was constant alternation between the two opposing tendencies. For a few minutes they sat rigidly without moving. Then for some minutes they were very restless, grimacing, groping around their bodies, moving head and trunk repeatedly. The movements were grossly distorted. These patients seemed to pass through akinetic and hyperkinetic phases in extremely short periods of time.

The combination of hyperkinetic and akinetic phenomena continues in the defect stage which becomes more prominent with each successive phase. Generally besides **diminished incentive** there is **increased impulsivity.** Most frequent are grimacing movements which in contrast to medication induced dyskinesia affect the upper half of the face more than the lower. Moreover there may be tics of various parts of the body. Occasionally iterative grasping and rubbing of the hands may be observed. More often there are impulsive acts: for example, patients may suddenly stand up during the examination and turn towards some event going on about them. Some patients tend to answer questions impulsively; they do not think, instead they say whatever comes to mind. They seem to talk past the point ("Vorbeireden"). Such disconnected talk does not have to be a cataphasic trait. Cataphasics tend to reply to questions, albeit in an unclear fashion. In contrast catatonics reply senselessly and then are silent. Sometimes they simply repeat a word from the question. Speech impulsivity may also take the form of certain unexpected remarks. The patients may even begin to tell stories; nevertheless, lack of drive remains clear in their lame speech.

Sudden excitement in periodic catatonics after quiet conduct may be partly understood in terms of a potential for impulsivity. Severe **aggressiveness** indicates the presence of severe **affective tension.** Such aggressive outbursts are found in a similar form in negativistic catatonia of systematic schizophrenia. Affective disturbances occur also in the form of **anxiety** and **ecstasy.** Anxiety may even be the most noticeable feature in acute phases of periodic catatonia.

A special kind of impulsive reaction came to my knowledge rather late. Patients with periodic catatonia who previously may not have been particularly noticed for their facial expressions occasionally break out in **exaggerated laughter** without any or sufficient reason. The laughter may be loud but more often is nearly silent, with exaggerated laughing facial expressions sometimes involving the entire body. It then seems as if the patient is literally doubled over laughing. It is reminiscent of compulsive laughter. I thought at first it might be due to medication; however I not only found it occurring without medication but also in milder form in supposedly healthy relatives. We found that the "healthy" father of a patient accompanied his report with continuous inappropriate laughter. On another occasion, while we spoke with a patient, in the background we heard the continuous inappropriate laughter of his brother, who had once been under psychiatric treatment, but now was supposed to be well.

As more phases occur, the nature of the **defect** becomes clearer. Actually the lasting characteristics of the deficit are often minimal; many phases do not always imply severe defect. In some cases the disorder does progress to a severe defect. Akinesia seems to go deeper and lead to defect more rapidly than hyperkinesia. The lack of drive which marks the end state may however develop from hyperkinetic phases as well. In the mildest cases only a certain **apathy** is found. Drive is diminished, movements are initiated with difficulty and are carried out slowly. Affect is flattened, thought processes are slow, and thinking tends to be short-circuited. There are many gradations between these mild states and the more severe pictures. Eventually lack of drive my result in severe **blunting** so that the patients have to be made to perform the most primitive functions. In all degrees of defect there may be **irritability** – not associated with disturbed ideation as in affective paraphrenia but unpredictable in its discharge.

Because of their impulsive excitability and aggression periodic catatonics are generally kept under **continuous medication.** Hence it seems that poverty of affect accentuates the diminished drive. Diminished initiative is always accompanied by a certain flattening of affect, but the latter becomes more marked under medication. When patients answer disinterestedly with low voice and without modulation, they are reminiscent of hebephrenics; the differential diagnosis may become more difficult. Not only the present picture, but also the course of the disorder should be taken into account. The course is not at all like that of hebephrenia: in catatonia it is periodic, with both hyperkinesia and akinesia and nearly always with episodes of impulsive excitement, even if the latter is subdued by medication.

In describing periodic catatonia I have emphasized the specific symptoms which are essential for the diagnosis. There are often variations from the typical picture. Particularly in the early phases, when the defect is not yet clearly present, the hyperkinetic and akinetic phases may resemble the phases of motility psychosis. Pictures of the other cycloid psychoses, such as symptoms of incoherent excitement, perplexed stupor, anxiety, and ecstasy occur. Furthermore traits of cataphasia and of affective paraphrenia are also found. Thought disorders, delusions and hallucinations may be like those found in related schizophrenias.

The following are descriptions of several Frankfurt patients where in the families there were other cases of endogenous psychoses. The clinical pictures are those seen before neuroleptic treatment became available.

Case 56: Lina Wei, born in 1892, became suddenly excited in 1921. She howled, screamed, raged, scolded, threatened to jump out of the window, and sang religious songs. She was admitted to the Frankfurt Hospital, where she ran around, sang incomprehensible syllables, pulled her hair, clapped her hands, grabbed at things, stripped her bed, bored a hole in her mattress, and lay down in bizarre postures. In the bath she gesticulated with arms and legs and scolded loudly in a stereotyped manner, giving no answers. In contrast she became akinetic, lying down rigidly, omitting not a sound. She assumed strange positions with her head thrown back to the side. She lay with her legs apart while her fingers and hands performed playful movements. Statuesque postures and grimaces always followed excitement. At times, apart from otherwise akinetic behaviour, she repeated the same sentence over and over for hours. After three months she was transferred to Institution E. There she was impulsive and violent, and was described as giving "the impression of manic excitement, but with confusion". Later she became calmer, seemed numb, and was discharged. At home she was frequently moody and paranoid in regard to her husband. In 1928 her distrust became so severe she had to be readmitted to the hospital. There she claimed that her husband had two illegitimate children, that he was secretly married to someone else, and he flew around in an aeroplane with his other wife; he smelled like wine and cigars and had pockmarks on his face. She believed people were talking about her and that her husband's office employees had made peculiar dancing movements. During the examination she suddenly lay down, remaining stiff and unresponsive. Otherwise she alternated between perplexed anxiety, akinesia, or excited talkativeness. Her remarks continued to be perplexed and often quite incomprehensible: "Everything is green with poison, verdigris"; "now comes the Caesarean section"; "they have sucked all my blood"; "here comes Queen Luise"; "everything is burning here." After two months she was transferred to Institution E., where she was mute, negativistic, and sat stiffly with her head bowed. Her posture was stiff, she made peculiar movements and grimaces. She was discharged against medical advice. During the following years at home she was often argumentative and jealous. In 1940 she again became extremely excited and was readmitted to the hospital. She ranted and raved in a sing-song voice and expressed ideas of reference. She developed a high fever which was assumed to be due to septic sore throat. At the same time she was quite unapproachable, she lay stiff in bed, rubbed her stomach continuously, and shook her head back and forth. She died of circulatory collapse. The autopsy revealed a subdural hematoma, which had probably occurred during her excitement at home.

The **sister** of the subject, Johanna Uff, born in 1878, began hearing voices in 1926. The voices insisted that she had had relations with dogs and with her son, and that she would be executed. She was admitted to the Frankfurt Hospital, where she complained of all the voices. She sat motionless with her hands folded and her mouth open and gave only very hesitant answers. At other times she was anxiously excited, cried, and trembled. She kept on repeating her complaints about the voices. After five weeks she was transferred to Institution E., where she cried and moaned in a low voice. After she had calmed down somewhat she was discharged. At home she neglected her housework, was messy and personally untidy. She was reported as being "dull, uninterested, and phlegmatic". At times she was confused; she burned things, sawed off the decorations on the furniture, and continued to hear voices. She died from an accidental fall from a window in 1936.

These two sisters demonstrate the polymorphic extent of periodic catatonia. The original subject during her second admission showed for a while a reference syndrome with peculiar ideas. Above all the mixture of akinetic and hyperkinetic symptoms characterize the subject as a case of periodic catatonia. For example, though she lay stiffly, she assumed bizarre postures and grimaced, during her akinesia she repeated the same sentences at length. From the hyperkinetic distorted movements and the lack of appropriate expressions the difference from motility psychosis is easily recognized. The sister's akinesia indicated the probable presence of periodic catatonia. For a while she had the symptoms of anxiety psychosis, but most unusually with many hallucinations. Eventually the dull defect state was compatible with the diagnosis of periodic catatonia.

The next patient to be described as well as her sister illustrate the overlap of the symptoms of the two poles, particularly because of their impulsive excitement.

Case 57: Pauline Ha, born in 1905, began to act abnormally at the age of 14–15 years; she forgot things, and talked to herself. In 1923 she again became ill, stood around and no longer went out of doors. A bit later she became extremely lively and loquacious. Then she again became abnormally quiet and at that point, in 1924, was admitted to Institution K. Here she was very reserved, spoke very slowly and stood around. Later she showed some improvement and she was discharged. At reexamination in 1934 she appeared almost stuporous and gave irrelevant answers. In 1939 she became increasingly excited, evidenced by her scolding of strangers and her claim that she smelt diapers. She was admitted to the Frankfurt Hospital. Here at first she was irritated and negativistic; later she became rigidly akinetic with an anxious expression. She often repeated in whispers what had just been said. At other times she screamed over and over again stereotypies such as "I belong to my mother" or "I am not stupid". Eight weeks later she became calmer and was discharged. After a year at home she again became excited, began suddenly to scold at night, and had to be readmitted to the hospital. Here she was nearly motionless, but again screamed in a shrill voice. She responded negatively to attempts to influence her. After three months she was taken home. In 1944 she bothered people on the street and wanted to walk with them, consequently she was brought to the hospital by a stranger. Here she spoke vivaciously, stating that men had talked to her. She accompanied her speech with stereotyped movements of her hands. On examination she gave illogical answers: (Stream/pond?) "There is a cakedough [dough = Teig; pond = Teich – trans.] and there is a pond where the ducks swim." (Error/lie?) "There you can look at a document." (?) "One does not need to lie, one can speak the truth." (Hunger is the best cook?) "If you have not yet been abroad you have to get used to it." (Make a sentence of hunter – rabbit – field) "The hunter shoots the rabbit." (Field?) "I am supposed to use three words ..." She went no further and looked about perplexed. She continued to be either restless – running around the room anxiously, pushing at doors, talking confusedly under pressure – or she lay motionless in her bed. Then she suddenly jumped up, ran through the ward, and just as suddenly stopped, looking puzzled. Occasionally she let out a loud scream. She developed diarrhea, deteriorated physically, and died in 1946. Autopsy could not be performed.

The personality of a **sister,** Elisabeth Ha, born in 1910, changed in 1931. She was given to brooding, believed she had a lung disease, suspected poison in her food, claimed people on the street greeted her strangely and the newspaper wrote about her. She was admitted to the Frankfurt Hospital in 1932, where she continued to feel that everything was strange. She felt influenced and that people were talking about her. She was very perplexed: "I tried to find a connection." She felt as if the priest was her father, that he wanted to make her "his own". People in the office were coming and going strangely, everything referred to her. The nurses were trying "to harm her somehow". The doctor was perhaps in disguise. She felt herself to be in the center of many people. Her affect was labile and she demanded her discharge in a stereotyped manner. After four weeks she was taken home. There she simply sat, indifferent and inactive. Again she began to refer everything to herself and in 1933 she was readmitted to the hospital. She looked around with astonishment, all things seemed strange to her, her own face had changed, at one time it was brown, then it became white, or she felt it was abnormally wide. She found significance in everything. She was limp and inhibited, but at times suddenly ran aimlessly through the ward. She became more perplexed, everything seemed completely different, the doctor had changed, he had different eyes. She also began to hear voices. Two weeks after admission she was transferred to Institution W. Here she for most of the time sat motionless, with mask-like facies. Then she suddenly jumped up and shouted: "I don't want to hurt him at all." During the examination she gave devious answers. (Greed/thriftiness?) "I can only tell you that greed is shorter than thriftiness, I mean the word greed." In 1934 she became excited, jumped about the ward, hit other patients, and spoke incoherently. The following year she usually lay in bed without moving, but was occasionally impulsively excited, and drank out of the bathtub and the toilet bowl. She developed decubitus, diarrhea and vomiting, and died late in 1935.

The subject most of the time was akinetic and negativistic. Above all she and her sister displayed impulsive excitement arising out of otherwise quiet behaviour. The subject was said to have been lying in bed without moving, but suddenly jumped up and ran through the ward, then she stopped without budging. Her sister was sitting motionlessly with mask-like facies, then suddenly she jumped out and shouted. It could not be determined from her impulsivity whether a cataphasic trait was present. The sister's reference and significance syndrome was striking. This confirms that the inhibition of periodic catatonia is not related to that of inhibited confusion psychosis. Against this diagnosis was the fact that though the inhibition was slight, ideas were plentiful.

The following subject with his many relatives, apart from some special symptoms, again shows the nuclear syndrome of periodic catatonia, with mixtures of akinetic and hyperkinetic elements. The copious family history is typical for periodic catatonia.

Case 58: Wilhelm Mün, born in 1898, became ill in 1915, and on examination at the Frankfurt Hospital exhibited the habit of stretching his body in a peculiar manner. He was transferred to Institution H. where he was excited for several months before becoming stuporous. In 1917 his condition was improved and he was discharged. In 1920 he returned to the hospital because of akinesia. He was again transferred to Institution H.; there although generally inhibited intermittently he scolded loudly. In 1921 his condition remained the same. In 1922 the institution report described him as friendly. In 1923 he was again stuporous, but assumed peculiar positions, grimaced, and was not clean. On one occasion he came out of his stupor, but soon fell back into it. In 1924 his mood was elevated and argumentative. In 1926 he was rigidly inhibited, he grimaced, and whispered to himself. Similar changes occurred during the following years. He was sometimes inhibited, displaying grimaces and stereotyped movements, at other times he was neutral. In 1932 he began to become excited with confused pressure of speech and severe hallucinations. In his good periods he remained apathetic. When SCHWAB and I reexamined him in 1937 he was akinetic, sitting bent over with empty facial expression. He had flexibilitas cerea and stayed in one position for a long time. Nevertheless his speech was compulsive and characterized by a high degree of uniformity: "Oh but, yes, yes, pretty garden, yea, yea, beautiful garden, yea, yes, ah so, yes, yes, I understood it right away, yes dictate, as the doctor is, ah, so, yes, yes, yes, yes, ah so, there are names, yes, Mün, Willi, yes, yes, correct, that will be written down, will yes or Willi, yes, nice man, yes, friendly, that will all be taken down, yes, yes, yes, also so yes, yes, yes, Willi, first name, yes Willi Mün, yes, yes, yes, yes, Willi yea, Willi yea, yes, yes, Willi, yes. Willi, one speaks so well, yes, yes, when one looks at life like that, yes, yes. ah so."

The **brother**, Adolf Mün, born in 1901, stated in 1919 that people were watching him strangely, and he expressed ideas of sinfulness. He was admitted to the Frankfurt Hospital where he stood around moaning. After four months he seemed well again, and although somewhat excited he was discharged. In 1928 he became ill again, he was restless and heard voices. He was readmitted to the hospital with incoherent pressure of speech, clang associations, and many linguistic errors. Then he was stuporous for a period of time during which he held his body in a rigid posture. Transferred to Institution H. in 1928, he was silly, talkative, in a cheerful mood. In 1929 he was smearing feces, indifferent, apathetic, sometimes irritated, and confused. In 1930 and 1931 he was apathetic and unclean. In 1932 he pulled the beds away from other patients. In 1933 he was noisy, aggressive and destructive. In 1934 he behaved similarly, and in 1935 he was described as excited and "totally confused". That year he died of heart failure and loss of strength.

The **sister,** Anna Mün, born in 1899, became ill in 1918 and was admitted to the Frankfurt Hospital. There she was excited and confused, she danced, laughed and cried. After five months she was discharged, but at home her behaviour was "unpredictable", and she had an illegitimate child. She was admitted to Institution K. in 1920, where she was described as silly, unrestrained, and apathetic. She was transferred to Institution H. The reports for 1921 described her as being mute, unclean, smearing feces in her face and ripping her clothes. In 1922 she hid under the bed covers, but then suddenly jumped out of bed and knocked over people who got into her way. She displayed stereotyped movements such as putting her left hand on her mouth. In 1923 she was still described as destructive and unclean; she died of heart failure.

A **mother's stepbrother's daughter,** born in 1891, had been confined to Institution B. since 1915. At first she appeared religiously ecstatic and spoke slowly. Later she became anxious and heard the devil's voice. In 1916 she was akinetic, but showed impulsive excitement. During the following years she fluctuated between primarily akinetic states and periods of confused excitement. From 1928 on she was generally negativistic, autistic under the blanket, unclean, and impulsively excited.

A **sister's son,** born in 1915, became suspicious, and was admitted to the Frankfurt Hospital. He was taciturn, off-putting, slow and apathetic. Previously contrariwise he was said to have been excessively demanding. He was discharged. The depressive symptoms diminished, and when observed in an examination three years later his energy was low and his affect flat.

Subject 58 presented a very good example of periodic catatonia. In the periodic course of his disorder he was generally akinetic and hyperkinetic as well. Of particular interest is the subject's linguistic behaviour. In stereotyped and repetitive manner nonsensical words were linked to one another. In general the words were only fillers which revealed no content. Apart from the akinesia there was noticeable verbal impulsivity. The subject and his brother showed the kind of excitement seen in confusion psychosis. Their sister was severely disturbed with negativism, aggressive excitements, destructiveness, and gross uncleanness. All three and their cousin as well progressed towards apathy, whereas their nephew ended up with only mild defect, which actually was already present after the first episode.

Thus I have presented both the symptomatic picture and the course of **periodic catatonia.** While there are some connections between it and the phasic psychoses, especially motility psychosis, there is no connection with the **systematic forms** of catatonia. The truth of this will become clear when the systematic forms are presented and the hereditary relationships are revealed. Repeatedly we find a periodic course and a tendency towards the two poles, particularly in acute and severe attacks. There was rarely ever a picture of insidious progressive schizophrenia, and no overlap. For this reason it is unfortunate that periodic catatonia, as well as affective paraphrenia and cataphasia, have been lumped together with the systematic forms under the catch-all name of schizophrenia. If a more precise diagnosis is to be made the overall concept of schizophrenia will become more and more unnecessary. The differentiation of periodic catatonia from other forms of schizophrenia is readily facilitated by the fact that heredity in this psychosis is unusually high. This fact shows its particular position, as we shall see in the statistical findings in more detail.

Summary

The course of periodic catatonia involves both hyperkinetic and akinetic phases. These are rarely present in pure form, rather, symptoms of one pole are mixed with those of the other pole. In the presence of akinetic traits hyperkinesia displays a certain rigidity. Movements occur with stiffness and jerkiness, and natural harmony is lost. From this change the original significance of the movements can no longer be detected. **Reactive movements** are no longer recognized as such; **expressive movements** lose their content to a great extent. Gestures turn into undefined poking movements, facial expressions turn into grimaces. This distortion of natural activity gives the excitement of periodic catatonia a **parakinetic character.** Even more clearly the opposite pole may be recognized as it impinges on the akinesia. Despite general rigidity of posture and facial expression aimless movement of an extremity may occur. The movements becomes uniform, either as a stereotype or iteratively. As well stereotyped postures develop in this manner; despite poverty of drive certain postures are repeatedly actively assumed. Another form of the admixture of hyperkinetic traits in akinesia is the occurrence of impulsive actions, often associated with aggressiveness. Furthermore some patients, with little provocation, suddenly burst out with exaggerated laughter. Negativistic behaviour may also imply the addition of a motor element to the akinesia.

Remissions regularly follow the acute phases. Patients with hyperkinetic states have, in fact, a relatively good prognosis. Even after several attacks, only a relatively small defect may remain. However akinesia is more often followed by permanent defect. If the defect is only mild it is characterized by general **weakness,** primarily psychomotor, but also involving affect. In more severe cases it is rather a matter of **indifference.** In any case a certain irritability exists which may easily lead to aggressive outbursts. In the **final phase** symptoms from both poles of the disorder still occur side by side. Besides lack of drive there may be grimacing facial movements, impulsive acting out, as well as verbal impulsivity, giving rise to "Vorbeireden" (talking-past-the-point).

Mild cases of periodic catatonia may temporarily imitate the pictures of motility psychosis; other polymorphic phasic psychoses, as well as the other two unsystematic schizophrenias may present similar relationships.

The Systematic Schizophrenias

The **sharpness of their symptomatology** differentiates the systematic schizophrenias from the unsystematic forms. Whereas in periodic catatonia, cataphasia, and affective paraphrenia we always drew attention to their symptomatological polymorphism, precluding strict delineation, in the systematic forms we find sharply circumscribed pictures. Apparently this parallels what we have observed in the phasic psychoses where alongside the polymorphic forms we had pure forms with their firm symptomatology. Probably the parallel exists because both the pure phasic forms and the systematic schizophrenias involve specific functional areas, the limits of which produce the sharp borders of the psychoses. No further relationship should be assumed because the phasic psychoses may be remediable while the systematic schizophrenias show permanent defects. The functional areas affected by the two sets of disorders are not only different here and there but are in fact of a very different nature. Pure depressions, euphorias, and manic-depressive illness are disorders of the emotions in close association with the vegetative nervous system, whereas the schizophrenias clearly imply disorders of higher thought and of willpower. Even the affective flattening of the hebephrenias does not suggest a disorder of the basic emotional system; rather it is at a higher level that affectivity is changed, while the more bodily types of feelings and instinctual drives are maintained. As I have discussed precisely elsewhere (LEONHARD 1970a), in the schizophrenias the highest, and at the same time the phylogenetically youngest human functions of the psyche are involved.

Because the higher human psyche is extraordinarily differentiated, it may be assumed that in the systematic schizophrenias a large number of functional units are involved. When it is shown that defect diseases, like the schizophrenias, can lead to disorders of individual functions, the thought occurs that here there may be a resemblance to neurological systems which also comprise functional units. This is even more likely since the entire nervous system is made up of "systems", i. e. networks of neurons with their axons.

Thus I believe, as KLEIST had always assumed, that schizophrenias are based in system diseases. However, while KLEIST applied this assumption to all schizophrenias, I restrict it to the systematic schizophrenias. The unsystematic forms, or the "atypical" forms, as I once called them, must be considered separately. Individual functional areas are generally affected selectively by the unsystematic forms of schizophrenia, but the cause of the disorder lies outside the functional area and can therefore affect other areas. In systematic schizophrenia, the disease process is primarily found within the affected system. The systematic schizophrenic pictures, which differ sharply from one another, are found in the end-state of the disease. In early stages accessory symptoms are associated with the schizophrenic process. Anxious or euphoric moods, ideas of reference, and hallucinations may characterize the onset of every form of schizophrenia, and are the manifestation of an unspecific effect of the process. As well, psychologically comprehensible reactions to the subjectively experienced onset of the disease play a role, including the feeling that a catastrophe is occurring. In every case of systematic schizophrenia this process invades the personality centrally; it is no surprise that at the beginning the still preserved personality reacts in a multitude of ways, especially with dismay and rationalizations. But often in the early stages the defect syndrome with increasing intensities is in the foreground, and although at first it may be minor, it does permit a precise diagnosis. Occasionally as FAUST and I have shown (FAUST 1953), the specific pictures may even be traced back to forms of latent schizophrenia, i. e. when the patient is not grossly psychotic. I will show that the symptomatic pictures do not apply only to defect states, although they may become clearest there. That it is a systematic form can generally be determined by the insidious onset which lacks many process symptoms. I do not dwell on the process symptoms because they do not contribute greatly to differentiating the individual forms. They are however presented in the various cases histories.

Simple Systematic Schizophrenias

In the majority of cases only one of the psychic systems is involved. The psychopathology represents the basic syndromes of systematic schizophrenias. Even if their dependence on certain systems is questioned, their significance is evident. The basic syndromes would still demonstrate that the nervous system in these schizophrenias is affected in such a way that there is not an indiscriminate mixture of single symptoms, but rather there are strict syndromes, or, as C. SCHNEIDER expressed it (although in a quite different context), strict symptom combinations.

In the following discussion I will continue to illustrate the disorder with cases in which there was a family history of psychoses. Due to the low incidence of heredity in the systematic schizophrenias there were not many cases of this sort, but the few we found suffice because their medical histories, at least in so far as the terminal state was concerned, were so similar. I first developed my system of classification on the basis of defect states observed in the Gabersee Institution, and published it in my book on "defect-schizophrenic clinical pictures". Later, in Frankfurt KLEIST taught me a great deal more about the classification of schizophrenia. At that time he accepted my separation of atypical, i. e. unsystematic, from typical, i. e. systematic forms; this made a more precise description of the systematic forms possible. There is always a combination of symptoms – this is to be strongly emphasized. I do not accept a "phonemic schizophrenia" just because phonemes are in the foreground of the picture. I accept it only if the voices are of a certain type and if the characteristic mood and thought disorder are present. Phonemic and hypochondriacal schizophrenia differ in relation to the presence or absence of bodily sensations, but the basic mood of the two forms is different. For example, in speech-prompt catatonia patients not only talk past the point ("Vorbeireden") but at the same time are anergic, autistic, and their facies are peculiarly expressionless. Thus the delineation of the single form is always established when there is a strict combination of symptoms, permitting the correct diagnosis to be made. This will be elucidated in the cases to be described.

Catatonic forms

The systematic nature is clearest in the catatonias. There may be "peripheral" manifestations, reminiscent of neurological diseases, particularly Chorea and Parkinson's Disease. Here and there similar symptoms are found, but always at a level higher than in the neurological diseases. This in itself is extraordinary evidence for the systematic nature of the catatonias. In Chorea and Parkinson's Disease (no matter of what etiology) specific systems of the nervous system are affected. In the "peripheral" catatonias the systems involved point to higher levels, i. e. systems affected are no longer purely neurological, but take on a psychological significance. Opposing symptoms of the catatonias are also evidence of a systematic origin, since in the nervous system one regularly finds that one function corresponds to another and the two balance each other. This antagonism occurs in catatonias which are not so peripheral, particularly in negativistic and proskinetic catatonia. It is not a coincidental phenomenon; we find it again in the hebephrenias and hints of it in the paraphrenias.

Parakinetic Catatonia

KLEIST designated the term **parakinetic catatonia.** He used it not only for systematic cases, as I do, but for all cases of catatonia with parakinesia in the foreground. Since his cases were, in the main, acute disorders they were generally **periodic catatonias** in which hyperkinesia, as we have seen, can be clearly parakinetic. The systematic form, which I originally described as "foolish", rarely manifests parakinetic movements in the phase of acute excitement; rather, parakinesia develops so gradually and inconspicuously that care must be taken not to overlook it. I have described periodic catatonia as such sufficiently, and have regarded acute parakinetic excitement as one of its manifestations. Therefore I am safe in limiting the term "parakinetic catatonia" to the systematic form. KRAEPELIN also recognized this condition, calling it "manneristic dementia"; under it he described the end-state of parakinetic catatonia. The illustration which accompanied his text shows a patient who was hopping and whose posture confirmed the presence of the parakinetic form.

At the onset of the disorder, no thought should be given to the picture of gross parakinetic excitement, otherwise the small changes will be overlooked. A fleeting grimace, a shrug of the shoulders, a slight twisting of the arms or of the trunk, a brief rocking of the body, and similar phenomena may be hardly more noticeable than in normal people who are embarrassed. If these fleeting parakinesias appear, without the rest of the behaviour revealing any embarrassment, the systematic form of parakinesia is more likely to be present than if severe parakinetic excitement occurs soon after the onset of the disorder. Often in these early stages the correct diagnosis is made from the kinds of verbal expressions which, as will be described, are somewhat sudden and jumpy.

Even after many years the picture may still be relatively mild. The close observer can no longer overlook the parakinesias, but they often go unmentioned in the progress notes. In the face movements resembling expressions – really distortions – are more common here. The corner of the mouth may stretch to the side a bit, or the cheek may be raised; the eye may wink or the forehead twitch; there is a movement of the hand, the fingers spread apart fleetingly, the shoulders jerk forward, the upper trunk jerks briefly; one leg is lifted off the ground for a moment, and so on. Frequently the movements seem to have a purpose such as reaching for an object or a turning of the head, but the movement is not carried out naturally.

If the parakinetic disorder is recognized in the severe forms which develop in the end-state, it will not be overlooked in milder forms. The picture, by then, is impressive but apparently hard to describe since progress notes, even at that stage, do not usually go beyond general terms such as "manneristic" or "eccentric", and thus rarely give the impression that the disorder was a severe parakinetic catatonia. Patients' voluntary movements are changed; they lack the fluidity of normal movement. Involuntary purposeless intermittent movements further distort the voluntary movements, becoming stronger when the patient is excited. Moderate stimulation may precipitate a brisk parakinetic restlessness; such movements appear to be **reactive.** The patients turn toward everything around them with jerky movements, look here and there, grab oddly at other objects, feel them, and speak about the things they see, with short factual remarks, but not expressing opinions. They may run impulsively through the ward, incited by the appearance of the doctor or disturbed by other patients. When their attention is seized by new surroundings, reactive movements predominate. In surroundings to which the patients are accustomed, **pseudo-expressive movements** predominate and create the impression of "foolishness". Gestures of threat, greeting, astonishment, love, superiority, and many others appear in grotesque distortion. Facial expressions reflect a great variety of emotions. Laughter exaggerated by parakinetic distortion looks more like grinning. To a certain extent the grimaces are dependent on the actual mood of the patient. But the connection is often lost.

Movement disturbances can also be recognized in speech – rounding off and modulation are absent, individual words are produced rapidly and jerkily; after every short sentence there is always a pause, with the next sentence following abruptly.

Constructually the parakinesias differ from voluntary movements only in the form of execution which is choppy; at times the movements are simpler and nearer to choreatic movements. Perhaps the latter are fragments of pseudo-expressive movements. A waving of the arms, a brief shrugging of the shoulder, shaking of the head, rocking of the upper body, and other movements are reminiscent of chorea, but are not so fleeting and above all do not return as rapidly to a position of rest. A movement starts, continues for a while, is frozen for a short time, then is immediately chopped off and is replaced by another movement. The grimacing of foolish catatonia is reminiscent of Huntington's chorea. It need not always represent a certain mood; indeed, the most varied facial distortions appear behind which no content is recognizable.

The above is a depiction of parakinetic catatonia in a very severe form, in order to clearly demonstrate its essence.

The longer the defect lasts, the more uniform the movement disturbances become. The simple choreiform parakinesias lose all variants, although the movements are not particularly clear-cut in all cases. Pseudo-expressive and other complicated movements are affected more. If a patient is seen only once, the movements may deceptively appear quite varied, and because of their individually different character appear new to the observer during the first examination. Prolonged observation reveals that the same patient always produces the same series of movements, some assuming even a stereotyped character. In order to emphasize their special character it is fitting to label them as **bizarre,** differing from the stereotyped movements and mannerisms of other catatonics. I present a few **examples:**

A female patient holds a spoon in her finger tips and brings it to her mouth in a twisted motion. Another greets the doctor with a grimace of great astonishment and with uniform gesticulations. Another parades with grotesque posture and exaggerated grimacing in front of the doctor. Another tries to run out of the door while making all kinds of grunting noises and contortions. Another keeps on giving strange compliments. Another grins repeatedly while moving her hands with parakinetic restlessness. A male patient grimaces and produces unarticulated noises. Another jumps over the cracks in the wooden floor. Another from time to time jumps from one leg to the other. Another on his way outside jumps a few feet to the left and then to the right. Another kicks his slippers off his feet and then retrieves them.

Such mannerisms could be partly described as non-mannerisms that may be prevented by suitable industrial therapy. But then the disorder leads to other types of parakinetic features. The characteristic picture itself does not change.

It is perhaps related to the inability to produce words other than in isolated outbursts that the patients only construct **short sentences** which are often ungrammatical. The following remark about another patient, produced in agitation, is very characteristic of foolish catatonics: "That one must leave! I will kill him! Don't want to see it! Need knife! Must be killed! Won't put up with it! I don't care! Men are here! Out, out!" The contents of patients' remarks present themselves very differently. Besides appropriate remarks which relate to the situation, there are **digressions** appearing nonsensical. Thus, with no cause or relevance, I was asked by such a patient, "Have you no dog? Don't you ride a bike? Are you here today as a straw man? Have you been on Phoebus or in the Cafe Wien?" Another patient greeted me with the remark, "The battles of the World War!" The impression is gained that these patients respond to disorganized speech impulses.

In some patients verbal expressions are common, others hardly speak at all. It seems to depend on individual inclinations. How far thought itself is actually disturbed cannot be determined from such remarks. Most noticeable is the abrupt jumping from one theme to another, i. e. **jumpiness of thought.** Despite this the disturbance of logical thought cannot be very great, since the patients observe very closely what goes on about them, as their relevant remarks, mixed into their speech, demonstrate. They are well oriented and do not fail to take note of current events. When their intelligence is questioned they may digress from the theme but usually are able to give useful answers. An important observation is that the tendency to short sentences also appears in the written expressions, which can be only partly explained by difficulties of articulation. It can be assumed that thoughts change as impulsively and quickly as parakinetic movements. In the second half of the following letter of a foolish catatonic, the disconnected thought is very clear.

„Dear Mother! Thank you very much for the nice package and the letter with it. Have you already got used to Munich? I would like to go there again, as Munich is a beautiful city. Best regards to you, I want to come home again. Regards also to relatives and friends. How are you all? I hope, well. W. was so nice in giving me bread and butter, I am satisfied. How is Karl doing at his new job? I hope I will be able to visit the plant sometime. To be on board or on a ship is very nice. If you are coming back, please visit. Bring a lot with you, dress nicely. There is money in the institution, as much as you want. Regards to Papa and Mama. I am always worrying. Besides come with car, horse and wagon. And especially have patience with the patients. And when are you arriving, Sir? Soon, I hope. Many kind regards until we see each other again in ten weeks. Happy Easter. Springtime. Ditto Goodbye. Send the letter home. Bring package and something with it. Or with Berlin or Sanatorium Nazareth? How is it with the car Berlin? Well! Is never spun. Besides to a big city or a trip abroad. Mainly to get enough sleep. Regards from Luise M. Have a visit. Luise and otherwise everything with the old and new. Yours Elsa.“

Foolish catatonics still seem to function well affectively. Female patients are often quite erotic and, as a consequence of their general lack of inhibition, with their expressive movements openly maintain the eroticism. They keep in contact with their relatives better than other catatonics and they express the greatest joy over a visit. The basic mood is generally carefree and satisfied, but they may become quite angry. Emotions are abnormally strong and are manifested outwardly through the expressive parakinesias. Their motivation is just as inconsistent as their thought and speech. They initiate all sorts of projects and express interest in everything, but they never bring them to completion and cannot complete any useful task. Delusions and hallucinations are not present.

Cases with psychoses in the family of patients are relatively frequent, more so than in other forms of systematic schizophrenia. In earlier editions I described all such cases from my investigations in Frankfurt, because I wanted to show that when there are several cases in the family the clinical picture is always the same. Here I present one such family.

Case 59: Friedrich Dob, born in 1890, the son of a factory owner, became ill in 1909 while continuing his education abroad. In a state of excitement he suddenly began to talk loudly in a theatre, exclaiming that his neighbour had the head of a devil. In the following days he went to church frequently and claimed to have seen Jesus alive, to have fought with the devil and to have won. He then became anxious and felt he was being watched. He was brought home and admitted to Sanatorium N. on June 26, 1909. There he spoke with a roguish smile, claiming that the devil had appeared to him in the form of a dog. Otherwise he was not very conspicuous. On one occasion he appeared to be very anxious, saw the ceiling on fire and many crosses in his hand. Generally he showed little interest and smiled stereotypically. Sometimes he spoke "a lot of nonsense, about the war". He was discharged on August 13, 1909. At home he was usually happy and wanted to have "fun". On November 2, 1909 he was admitted to the Psychiatric Hospital M., where he spoke cheerfully about his failures in school and made jokes. On November 15 he returned to Sanatorium N. He continued to joke, but at times became irritated. He disturbed other patients with his naughtiness, often gave senseless answers to questions, and behaved impulsively. He was discharged on February 15, 1910. At home he often spoke nonsense. On July 8, 1910 he returned to Psychiatric Hospital M. where he made jokes, laughed, and spat. He was discharged in September. At home he was very demanding. He was readmitted in January 1911. He had a mischievous smile, digressed continuously and grimaced while speaking. It was reported that his movements were manneristic, his behaviour was "highly mischievous", he made many puns, and often spoke "complete nonsense". In March his manner of expression became bizarre and his speech was described as "disjointed". He was again discharged in May. At home he was occasionally excited, spoke about hearing voices, and in August he was readmitted. He wanted to smoke during the examination and wanted a driver's license. He still had the sudden outbursts of silly laughter. On November 17, 1911 he was transferred to Institution E. There he spoke of visions, told of seeing the spirit of his soul, but it was not clear whether or not he was still making jokes. In 1912 he was more often irritated and claimed to be the guinea-pig of evil spirits. In 1913 he spoke of voices and electrical currents. In 1914 he dug holes in the earth because the voice clown ("Stimmhansel") was allegedly there; again, as a response to other patients' utterances, he may only have been joking. At times he was inactive, at other times mildly excited and he often made "all kinds of senseless movements", which he could repeat for hours. He also talked a lot. In 1915 he made further "manneristic movements", but was otherwise dull. The medical record was very sparse owing to the war time situation, but in 1917 it was stated that he grimaced, carried on conversations with himself, at times was excited, scolded confusedly, and became aggressive. In 1918 he was described as displaying distorted postures and raising his eye brows. In 1919 he tried to strangle an attendant and was very angry. At times he limped around curiously or kept one hand in his pocket and the other behind him. He took food away from other patients. In 1920 in response to a greeting he made peculiar movements with his legs and touched himself with his arms. In 1921 he ate everything he could reach. In 1922 he was "extremely manneristic and grimaced exaggeratedly"; he was scatter-brained and his basic mood was cheerfully silly. In 1923 and 1924 he was "continuously silly". In 1926 and 1928 his movements were again described as "peculiarly manneristic". In 1929 he was "silly, stupid, cheerful, and full of stereotyped grimaces and gestures". His affected movements, grimaces, and gestures were mentioned consistently in the years that followed. Otherwise he was described as dull, with a constant silly smile. B. SCHULZ and I examined him in 1939, noting the following: Continuous parakinetic restlessness, with repetitive series of movements; he reached with his left arm which he held peculiarly twisted up to his mouth and back, sometimes licking his fingers with his tongue. For a while he made the same movement with his right hand. A second series of movements was partly associated with the first; it consisted of rocking back and forth of the trunk and slight raising up from the chair. Other movements included: unnatural turnings of the left hand, at times associated with slight raising of the left forearm; nodding of the head, along with exaggerated grimacing smile and stereotyped "yes, yes". Many other movements in different forms were combined with these stereotyped forms. They also varied according to what was being discussed and to a large extent represented an attentiveness to the outside world albeit via exaggerated expressive movements, i. e., he looked at a picture with a questioning, astonished expression, and he listened with an expression of exaggerated surprise to the doctor reading from his medical history. He then made a deprecatory gesture with his hand, his grimaces expressing contempt. Most characteristic was his exaggerated laughter, often erupting in association with other exaggerated movements. If stimulated more, his laughing shifted to intense grinning. His movements were choppy and distorted, so that on the whole a very strange picture was generated. His verbal output was always short and nearly always ungrammatical. Often he did not finish words, stopping in the middle, or adding an unarticulated sound at the end. The series of sounds always broke off, never was accelerated. His answers were generally incomplete and at times tangential. (Child/dwarf?) "A child is small and a dwarf is also small." (Difference?) "That it just pleases one."

A **brother,** Alfons Dob, born in 1888, a merchant, had always had a hard time making friends, had had no relationships with women; he spent money buying whatever pleased him (for example, 50 postcards at a time). He often talked a lot without making full sense. Now and then he made quick senseless movements and greeted strangers on the street whom he did not know, simply because it gave him pleasure. As he became more and more abnormal his relatives brought him to Psychiatric Hospital M. on April 12, 1913. There his speech was described as "crazy, disconnected in an affected manner". His answers to questions were at first rational, then digressed: his relatives wanted to poison him and had turned his father against him. Generally silly and cheerful, he was at times angry and lachrymose. His remarks were confused, for example: "Self-knowledge a means leading to improvement." Neologisms that looked like puns appeared: "Idealmaterialism", "magnetic-theoretical", "Thicktic" (instead of thicket), "coffeologists". On October 6, 1914 he was transferred to Institution E, where he was carefree and cheerful, but confused in his speech and affectedly ceremonial in his movements. Because of the war there were hardly any other notes in his hospital record. In 1918 he died of pneumonia.

A **sister**, Marie Dob, born in 1892, was a quiet, calm girl who read a lot. She began to change in her 16th year. She became suspicious, argumentative, unstable, and had an illegitimate child. On June 20, 1922 she was admitted to Psychiatric Hospital M. The record states: "The diagnosis is not clear since the signs of schizophrenia (delusions and hallucinations) are absent." In the progress notes it was reported that "the patient speaks with great haste, begins sentences without ending them. She continuously grimaces with her eye brows. She laughs in a silly and embarrassed manner. She speaks circumstantially, her answers always miss the main point of the question, and she breaks off in the middle of sentences to start something new." She was discharged on July 22, 1922. At home she did not get along well and had many boyfriends. She was readmitted on October 24, 1926. The record states: "She speaks haltingly, with individual words and parts of sentences sharply interrupted with breathless pauses. There are no complete sentences; she repeatedly starts again, but then suddenly stops." Her speech is also described as "manneristic and jerky". Grimacing is mentioned frequently, a silly smile, her "flamboyant theatrical pathos", and stereotyped licking of her lips. Her entire nature is described as unnatural, remotely friendly, and manneristic. On November 12, 1926 she was transferred to Institution H., where she remained eccentric and disorganized. She was discharged on November 21, 1926. She was back in the institution from October 7, 1931 to July 6, 1932. The record described vivid gestures, various facial expressions, as well as affected manner of speech and disconnected train of thought.

Another sister, Josefine Dob, born in 1899, was admitted to Institution E. on August 2, 1916, after having become anxious at home and having talked nonsense. In the institution she spoke with a halting voice, was generally calm but occasionally excited. She was discharged on April 15, 1917. On August 17, 1917 she was readmitted, having begun to buy useless things and to flirt with men. In the institution she was erotically importunate and at times irritated and aggressive. On December 15, 1918 she was discharged again. Twenty years later the patient was examined in Psychiatric Hospital M. because she was to be declared mentally incompetent due to profligacy and mental derangement. She had spent money unnecessarily and had run away from rest homes, where her family had apparently tried for years to keep her. She had entered into erotic relationships. Six years before a doctor wrote that she followed every sudden whim. She had, for example, suddenly travelled to Starnberg in order to be treated with rays. The records also state: "Because she had to have everything she wanted, she acquired a large number of toys and other junk, with no expenses spared. On one occasion she walked through the city wearing an absurd night cap. The examination for incompetence used the following descriptive notes: "The psychological findings are difficult to state. In a short meeting, D. may appear to be inconspicuous, although the manner of her appearance is very peculiar. All her movements are somehow disconnected and choppy, her presence is somewhat awkward, and occasionally a certain affectedness of individual movements is observed with which she accompanies speech, along with mild grimacing." Furthermore her thoughts about her personal progress last only from today to tomorrow. Apparently she has no goals which she would want to pursue any further." Her affective behaviour is described as "jumpy, explosive, unpredictable." After mentioning her poor concentration the record describes her speech as follows: "The way she talks has a special coloration due to the affectedness of her remarks, which at times become very stilted due to all the verbs being used as nouns." Finally, interruption of thought is mentioned.

In reference to the **mother** of the Dob siblings, the medical history of one of the children states: "The mother speaks affectedly, circumstantially, never gets to the heart of the matter, cannot give correct information, makes an unintelligent impression, continuously praises herself, and grimaces."

Subject Dob displays clear parakinetic catatonia. At the onset he apparently was mildly ecstatic and anxious, but later the picture became more typical. Early on, silly laughter played a role which was described as "compulsive" and was referred to in connection with his grimacing. The laughter was still present when I examined him and had a parakinetic nature which had probably existed long before. The often mentioned cheerfulness was probably derived from the laughter, which, however, need not necessarily have corresponded to a cheerful basic mood, but rather could have represented the parakinetic transformation of the facial expression of a momentary mood. The basic mood does seem to have been carefree and satisfied as is often the case in parakinetic catatonics. Dob's impulsive actions were noticeable very early, occurring as a reaction to external events. The "manneristic" manner of the movements cannot be missed. There were also his tangential and senseless answers. Thus I suspect that the picture of parakinetic catatonia was present from the start, but in the mild form I have described above as typical at the onset. Later parakinetic features became more obvious in the "senseless movements", "manneristic movements", and "eccentric postures".

The brother Alfons Dob unfortunately was not observed for a long time, nevertheless clear elements of parakinetic catatonia were present. The disorder is seen in his affected manner of speech, his ceremonial nature, and his senseless movements. There was also the characteristic abnormal thought process and the playing with words which I found in a very similar form in other patients with parakinetic catatonia. It may be explained by the disconnected thoughts producing all kinds of ideas, as well as by the prevailing cheerfulness. Friedrich Dob made jokes, too. Both he and Alfons acted impulsively in reaction to accidental stimuli of external origin. Furthermore, since motor excitement is described while paranoid phenomena are absent, I do not doubt that the diagnosis of the subject's brother was parakinetic catatonia as well.

The sister Marie, too, surely suffered from parakinetic catatonia. The description of her speech, with its haste and haltingness, is very characteristic, and may lead to loss of articulation. The affectedness of all her movements is described. Since parakinesia is clearly recognizable in the medical history, I want to emphasize that it was certainly present to a marked degree. In my experience I have seen that the description of clearly parakinetic restlessness is overshadowed by remarks about affected and manneristic behaviour. Paranoid phenomena are generally given more importance. Since these are absent, the presence of a schizophrenic disorder is often doubted at the onset, although parakinesia and the personality change are quite evident.

Likewise, the parakinesia of the sister Josefine is clearly described in the form of affected movements and grimaces. Since her records describe disconnected thought, sometimes taking the form of abrupt stoppages, and impulsive actions based on momentary impressions, and since there was a certain motor excitement and an absence of paranoid features, I feel I am justified in assuming the presence of parakinetic catatonia in a somewhat milder form. I also emphasize the erotic impulsiveness of the two sisters; uninhibited eroticism often appears in female parakinetic catatonics.

Thus the diseases in the Dob family were very uniform. I have good reason to diagnose parakinetic catatonia in the subject and his three siblings. In none of the siblings do I find symptoms of any other condition. The presence of the disorder in so many individuals in one set of siblings is remarkable. It may be explained by hereditary loading from both the mother and the father. The mother herself may have had mild parakinetic catatonia, as may be deduced from the brief remarks about her mentioned above.

Summary

The hyperkinesia of parakinetic catatonia, which in its early stages may develop only after stimulation, draws its characteristic form from the nature of the movements. Voluntary movements are unnaturally jerky, involuntary movements even more so. Fluid connection between movements is absent: a movement occurs, stops for a moment, and then the next one suddenly begins. The motor process is jerky and grotesque, even when individual movement is accelerated. In terms of content, involuntary movements are characterized by bizarre reactive and distorted expressive movements, with facial expressions appearing vividly in the form of grimaces. Simple movements resemble chorea. In the defect state parakinesia may still be varied, but certain forms are continually repeated as bizarre gestures. The peculiar movement disorder is also expressed in verbalizations. Words are chopped off, produced in isolated jerks, and generally only short, non-grammatical sentences are formed. Oral as well as written utterances of the patients are partly relevant, partly irrelevant, and often non-sensical remarks are found. Additional cases in the families occurred relatively often and were usually also characteristic of parakinetic catatonia.

Manneristic Catatonia

Earlier in my career I characterized manneristic catatonia as rigid. At that time I was referring to very severe cases and had not yet realized the importance of mannerisms in this type of catatonia. The motor aspects certainly do, even in milder cases, contain a rigid element, but the term "rigid catatonia" suggests akinetic forms with which this disorder should not be confused. What KLEIST calls stereotyped or manneristic catatonia corresponds to this subcategory of mine. On the other hand it is my opinion that his akinetic form really belongs to periodic catatonia. In severe cases motor rigidity dominates the picture, but in milder and beginning cases mannerisms are more prominent. At first they may resemble compulsive actions; it is particularly important in that instance to observe motor activity in order not to miss the onset of catatonia. Emma FAUST (1953) described relatives of patients with manneristic catatonia as having "compulsive neuroses", not yet recognized as catatonic, but who according to the family picture certainly represented latent manneristic catatonia.

The mannerisms of my patients included stereotyped kneeling, touching the floor, touching other patients, turning the body before passing through a door, holding a spoon peculiarly, putting the fork down after every bite, collecting worthless objects, pushing rocks and papers off the side walk, refusing certain foods, refusing all nourishment necessitating tube feeding, and mutism, at times accompanied by sign language.

The mannerisms continue so long as the rigidity has not progressed too far. Otherwise they are only present as stereotyped "omissions", which have to be distinguished from general lack of motility. Alongside the **"movement mannerisms"**, these **"omission mannerisms"** always play a role, especially with such symptoms as refusing nourishment, mutism, and standing in a fixed place. When some patients in severe end-states hardly move at all these omission mannerisms are generally quite significant. This can be prevented by continuous occupational therapy. Movements which the patients must continually perform remain preserved even if the patients are otherwise quite rigid. For every daily necessity there is a certain series of acts one following the other; all other movements are absent. In a sense the whole course of the day becomes a rigid mannerism. If the set course of movements is externally interrupted no other activities take the place of what has been lost. I observed a patient who was prevented from entering the dining room from a side opposite that of the other patients as she had been accustomed to doing – she stood still and did nothing.

The peculiar rigidity of these catatonics, on the basis of which I gave them my earlier term, is due less to an absence of voluntary than to the disappearance of involuntary movements, particularly movements of expression. If a person is completely motionless, but in a natural posture he is not rigid. On the other hand if a person is rigid while moving voluntarily, the movements are somehow unflowing, tight, and "wooden". This phenomenon occurs in Parkinsonism with which I have compared catatonia. Differentiation from neurological rigidity, so often seen nowadays due to neuroleptic treatment, is quite easy, since in catatonia only expressive movements are affected, while voluntary movements take place at a normal speed. Of course catatonic rigidity can no longer be identified as such if it is accompanied by Parkinson's disease.

Patients with manneristic catatonia stand stiffly and walk with choppy steps. The arms do swing back and forth, but the finer adjustments are missing and so the patients appear wooden. In speech, too, the finer modulation is absent. It is not softer or louder, higher or lower, faster or slower, in harmony with the expression relevant to the content; rather the speech is monotonous, although its average speed or loudness need not be abnormal. There is also impoverishment of facial expressions. The finer facial movements shown by healthy persons even in unimportant conversations are absent. The rigid tone of the muscles of the face makes it look quite different from the face of a normal person. When all these described traits are mild and voluntary movements are still uninhibited, rigid catatonia may still be assumed with certainty if the presence of mannerisms is demonstrable.

As the catatonia progresses spontaneity becomes more and more limited. The worst impairment appears only if without occupational therapy the most important movements are not practised. Even with continued occupational therapy patients may stop speaking, no longer perform the simplest tasks, and unless stimulated stand or sit like wooden blocks without moving. Impulsive actions may occur even in otherwise immobile patients. Such catatonics may suddenly hit another patient. With careful observation it may be discerned that this sort of aggressivity occurs if the patients' mannerisms are disturbed.

The **patients' affectivity** is relatively well preserved. Even in mild cases some of their affect is lost because they are generally fully preoccupied with their mannerisms. As their motility is increasingly impoverished this phenomenon increases. However even in severe cases deep affective changes may still be observed; for example, if one speaks tactlessly of their disease the patients will be offended.

In milder cases **thought** seems not to be disturbed. In severe cases, intellectual activity decreases somewhat, presumably because thought processes are inhibited as well. Paralogical disturbances do not appear. With appropriate guidance patients are still able to perform intellectual tasks. Even in very severe states, when patients stop speaking, they continue to be sensibly concerned with the events in their surroundings. This can be seen in their appropriate facial reactions to given situations. Despite the rigidity affecting facial expressiveness, the expression is still comprehensible to the observer. The face may somehow reveal sorrow, anger, and the like. More often I noticed a smile which was not senseless and occurred when appropriate. It has an understandable, even roguish character, apparently implying that more is being assimilated internally than can be gathered from the external appearance.

The following is a case from my Frankfurt series.

Case 60: Viktoria Nied, born in 1891, had been ill for years before she first came to our hospital. In 1929 she wrote letters to the welfare department in which she referred to herself in the third person and reported fantastic experiences. She claimed that the police wanted to take her pension away and that she had felt as if someone had put an axe on her chest. Someone had pulled her breast three times during the night, a man stood naked in front of her and had pumped her stomach. She had had sexual intercourse and felt as if her body would be torn apart, and as if a man's body were inside her. Two men were tight together and hung by their buttocks. She had various animals in her body, while the doctor was full of cats. Two doctors tied up were thrown into a court room. In 1930 she was admitted to the Frankfurt Hospital. She spoke of experiments and made mysterious allusions. She exclaimed stereotypically: "Mrs. Nied wishes to be released." She was transferred to Institution W., where she continued to make the same remarks. She neither ate nor spoke for a while and communicated only by nodding or shaking her head. Sometimes she stood motionless in the ward, her eyes closed. In 1931 she ate only cookies and drank milk, and eventually had to be tube fed. She generally sat in the same place with her eyes closed. She expressed her desires only in writing. In 1932 she displayed a rigid, stereotyped posture and kept her eyes closed. She sometimes gave the doctor a note in which her complaints were described stereotypically: She was a "fortified oath person" and the persecutions should stop. In 1933 she stood stereotypically at the window, her eyes tightly closed, her hands stuck into her belt. In 1934 she was incontinent of urine every night, and when reproached she gave no answers, only pretended to write. She tilted her head stereotypically to one side, holding it in her hand. In 1935 she displayed the following stereotypical movements: She rubbed her left cheek a few times, held her chin, then made some hissing sounds. At night she regularly urinated in the sitting room. In 1936 she displayed a catatonic posture with the left hand pressed to the face. She like to walk backwards with eyes closed. Her hissing mannerism continued. In 1937 she let herself be occupied in sewing. She responded correctly to questions, but only in writing. Her spontaneous written expressions were nonsensical. SCHWAB and I examined her in 1939. In 1940 she died of enteritis. The results of the 1939 examination follow: She did not speak at all, standing in the same posture with the left hand on her cheek and the right hand under her apron. She did not sit. She stepped back when pressed and shook her head vehemently. When pressed further her face became pained and disapproving. According to the nurses she never said anything but understood everything and gave written answers. She wrote with her hand in the air or on the wall. She always sat in the same position, but otherwise behaved sensibly and took care of herself.

A **sister**, Johanna Kon, born in 1887, had had a good sense of humour, had performed well in school and as a domestic servant. During the time of the World War she suffered from nocturnal fears. In 1918 her condition began to worsen. She sometimes mistook men for her deceased stepfather. In 1923 on one occasion she suddenly became excited and wanted to throw her children out of the window. On July 16, 1923 she was admitted to Municipal Hospital A. She barely gave any relevant answers, but declared stereotypically, "You just can't tell what's happening anymore." Later she added the phrase, "I believe that," to all answers. She spoke somewhat slowly, displayed catalepsy and "light feathery resistance" to passive movements. She was transferred to Institution K. where she was apathetic and claimed that she heard her children. On September 28, 1923 she was taken home against medical advice. At home she occasionally burst into tears and expressed jealousy. In 1925 she became excited and broke windows. She was readmitted to Hospital A. Here she did not speak, or at least had long "blocks" before answering. She appeared depressed or non-participatory. After a few days she was again taken out, became excited at home, and was brought back ten days later. She was depressed, at times laughed senselessly, and then appeared blocked. She was transferred to Institution K. She gave information reluctantly, kept turning her head to the side, and appeared affectless and dull. She was again taken home after six weeks. Four months later she returned to Hospital A., after having again become excited at home. She was again "blocked", hardly gave any answers, and her behaviour was reported to be "unnatural, perverse". She occasionally lay down in bed backwards, and threw her bed jacket under her bed. She was transferred to Institution K., where she was resistant, answering stereotypically, "I won't stay here, I won't be locked up." On one occasion she claimed that her husband had been switched during the night. In 1926 she was inhibited, gave no answers, and refused to get out of bed. She was discharged in January 1927, but returned to Hospital A. in March. Her behaviour was silly and eccentric, and she was transferred to Institution K. There she totally refused food and had to be tube fed. She broke windows in a sudden outburst, and then became quiet again. She spat in her hand and on the floor. In 1928 she did not reply to questions, but only nodded or shook her head. In 1929 she occasionally made silly remarks, but otherwise was silent. In 1930 she was "full of eccentricities, gave no answers, and stared blankly ahead of her". In 1931 she was silent, but did what she was told. She often smiled "cunningly". She displayed mannerisms while eating: unless she had a certain dish she refused to eat at all, but ate the bread from the garbage pail. In 1932 her peculiarities continued. She would not, for example, take a dish from the nurse – it had to be placed on the floor first. She also grimaced. In 1933 she smiled a lot and displayed other peculiarities. In 1934 she produced odd sentence constructions, in one letter for example, "You can certainly consider everything more profitably if you are the reward of your will." In 1935 she spoke to herself at times. In 1936 she was busy and had a silly smile. In 1937 she displayed many oddities, washed herself with cleaning rags, swept the floor with her dress, and hid some objects. Her speech was eccentric and she missed the point. In 1938 she was eccentric and displayed mannerisms while performing house work. She remained in the same state in 1939 and 1940. She was irritated during menstruation. In 1941 she became a victim of the cruel "euthanasia".

The subject Nied illustrates a severe form of manneristic catatonia. For years she assumed stereotyped postures, was often rigidly immobile, hardly spoke, sometimes refused food and had to be tube fed. In the end-state her only movements were either mannerisms or simple occupations to which she was led. The poverty of movement itself was codetermined by the mannerisms. Thus she was completely silent, although she could express herself in writing. The association of rigidity with mannerisms proves the presence of rigid catatonia.

The fantastic phenomena which she displayed prior to her institutionalization are noteworthy because they hardly ever occur in manneristic catatonia. A combination with sluggish catatonia might be considered; during the onset of the latter fantastic confabulatory phenomena are common. Since otherwise the patient did not express anything when hospitalized, although she had recently written about them, I consider it possible that they were only erotic fantasies which she elaborated in a manneristic fashion. Sadistic and masochistic tendencies appeared in gross forms. Her extreme erotic attitude was independently mentioned in the progress notes.

Her sister had manneristic catatonia as well. At first her excitement was rather noticeable, but it occurred only at home and was probably of a reactive kind. Furthermore it seems to have been at least occasionally based on menstrual irritation. Otherwise stereotypic behaviour is described from the onset, as well as poverty of movements with "blocking". Like her sister she later had mannerisms, not speaking but communicating with movements of her head. When later she performed tasks her activity was full of mannerisms. The "cunning" smile is also noteworthy, since it is characteristic of rigid catatonics. Thus I take it that both sisters had the same form of systematic catatonia.

Summary

In manneristic catatonia we find increasing impoverishment of involuntary movements with the development of rigidity of posture and movement. There are also mannerisms which at the onset are often more prominent than rigidity. In the further course movements become more narrowly limited to stereotypes. Mannerisms may continue all day long. Impoverishment of motility becomes even greater when the "movement mannerisms" recede and "omission mannerisms" become dominant. Then the particularly severe lack of movement develops and the patients stay in the same spot with stiff positions and facial expressions. Appropriate occupational therapy may prevent these severe states, as movements that are constantly practised may be preserved.

Proskinetic Catatonia

In proskinetic catatonia there is an abnormal tendency toward automatic movements which occur reactively to external stimuli. Patients are interested in external events; they hold on to objects ("Nesteln" in German), go along with other people ("Mitgehen"), and make responsive grasping movements ("Gegengreifen"). Furthermore the speech of these patients has a peculiar mumbling character. When the patients appear interested and friendly and do not yet display the marked lack of initiative that they later develop, they may not be recognized as catatonic at the onset. A few ideas of reference may be present, which are seen in many cases of early schizophrenia, giving the impression of a paranoid form. At this stage the holding on to objects is still generally absent and speech, although soft, is still comprehensible. Yet during this period the characteristic grasping and going along movements may already be obvious. At this point patients should be given appropriate tests. As the condition progresses initiative decreases more and more, permitting the diagnosis of catatonia to be made from the lack of motivation alone. Speech becomes quieter. The holding on to objects is quite noticeable. The total picture in the end-state is very characteristic.

Patients with proskinetic catatonia respond to **being addressed ("Zuwendung").** They look up when one talks to them. They do not become tired of responding even if the conversation is prolonged. The face reflects little other than a recognizable friendliness and interest. At the same time verbal expressions are common. The lips begin to move and **mumbling sounds** appear. As long as the observer occupies himself with the patients they speak, sometimes louder and sometimes more quietly, with short pauses and not always comprehensibly. If they stop talking, usually only a simple nod or an encouraging look induces them to continue their mumbling. It is not a whispering but talking softly with a monotonous voice: hence, a mumbling. When from their facial expressions it seems they are interested, it may be at first assumed that they want to communicate something. However that is not at all the case. By listening very carefully one can understand enough to judge the content. But it is mostly a stereotyped repetition of certain expressions in short sentences, referring to events in the patients' surroundings, but mostly nonsensical. Sometimes these are repeated several times, sometimes they alternate with similar expressions. In the mumbling of one patient, for example, I could hear the following verbigeration: "The stolen women, the stolen women must go to prison, the stolen ..." Another patient's verbigeration: "Do you have an apron, apron, do you have an apron, do you ..." If the onset of the disorder was not too far in the past one can understand considerably more, especially when the patients are encouraged and stimulated. In such a case I recorded the following, which clearly reveals the stereotyped repetition:

(How old?) "Four years younger than my sister, four years younger than my sister; yes, four years younger than my sister." (Go on) "Go on, what more? If I am four years younger than my sister, four years younger than my sister, and now, what is? And now, what is? And now, what is being bad? To be bad, to go to bed, are you bad too? And now, and now." (When were you born?) "When were you born, when were you born, when were you born." (You!) "I am not born, I am not born, not even born smart, so what?" (What is your mother's name?) "She is not born, born, is not born at all." (What is your mother's name?) "She is not born at all, she is not born at all." (What is your daughter's name?) "She does not go to school, she is stupid, when she goes to school she is very stupid, stupid, very stupid, not born at all, not been at all."

When patients converse in whispers with their voices they look distractedly here and there; but when they are spoken to they look at the speaker directly as if they have something to say to him. When asked questions their mumblings become louder and even more so if the questions continue, whereas hallucinatory whispering is most obvious when there are no external disturbances. Thus the differentiation presents no problem.

While mumbling follows after minimal stimulation holding on ("Nesteln") usually occurs only after patients feel strongly challenged. They reach for nearby objects and handle them, or they turn and twist them in their hands; they may play with a button or their clothes, they may pull threads, and the like. In many (though not all) cases, the symptom is best demonstrated by placing an object in view of the patient and saying nothing. After a short time the patient will start his purposeless handling of the object. This may be similar to what some patients with Alzheimer's disease do.

If proskinetic catatonia is suspected appropriate tests usually confirm or negate the diagnosis very quickly. If a hand is held out to the patient he reaches for it. It is not unusual, the first time, for it to appear a matter of shaking hands. But if the hand is held out ten or twenty times in a row the patient continues to reach tirelessly. If asked, "why are you doing that? You can see that I don't want to shake hands" implying he should stop, the patient may hesitate for a moment, but one word of distraction suffices to allow the tireless responsive grasping to reappear. A certain shortening of the movement may occur in the wake of frequent repetition of the experiment. Whereas at first the patient grasps the observer's hand firmly he later touches it quickly without really grasping it. Otherwise the phenomenon does not change. The **responsive grasping** occurs after the 20th or 30th time with no more hesitation than after the first.

If the patient is blindfolded touching the hands precipitates no grasping which proves that the phenomenon is not a grasping reflex. The holding on to the hand commonly seen as part of a grasp reflex is not a feature of the responsive grasping seen in catatonics. It is not the matter of a reflex but only of an automatic movement. It is clearest when opposite tendencies are stirred up in the patient. If the patient is told that he should not grasp the hand that is held out to him, **ambivalence** is created. He may not return the grasp at first, but failure to do so does not take the form of a simple omission. Rather he holds his own hand visibly cramped on the side of the body, or even behind it. If the examiner now holds out his hand again, the two tendencies can be seen to exist simultaneously. On the one hand the hand is held back, and on the other it is stretched out somewhat as soon as the patient is slightly distracted from his intention not to reach out. This behaviour shows how the higher will may suppress the movement if only momentarily.

The **going-with reaction** ("Mitgehen") of patients with proskinetic catatonia supplements the responsive grasping. If the patient is grabbed under the arm and a mild upward pressure is executed the arm begins to rise immediately. The lightest touch brings the arm into any position, even an uncomfortable one, without the patient's arm being moved by the tester; rather the movement is carried out by the patient. The test can be applied to many different parts of the body. If the calf of a seated patient is touched the leg goes up as far as it is mechanically possible. If the forehead is touched lightly the head goes back. When the head has moved back as far as possible, and if the observer continues to touch the forehead, then the whole upper body begins to bend back until the patient threatens to fall over. A mild touch at the back of the head causes a forward bowing until with knees bent the hands touch the ground. If the patient's shoulder is lightly tapped he begins to revolve. Thus he can be rocked to continuous rotation around his own axis merely by the examiner's light pressure on the shoulders, the patient turning himself with his own energy. Admonishments against these curious movements will interrupt both rotations and responsive grasping for a short period, but they begin soon again after a few words of distraction and new physical stimulation. In any case strong **countersuggestions** must be given, so as not to confuse a patient's cooperation with the doctor's requests for the symptom itself.

Responsive grasping and "going-with" certainly have the same foundation; they can be grouped under the term **"impulse-automatism"** ("Anstoß-automatie"). As impressive as the symptom may be, I was unfortunately incorrect in believing it to be present only in proskinetic catatonia. Because I discovered it relatively late in my Frankfurt studies, I did not have the chance to examine its occurrence in other catatonics. In my Berlin series I determined that, even if on another basis, it is also present in negativistic catatonia, which is otherwise quite different from the proskinetic form. It also appears, albeit less markedly, in the other two catatonias to be discussed below. Impulse-automatism can help in making the differential diagnosis only in reference to parakinetic and manneristic catatonia, where it does not occur. Parakinetics can, of course, be similarly stimulated to move, yet they do not simply grasp or go along with others but rather playfully transform the symptom with all sorts of secondary movements. Nor have I ever found impulse-automatism clearly in periodic catatonia. Patients may like to keep on grasping and be induced to make rotating movements; however, as soon as they are requested to stop they no longer grasp and begin to resist pressure on the shoulder. With proskinetics even repeated commands to stop stretching out their hands or turn their bodies only result in short interruptions of their behaviour.

The extent to which the **thinking** of proskinetic catatonics is disturbed cannot be deduced from their soft speech. Their utterances are probably only a motor automatism. Just as external stimuli lead to "holding on", "responsive grasping", and "going-with", an external stimulus, or a gesture encouraging speech may lead to automatic speech movements, even when there is no content. For this reason the same content keeps on being repeated automatically as a verbigeration. As well the lack of volume in speech indicates only an automatic action, behind which there is no active will or thought.

Thinking is surely better preserved than verbigeration would imply. When patients are urged to perform a task their behaviour is relatively understandable. Sometimes they give an answer independent of their mumbling; it may then appear to be sensible. In earlier stages of the disorder conversation with them may still be quite appropriate. Though they often do poorly in intelligence tests illogical thought processes are not found. Their mumbled verbigerations may contain some sensible answers. In later stages actual answers are spoken out somewhat more loudly. In one of my female patients this double manner of speaking was quite impressive. She gave sensible and clear answers to questions, but between answers she mumbled to herself with verbigeration; for example she repeated, "Where is the butcher?" twenty times.

Despite their abnormal motility proskinetic catatonics generally appear to show poverty of movements. Their **initiative,** that is, inner drive is diminished. Without stimulation they are inactive all day long. They are able to perform routine tasks well, carrying them out in the automatic manner characteristic of all their movements. One of the patients liked to iron but she had to be watched because she sometimes forgot to move the clothes and kept ironing on the same spot. The patients appear to be **affectively** distant and do not display any of the deeper emotions. Their flat affect gives the impression of carefree contentment which contrasts strongly with the frequent irritability of patients with negativistic catatonia.

Excitement is rare in the defect state of proskinetic catatonia but occasionally occurs in the form of brief scolding or hitting. It is apparently caused by moods that interrupt the patients' contentment. In acute phases proskinetic restlessness may develop into excitement in which the patients run around, interfere in everything, climb on chair, bother other patients, try to get out of the door, and so forth. The proskinetic character of such restlessness is observable in the patients' heightened interest in their surroundings.

The next case to be described again shows a family history.

Case 61: Martin Ile, born in 1890, was first institutionalized in 1918, but had apparently been gradually developing mental changes for years. In 1913 he had expressed hypochondriacal ideas and made a suicide attempt. Despite this he served during the war. In 1917 he came to the attention of authorities by being absent without leave. In prison he saw faces and heard voices speaking out of his chest. The military doctor's examination found him to be apathetic and he displayed both echopraxia and flexibilitas cerea. In March 1918 he was admitted to Institution H., where he was indifferent and lacked interest. Later he scolded a lot and wanted to have an operation, claiming he had either a beetle or a tapeworm in his stomach. He ripped up his shoes and threw his clothes out of the window. In 1919 he was dull and gave only short answers. In 1920 he sometimes ran purposelessly to a window, or laughed mischievously, and smeared feces on his bed. He gave short, matter of fact answers to questions. In 1921 he let other patients take his food. In 1922 he urinated in his clothes. In 1923 he was dull and often spoke softly to himself. In 1924 he assumed strange positions; another patient amused himself by ordering him about, and he automatically obeyed. He wandered around wringing his hands, mumbling to himself. Sometimes he screamed. In 1925 he defecated in his trousers. In 1926 he spoke to himself and rolled small pills from his feces. In 1927 and 1928 he was dull and only responded to questions by speaking to himself. In 1929 he continued to mumble all sorts of incomprehensible things to himself. In 1930 he gave no answers but spoke only to himself. If someone bothered him with questions he jumped out of bed and hurried to the window. In 1932 he responded correctly to simple questions, but with a "monotonous, barely comprehensible voice". He continuously rubbed one hand with the fingers of the other. He is repeatedly described during the following years as dull, unapproachable, and mute. B. SCHULZ and I examined him in 1939, but he did not answer any questions. After considerable urging he gave his name and place of birth, the latter as a response to being asked for his date of birth. Instead of answers he produced only a mumbling sound, partly as "Hm, hm". If stimulated a little he moved his hands rubbing them together repeatedly, cleaning a spot on his jacket, rubbing his thigh, or plucking at himself. If left alone he became calmer and eventually sat nearly motionless, looking around. The beginning of dictation sufficed to precipitate all sorts of finger and hand movements. When addressed he always looked up, his face showed some interest, but at the same time his facial expression was quite vacant. There was marked responsive grasping. If a hand was extended to him he repeatedly reached out in return. Only the smallest pressure was necessary to create a "going-with" phenomenon. On command he repeatedly stood up and sat down.

A **sister** of the patient was mentally ill, but unfortunately I could not obtain her medical record. I did however receive the record of a **cousin** (father's brother's son). Max Ile, born in 1896, developed normally but was not very active. In 1918 he attempted suicide for no apparent reason. Following this he did not appear abnormal. But in 1928 he became disturbed. He spoke of voices and felt persecuted, broke a window, and was admitted to Institution E. He answered questions in a quiet, monotone voice, kept his face expressionless, and showed no initiative. His mood was by turns cheerful and anxious. He was discharged after three months, but readmitted several weeks later, having smashed a closet door at home. In the institution he again spoke of persecutions, and of wanting to form a political party. He had a forced friendly manner and had a rigid look. He responded to questions with general phrases and spoke softly. Later the record states, "When speaking he hardly opens his mouth." In 1929 his eyes moved restlessly back and forth, and he gave evasive answers. He was discharged but returned in 1931 when he was described as having a "frozen smile" and running restlessly up and down the corridors. In 1932 he occasionally spoke of a "sexual need" and demanded help. He claimed his illness was due to lack of sexual intercourse. His remarks were stereotyped and presented in a "soft, secretive voice". In 1933 he was frequently irritated, claiming that strange thoughts were being put into his head. He had little facial expressiveness and spoke in a soft voice. In 1934 he was obedient and modest. He greeted the doctor with a bow and watched him expectantly, but had nothing to say. In 1935 he was again frequently irritated. In 1936 he occasionally scolded. In 1937 and 1938 he lacked initiative and spoke little. In 1941 he did some basket weaving but was otherwise "rather stiff and tight". In the following years he was occupied with simple tasks and was described as unresponsive. In 1946 he remained dull and lacking in interest. He died that year of lung tuberculosis.

There is no extensive description of the early years of the subject Ile's condition. That he served during the war despite his disorder is explained by the obedience of proskinetic catatonics. As described above, the catatonia is often not obvious in the early stages. When in 1918 he ripped up his shoes and threw his clothes out of a window his actions were probably signs of proskinetic restlessness. Later the rubbing and the finger movements are repeatedly mentioned. Perhaps his rolling pills from feces points to the increasing apathy. From 1923 on his soft speech is noted regularly. In 1924 his considerable automatic obedience is obvious in the description of his obedience to the orders of another patient. In my 1939 examination he showed all the signs of proskinetic catatonia: finger movements, mumbling, going-with and responsive grasping ("Mitgehen" and "Gegengreifen"). The onset of the cousin's disease was also so gradual that for a long time institutionalization was not necessary. Throughout the medical history it is mentioned that he spoke with a peculiarly soft voice. When he was not upset he behaved obediently and modestly. At the same time he was dull and without initiative. As with his cousin Martin Ile, the combination of symptoms indicated the presence of proskinetic catatonia even though paranoid symptoms were mentioned here and there.

Summary

Proskinetic catatonics respond with facial expression to being addressed and despite unresponsiveness display a certain interest in the examination. At the same time or after a certain further stimulation the patients begin an incomprehensible mumbling, which turns out to be a **verbigeration** of isolated forms of speech. In milder cases speech is soft but still comprehensible. Sensible answers may still be given although they may contain several repetitions. More stimulation leads to grasping, **handling** and holding on to objects, plucking clothes, rubbing of the skin, etc. Appropriate tests reveal the presence of **"impulse-automatisms"**. The patients tirelessly grab the repeatedly offered hand; they are easily stimulated to active movement by the mildest pressure on any part of the body, assuming any position the observer desires. If opposite suggestions are given the grasping and "going-with" may be temporarily interrupted, but begin again as soon as attention is distracted. **Initiative** of proskinetic catatonics is severely limited so that their total behaviour, despite willingness to move, seems lacking in motility. If encouraged externally they are able to work effectively. **Affect** is severely flattened and appears as carefree contentment. Some patients have bad moods with scolding and aggressivity. In later stages when the patients no longer speak at all but only mumble, **thought** processes can no longer be identified. Even though the patients still answer, their responses remain incomplete without, however, becoming entirely illogical.

Negativistic Catatonia

Negativistic catatonia is more or less the opposite to proskinetic catatonia. The essential symptom of this disorder, as described by both KRAEPELIN and KLEIST, is, as the name implies, negativism. In mild cases and at the beginning negativism is often only hinted at. Patients may still answer questions correctly, and carry out orders, but the tendency to resist can already be recognized from their behaviour. They may look at the doctor only when forcefully queried, but otherwise look indifferently to the side. The characteristic position in which the head looks in another direction than would be expected from the body's stance is often recognizable early. If irritated patients become more clearly negativistic. States of excitement may develop early, and if impulsive and severe in nature are evidence of negativistic catatonia.

If the negativism is not momentarily accentuated by an irritated mood it need not be accompanied by gross resistance. In pure form it is less a resistance than a simple abstention. If such a patient is approached he looks up briefly or not at all. If offered a hand he appears not to notice it. If the patient's hand is taken he lets it happen indifferently without reacting. If given an order he may not move at all. Questions remain unanswered. The facial expression may appear immobile. The more contact with a patient is attempted the more unwilling and resistive his attitude becomes, even at the onset. Gross forms of resistance occur at that point. If approached the patient now turns away. If a hand is offered he pulls his own back. If his hand is grasped he pulls it away. At the same time he may swear, run away, even become threatening. The facial expression shows great irritation. More characteristic is a hardened, hostile expression which precedes the negative reaction. The look of the face is not unnatural though not lively.

The patients often smile embarrassingly when attempts are made to contact them. This smile expresses the fact that they do not totally identify with their own negativism. This internal contradiction may be called forth by trying to bring the patients into the most friendly mood possible. If one is very nice to them, through flattery for example, it is possible to bring them to a friendly mood, unless of course they are already angry. They do not lose the negativism but now they resist with that characteristic **embarrassed smile,** which in a way represents an excuse for their resistance. The ambivalent smile and behaviour are better evidence of negativistic catatonia than simple resistance, which may be due to simple irritation. In the **ambivalence** negativism is clearly reduced. The patients look at the observer more frequently and turn away with less forcefulness; if they had been completely mute before, they may now at least give simple personal data.

I no longer saw such gross negativism later on. Apparently the patients I had described earlier stayed in constant tension from which they easily became quite irritated. Inappropriate treatment may have played a part, as will be seen when negativistic excitement is discussed. The use of neuroleptic treatment may have helped to make the patients more approachable. All the negativistic patients of my Berlin series answered questions even though only in monosyllables and without much cooperation. Usually there was no difficulty in obtaining answers when they were asked to give personal information; they responded correctly to simple questions about current events. But to less simple questions or if pressed they resisted more definitely. They did not answer at all or said "I don't know." Their ambivalent smile was still noticeable in my later observations.

The patients' **posture** is also characteristic. They tend to sit in a peculiarly contorted manner with the upper body at an awkward angle to the legs, the head turned in yet another direction, and one shoulder pulled up. The posture is probably due to movements stopping half way. In their ambivalence the patients half turn toward the observer with the head while the body remains fixed; or the body turns while the legs stay put. And events in the surroundings stimulate some reflex movements, which are again stopped half way by the negativism. Impulse-automatism (to be discussed below) and negativism may exist here side by side.

So far the characterization "negativistic" is appropriate. But another symptom, described in reference to proskinetic catatonia, I first noticed in negativistic patients in Berlin. Despite their negativism the patients display an abnormal tendency to move in the form of **impulse-automatism.** Negativistic patients may grasp responsively without tiring, and with slight pressure against the body actively go along even when asked to stop. It is difficult to understand how such different catatonics as proskinetic and negativistic can have the same tendency in their movements. Only finer differences may be noted. In proskinetic patients the going along occurs more closely in conformance with external stimulation, movements stop as the pressure is released. In contrast negativistic patients continue to move; with a short push against a shoulder they may make a complete turn. Sometimes they do not react directly, that is, they do not turn, but when pushed on the side of the upper arm they walk forwards with a few pattering steps. The pressure has more of a releasing effect than producing an appropriate movement. Going along, too, is not as complete. In response to mild pressure at the back of the negativistic patient's head the head may nod forward, but the body remains stable, while proskinetic patients with knees flexed bend so far forward that they touch the ground with their hands. This experiment cannot be continued indefinitely, for negativistic catatonics with continued pressure on their bodies become more and more irritated and eventually stop moving. On the contrary, resistance increases so that even with strong pressure no further movement results. With proskinetic catatonics this does not happen, they continue to move indifferently as long as they are challenged.

Despite the similarity the genesis of the syndrome is surely different in the two catatonias. In proskinetic catatonia there is an impaired ability to resist external stimulation, while in negativistic catatonia deeper motor mechanisms are abnormally involved, due to the general failure of the voluntary motor system. This abnormality is clearest when a single brief push produces continuous motion. The fact that the other two catatonias of low motivation, speech-prompt catatonia and sluggish catatonia, to be described below, also display impulsive automatism points toward an unspecific involvement of deeper mechanisms. If in descriptions of negativistic catatonia it is stated that the patients may not even shake the examiner's hand this does not contradict the responsive movements. In natural situations, when one simply wishes to greet them, they do not turn towards the greeter nor do they offer their hands. But when one demonstratively holds out one's hand again and again, it is a different matter. In this case the external stimulus is stronger. A further consideration is that in the examination situation patients' attention is much more drawn to the outside than when they are sitting apathetically in their usual place and are not stimulated by anything up to the moment of being greeted. Negativistic abstention is more at work here, automatic movements are not being brought into play.

Impulse automatism is related to **echopraxia, echolalia,** and **catalepsy.** These symptoms may appear in various forms of catatonia when higher initiative is greatly impoverished, depending on whether the patient subjectively takes the external stimulus to be a request to imitate a movement, to repeat a word, or to leave a raised limb in a given position. These symptoms are therefore not reliable for a more precise differentiation of the catatonias. With impulse automatism the subjective component is excluded by means of clear cut countersuggestions, i. e. by demanding or even calling out loudly: "Don't grasp" or "Don't let yourself be turned".

If patients are urged to do something they do not want to do they become increasingly irritated, and if the urging continues they become **negativistically excited.** For example, if a female patient does not want to go into the garden, attempts to convince her are useless and angrily she turns away. If her hand is taken to lead her out she tears herself free. If then attempts are made to force her out, perhaps with the help of several nurses, she will protest noisily. She defends herself with all her might; she tries to shake the nurses loose, hitting, kicking, scratching, biting or screaming. Such negativistic excitement may occur even for less reason. There may no longer be any connection with the original resistance. The patients may suddenly run through the ward, pushing objects over, throwing other patients out of their way, breaking windows, etc. The record of one female patient states: "Throws food, attacks an uninvolved patient, slaps the nurse, rips her apron." Generally when patients meet no resistance the excitement abates quickly. The disturbed behaviour is characterized by its acuteness, tendency to aggressivity, and short duration, but apart from excitement may also consist of other emotions. The patients may howl, jump, clap their hands, rotate, or turn somersaults. Any kind of mood may lead to motor discharges. At times excitement is so brief that a better term could be chosen – **impulse actions.** A sudden scream, an unexpected slap in the face, smashing of windows, followed by quietness may occur in negativistic catatonics. Among men excitement is less obtrusive, but may be more dangerous. The movements are generally hurried, sudden and jerky, but without real distortions.

Patients under neuroleptic treatment move more slowly, but the jaggedness of their movements is still recognizable. Excitement is less marked but may still break through in spite of the medication. The tendency to aggressivity remains if the medication is not given in strong doses. Improvement is gradual. With neuroleptic treatment excitement is no longer as strong as described above.

Affectivity in negativistic catatonics is largely lost. However a primitive instinctual interest remains and it is directed at food and, where opportune, at eroticism. Food is often taken away from other patients and devoured greedily. There is no **initiative** beyond simple instinctual activity. There appears to be little thought disorder. When patients agree to respond to simple test questions, the answers may be incomplete but not grossly illogical.

In the Frankfurt series of patients with negativistic catatonia one female patient had a psychotic cousin:

Case 62: Emilie Au, born in 1890, had always been sensitive, had taken things hard, and was never really happy. In May 1911 she became mentally disturbed and was admitted to Psychiatric Hospital M. She believed that she had been brought there to marry and that the Saviour had called her. She sat motionless and gave stereotyped answers to questions: "I swore eternal love and faith to my darling." She postured, occasionally displayed echolalia and automatic obedience. Sometimes she laughed and did not say why. She masturbated shamelessly. After five weeks she was transferred to Institution D., where she did not speak but often burst out laughing. During an attempted physical examination she slapped the stethoscope out of the doctor's hand. She then became excited, ran back and forth, screamed, laughed, and clapped her hands. Such states occurred repeatedly: she romped around in the garden, erotically approached another female patient who she claimed was a man, and several times became aggressive. Later she was again calm, negative, and silent, running away from anyone who addressed her, and often laughing to herself. In the beginning of 1912 she became more reasonable and was taken home, where she remained for three years. Exact information about her condition there was not available. In 1915 she changed for the worse, scolded loudly, and was readmitted to Institution D. She gave no answers, stared blankly, and sometimes broke out laughing. At times she suddenly ran to the doors and banged on them. In 1916 she mostly lay on the floor, and did not answer any questions. Intermittently she became excited, screamed, roared, banged on the doors, and ran back and forth. In 1917 she remained negative, generally lay under her blanket, looked sullen, and gave no answers. From time to time she scolded and hit other patients. In 1918 she remained under her blanket and "vegetated dully". Occasionally she was noisy, ripped her bed apart, and tried to push a table over. In 1919 and 1920 she spent much of her time squatting silently on the floor, did not respond to questions, but at times she screamed and became violent. In 1921 the periods of her excitement became more frequent, but were always short and accompanied by aggression. In 1922 she made brutal attacks with no external stimulus, on one occasion she suddenly jumped up, banged the doors, knocked over a heavy wooden bench, and swept the table top off the table. At other times she brutally mishandled other patients and attacked the nurse for no reason. In 1923 and 1924 she often had to be isolated because of her aggression. In her calm periods she was completely unapproachable. In 1925 she carried on a kind of battle in her room with a palliasse, constantly scolding it and throwing it around. In 1926 and 1927 she crouched in a stereotyped position and tended to impulsive violence. In 1928, influenced by work therapy, she sometimes sat down at a table, but her states of excitement, her screaming and her aggressivity continued. In 1929 she was less excited. In 1930 she was able to do some knitting work though she sometimes broke the needles. At meal times her behaviour was unappetizing. From time to time she was aggressive. Her behaviour remained the same for several years. In 1933 she broke a window. When excited in 1934 she lifted up her skirts. She was also incontinent of urine. – B. SCHULZ and I examined her in 1939. She let her head hang forward a bit, but at the same time turned to the side and kept her arms pressed rigidly against her body. She looked hostile and gave no answers. When her hand was grasped she tore herself away with curses and turned her back. Once she laughed out loud with no reason. When we attempted to approach her with a very friendly tone, her resistance became noticeably less intense; she had an embarrassed smile, no longer turned away, but could not be brought to any further cooperation. At that time I had not yet come across the impulsive automatism syndrome.

A second degree **cousin,** Georg Brum, born in 1886, was admitted to Institution M. in 1906, with the following report: "Has been ill with dementia praecox for two years. Periods of complete apathetic calm alternated with periods of excitement and compulsive movements. For the last few days he was in a severe state of excitement as never before; he smashed everything: windows, plates, tools, he also attacked people, and yesterday bit his brother's arm..." In the institution he sat dully, smiled and responded to being addressed, but gave no answers. When approached he withdrew. He kept his head down between his shoulders and, as was clearly stated, kept it "turned somewhat to the side". Sometimes he made jerky movements with his body. In 1907 he was more negative, resisted passive movements and turned his head away when addressed. Most of the time he sat with hunched shoulders and a shy smile. He was discharged home as unrecovered.

Subject Au had typical negativistic catatonia. At the onset, a few other symptoms which do not belong to the defect state were present, but the symptoms of this disorder dominated the picture. When I examined her she was negativistic and mildly irritated. I succeeded in bringing out the ambivalence. The cousin Brum is unfortunately described only briefly, but so characteristically that I cannot doubt the presence of negativistic catatonia. Both negativistic withdrawal and impulsive excitement are typical as described. The description of the severe state of excitement at home is very characteristic. Similar states did not occur in the institution, but occasional jerkings of the body were observed and these probably corresponded to such impulsive states. The characteristic posture of negativistic catatonia is described very clearly. The patient also displayed the embarrassed smile described as shy or silly.

Summary

Negativistic catatonia is characterized by resistance. If there is no irritation, gross forms of resistance may not appear, and negativism is expressed primarily in **omissions.** If patients are pressed they at first become moody and subsequently more clearly resistant. In trying to shake hands with them they pull back, tear themselves free, produce involuntary sounds, run away, or become aggressive. If attempts are made to overcome the resistance with force, **negativistic excitement** is precipitated which may become severe. States of excitement also arise without external causes. They are usually brief but associated with violence. The whole motor activity is somewhat impulsive and appears jerky and choppy. The posture of the patients is often peculiarly contorted. In combination with severe impoverishment of voluntary motility there is abnormal irritability of deeper motor mechanisms appearing as impulse automatism. Higher **activity** and **affectivity** is no longer recognizable in negativistic catatonia although patients maintain a lively instinctuality, observable in greedy eating and erotic propensities. Thought does not appear to be very disturbed.

Speech-Prompt Catatonia

My having described this peculiar form of catatonia as "speech-prompt" has occasionally been misunderstood to imply that the patients are very talkative. Actually the patients are only talkative when a relevant stimulus comes from outside. Left alone the patients hardly speak at all. Only in earlier stages when lack of initiative is still moderate does compulsive speech occasionally appear. Answers readily given to questions do not set off compulsive speech; on the contrary answers usually are short and often ungrammatical. Abnormal volubility is not an urge to speak, but the peculiar characteristic of giving a quick premature answer.

Non-sensical answers often occur during the onset of the disorder. At this period of time they are scattered among logical answers. Often the incorrect answers are not really devious but only rather meaningless or ill-considered. An awareness of this is necessary in order to recognize speech-prompt catatonia in the early phases. Autism, to be discussed below, is also present during the onset.

Some answers are still relevant even in later stages. If simple questions are asked, for instance about personal data, nothing unusual may appear. With further inquiries however, more incorrect answers are given. Thus a patient who has given his personal data correctly if asked how long he has been in the institution gives wrong or at least inprecise answers, such as "for a while". Attempts made to obtain a more precise answer usually fail. Tangentiality develops, which seems negativistic, and may lead to the patient's saying that he has been in the institution 80 years even though he has already given correctly his age as 50.

Vague answers and those missing the point of the question are both characteristic for speech-prompt catatonia. Some patients keep on giving only false or indefinite answers while from others much correct information may be obtained. Answers become imprecise and senseless most quickly when themes are touched upon which are unpleasant for the patient. The actual duration of their stay in the institution can rarely be obtained correctly from chronic patients. Talking-past-the-point ("Vorbeireden") may be provoked by speaking to the patient very briefly and concisely. The answers that follow are also very short and at the same time quite senseless. What the patient says is not only inwardly poorly thought out but outwardly is given very fast as soon as he gets into the rhythm of a question-answer game. At first verbalizations may occur with hesitation. The explanation for this is the patient's general lack of motivation, to be discussed below. This lack of motivation also explains why patients rarely say anything by themselves. In progress notes they are often described as mute because they have not been asked questions, which at first would have been answered hesitantly but later with promptness. I present a conversation I had with a speech-prompt catatonic patient in the Gabersee Institution.

(Where do you come from?) Patient gives place of birth correctly. (What was your last place of residence?) "Hier in Gabersee." (Before?) "At home." (Where?) "At our home."

(Where was that?) "It was at my place." (In which place?) "Gabersee." (Which place before?) "Before I can't seem to remember." (When were you born?) Correct reply. (What year did you come to Gabersee?) "Three years ago, three to four years." Actually she had come ten years before. (What year did you come?) "I don't know." (Around?) "1908." (How many years ago is that?) "Many years." (How many?) "Four." (Since 1908?) "Yes." (What year is this?) "1900." In reality the year was 1934. (And what was last year?) "Also 1900." (Two years ago?) "Also." (Is it always 1900?) "Yes."

It is a clear case of talking-past-the-point. The answers do not all look like active negativism. Often it appears that patients merely say quickly what comes into their mind. This impression is gained particularly when general answers are given which may not be entirely inaccurate. When patients quickly and thoughtlessly say whatever occurs to them, simple questions are answered correctly. If negativism were present it would have been just as clear in simple questions as in more difficult ones. On the other hand it does seem to be negativism when the patients become increasingly incorrect in their answers the more a line of questioning is pursued. In the above example it can be seen that it was impossible to discover the last place of residence from the patient and how she evaded the question again and again.

A further peculiarity of talking-past-the-point of these catatonics is not clear in the above example. Sometimes the answers are formed so absurdly that a deliberate wrong answer can no longer be suspected. With many verbalizations it is impossible to understand how they arose. That will be clear when I contrast the still comprehensible replies of the above patient with the extreme utterances of another Gabersee patient.

(How are you?) "Well." (What do you do all day?) "Nothing in deer." (What does that mean?) "I can't recognize that." (Where did you work today?) "On field prince." (What is that?) "To the board." (To the board?) "Yes." (Have you eaten?) "Yes." (What?) "Canvas roofing." (Canvas roofing?) "Yes." (How did that taste?) "Well." (What else did you eat?) "Holiday." (What day of the week is today?) "Garbage." (What date is it?) "88, 83." (Season?) "Sister Ludwila." (What year were you admitted here?) "I don't know anymore, 45." (What year were you born?) "That too I don't know any more." (What are your plans?) "To a worker's alley neighbourhood." (What do you want there?) "No, I don't know that." (Where are you from?) "The Ludwila Section." (Where is that?) "In Bohemia." (Are you Bohemian?) "Yes." (Where were you born?) "In Bethlehem." (Where is that?) "In the stable." (Are you Jesus?) "Among others." (Are you married?) "Yes." (To whom?) "Self in the prize." (What is your husband's name?) "Ludwila." (Everything is called Ludwila?) "Yes." (Do you have children?) "No." (But you do have a child!) "I don't know." (Where are you here?) "In the Ludwila alley." (And before?) "Before I was at home, that self and Staufen." (Did you live with your husband?) "I don't know anyone, was never married." (Not married?) "No." (But you were!) "No, I was never married." (But you said so yourself before!) "I was never married." (What was your maiden name?) "Ludwila." (What else?) "That's all." (What is your name now ?) "Schmidt." (The first correct answer!)

From this transcript it can readily be seen that to a large exent the will to give correct answers was absent. When other patients who were present laughed about the funny answers the patient became irritated and seemed to be losing her good will; but it was remarkable that she never refused to answer further questions. In her irritation her answers only came more rapidly than before. It was clear from the transcript that the patient often said whatever came into her mind, regardless of whether it was relevant to the question or not. When she answered "Yes" or "No" it was not from the matter being discussed, but rather depended on which answer corresponded better to the intonation of the question. With another patient whom I examined with SCHWAB we could predict the answer from the way the question was formulated in one sense or the opposite sense. Asked what she liked better, German or French, she replied "French"; but if asked, "French or German", then the answer was "German", because it was at the end of the question and was thus more obvious. A similar phenomenon was seen in the "Yes" and "No" answers of the above patient. KLEIST used the term "echological" to describe answers which simply repeat the question in a slightly different form. Such echologisms are frequent in speech-prompt catatonia. Also noted is the constant repetition of the word "Ludwila". Such repetitions are very characteristic. "Ludwila" was a nun who had been in the institution some years previously. Perhaps the name had some affective associations. It was however continually repeated because once it occurred it became available as an answer when no other was produced. One might even speak of perseveration in speech-prompt catatonics, although it is of a very different origin from the **perseveration** of organically ill patients. Short-circuit utterances of what came into their heads may be derived from external associations. Hence the origin of the remark: "Bethlehem lies in the stable".

Beyond short-circuit talking-past-the-point are quite incomprehensible answers like "on field prince", "Ludwila section", and "nothing in deer". Such peculiar verbalizations are also characteristic of speech-prompt catatonia. It should be emphasized that it is impossible to know what thoughts pass through a patient's memory, fragments of which are of the surface as questions are posed. If several ideas are present simultaneously they may produce incorrect word combinations. Here the notion of **contaminations** is limited since it is not a question of thoughts but words without sense. One can explain the **ungrammatical** nature of many answers as the result of the speech impulse bringing out whatever is available without the sentence construction first being put in the right order. Furthermore the brevity of answers is evidence of the effect of impulses which become exhausted as quickly as they arise.

Hence I believe that the talking-past-the-point of speech-prompt catatonics is based on a peculiar **"short-circuiting"** as a consequence of which it is not a thought that is verbalized but whatever word happens to become vocal. This explanation is given support by the outer behaviour of the patients. As soon as their lack of initiative has been overcome answers become so prompt that there is hardly any time for them to think out the proper reply. It is immaterial whether the question is easy or hard, whether it requires mental work or not. In response to the most difficult questions, such as the number of stars or the number of grains of sand in the sea, prompt answers are given; it may be any number, usually a smaller one as smaller numbers are more available than larger ones.

Thus it is understandable that simple questions are generally answered correctly. The correct answer is so easily available that it is called forth by the question without the intervention of any mental work. If simple questions are answered incorrectly there usually are affective reasons. In so far as the patients evade the questions negativistically it is most often a matter of the questions being embarrassing. Conscious or unconscious repression makes certain that unpleasant thoughts are not readily available. And if a false answer is given the correct one is rarely given by further questioning. Here too is clearly seen the affective refusal to admit that something was falsely answered.

Explaining speech-prompt talking-past-the-point as a short-circuiting of thought provides an essential parallel to another catatonic subform. We have seen how in **proskinetic catatonia** external stimuli lead to corresponding movements abnormally easily. In that condition action was immediately stimulated just as in speech-prompt catatonia verbalization was immediately stimulated. Thus there are two forms of catatonia with short-circuiting, one involving activity and the other speech or thought. Up to now we have seen opposites in the catatonias: between parakinetic and manneristic, and between proskinetic and negativistic. We must now make the observation that the forms on each side have something in common. In speech-prompt and proskinetic catatonia there is facilitation of will, a similar phenomenon in parakinetic catatonia. On the other hand in manneristic and negativistic catatonia there is an aggravation. The same thing is found in sluggish catatonia, to be discussed below. These regular associations clearly point out **that here very specific pathological processes involve clearly delimited functional areas.**

After close observation of the peculiar talking-past-the-point it becomes clear that this phenomenon is **quite specific and does not occur elsewhere.** It would be very wrong from the remarks of speech-prompt catatonics to infer a kind of "schizophrenic" thought disorder. In the past thought disorders of different schizophrenics were examined and were incorrectly ascribed to schizophrenia in general. Talking-past-the-point is exclusively associated with the disorder of speech-prompt catatonia. It is by no means clear that there is a thought disorder in the strict sense, since the patients apparently do not think at all before they speak. The disorder lies more in the processes of the will, thus indicating its catatonic nature. How thought is formed in these catatonics cannot not be determined readily. They behave sensibly when occupied with simple tasks and thus from that point of view there is no evidence of severe impairment of thought.

Other symptoms prove that this condition belongs to the group of catatonias. Although no actual anomalies of posture occur the patients do show conspicuous psychomotor stiffness. When they speak normal gestures are absent and the face remains remarkably expressionless. Mask-like it reflects nothing. It is not a rigidity as in manneristic catatonia or in metencephalic states, but rather an absence of expression. The patients can laugh but it is impossible to know if they really feel like laughing. They may give answers which seem to be deliberately false, but their facial expression reveals no negativism. It remains incomprehensible, strange, and impenetrable. Above all the expression of the eyes is absent, while in metencephalitis the eyes do have something to tell.

The **facial expressionlessness** of speech-prompt catatonics becomes even more noticeable when the patients turn toward the questioner and look him in the face. In this way a peculiar contrast arises between normal, purely external approachableness, seen in the readiness to answer, and an emotional unapproachability of the patient. **Interpersonal rapport** which depends strongly on unconscious facial expression is not disturbed to such an extent in any of the other schizophrenias.

Curiously the patients become attentive only when approached with questions. They do not react readily to their environment; on the contrary they appear carefree and untouched by everything. Neither do they join other patients, nor do they speak to them unless asked first. From their outward behaviour it seems that they attach no importance to the doctor's visits. They remain totally uninvolved during a conversation with a patient next to them. Their interest is awakened only when they are themselves addressed. Thereby another peculiar contrast shows itself. Despite their response to being addressed the patients classically display what is called **autism.** For example the patient from whom the second transcript above is taken liked to walk back and forth across the room, not looking to the side, nor looking up when someone stood in her way. Rather mechanically she stepped aside the other person and continued in her path. When food was distributed and all others were on their feet she wended her way through the others, not caring about pushing until she was called to eat. This autism is reminiscent of the autistic character of hebephrenics, to be considered later; but there the contrast to prompt answering is absent.

Considerable **lack of initiative** is part of the autistic lack of interest. Normal **will** and normal **affect,** which here can hardly be separated, seem extensively deteriorated. In the end state there is never the impression that patients are affectively tied strongly to anything. They never ask about their family, never express any wishes, nor ever undertake anything on their own initiative. They carry out their daily routine mechanically performing simple tasks given to them, but do not begin anything on their own. As stated above they do not speak unless stimulated to do so. Although the autism may hide something the deterioration of will and affect is certainly very severe in this catatonia.

Speech-prompt catatonics may become excited but not typically so. Sometimes the patients scold into the void and appear to hear **voices.** However they never explode during their excitement like sluggish catatonics may do, but rather may be calmed down by being spoken to. Occasionally attacks on the surroundings do occur which may perhaps be related to hallucinations. Other patients never become excited. Details about delusion or hallucinations cannot be ascertained from the patients. A female patient in Gabersee often called me her son, but it was never clear whether this was a delusion or talking-past-the-point. Probably neither delusions nor hallucinations play an essential role, since in the earlier stages of the disorder there is little indication thereof.

Among my speech-prompt catatonics there was one with a family history. I present the case here.

Case 63: Rudolf Stö, born in 1887, a law student, was normal at first, liked parties and did well in school. In 1909 he began to change gradually, his studies suffered, he became quiet and moody, walked back and forth in his room, and answered questions with irritation. In 1911 he wanted to sleep with the maid, he broke a window, and was admitted to Psychiatric Hospital M. There he stood like a statue, with mask-like face, displaying catalepsy and echopraxia. He gave monosyllabic replies to questions, accepted everything, and had a propensity to simply answer "Yes". He was transferred to Institution E., where he often laughed to himself and ran back and forth in the ward. In 1912 he was described as mute. The next entry in the progress notes was made in 1914 when he occasionally answered questions, sometimes laughed and at other times shouted to himself. In 1915 he gave some correct answers, grimaced, and carried on conversations with himself. According to the next notes in 1919 he lay in bed and was unapproachable. In 1920 he barely spoke and was unclean. In 1921 he played with his buttons. In 1922 he often tore his shirt and developed blisters on his feet from stereotyped pacing in the garden. In 1923 he generally lay stuporous under his blanket. In 1924 he repeatedly threw dishes in the ward. He carried on the following dialogue with the doctor: (How are you?) "Thank you, well." (Where are you?) "Bellevue." (What is the date?) "27 July 1918." It actually was 27 July 1924. (How do you like it here?) "Yes." (Would you like to leave?) "No." (How is your mood?) "Yes, good." In 1925 he was described as uninvolved and silly. In 1926 and 1927 he was stuporous, but occasionally excited and aggressive toward the attendant. In 1928 he mostly lay in bed. In 1929 he ran back and forth through the ward, hallucinating and grimacing. In the following years he was described as apathetic, autistic, occasionally excited and aggressive. In 1938 he was silly and unclean and laughed when addressed. – B. SCHULZ and I examined him in 1939. He gave prompt but brief answers which were sometimes correct and sometimes beside the point. Even when asked quite senseless questions he had a ready answer. In all his remarks his facial expressions remained empty. His forehead was always frowning, his hands were folded in front of his body. A raised hand remained up in the air for a while; with very ostentatious demonstrations echopraxia appeared. The dialogue was as follows: (How long have you been here?) "A very long time." (How old are you?) "30 years." (What is the date?) "1937." (How old are you?) "28." (Before you said 30!) "I give in." (How old is the institution?) "A very long time, a long time." (How long has the world existed?) "Longer." (How many stars are there in the sky?) "All." (How many?) "Two." (How many suns are there?) "Three." (How many moons?) "There should be none." (What is the difference between box and basket?) "Box is small." (What is the difference between tree and bush?) "Bushes, shrubs." (What is the difference between child and dwarf?) "Dwarf is right." (What does this means: 'Necessity is the mother of invention', in German: 'Not bricht Eisen') "That the need eats the iron." (No rose without thorns?) "Rose, that is Catholic." (Morning time is the best time?) "Good morning, morning time."

A **brother**, Friedrich Stö, born in 1886, also a law student, had until 1908 been calm, industrious, and successful in school. In 1908 he became apathetic, stopped eating, felt hypnotized, and heard voices. He was admitted to Psychiatric Hospital M where he was very stiff, assumed forced positions, displayed flexibilitas cerea, and sometimes laughed for no apparent reason. He complained about voices which berated him. The following dialogue was recorded: (Where are you?) "In the eye clinic." (Why?) "Because I have a problem with my eyes." (Why are you so depressed?) "Because I have to concentrate my mind on my eyes." (Why?) "Because otherwise it hurts, and I can't stand that, that pain." (What day is today?) "I don't know." (What is the year and the month?) "I don't know." On one occasion he returned the doctor's greeting with the following remark: "We won't play anymore. I don't want to remain in a fraternity." Later he declared, "I am thinking about my surroundings." (And to what conclusion have you come?) "To the circulus vitiosus." At times he was excited and exclaimed loudly, at others he hopped stereotypically on his bed. He was transferred to Institution E. In 1908 he was negativistic, gave no answers, and smiled mischievously. He was often excited, shouted, sang, stamped his feet, became aggressive, and hit his head with his fists. In 1910 he was under his blanket most of the time and emitted a strange grunting sound. When he got out of bed he paced through the ward with a peculiar gait. In 1911 he made inarticulated sounds and grinned. He walked with a peculiarly shambling and choppy gait. In 1913 he spoke softly to himself and turned away from others. In 1914 his shouting, singing, and running around were disruptive. He also became violent. In 1915 he was at times cheerful, talked a lot with a silly smile. In 1916 he carried on conversations with himself. In 1918 he was unclean, spoke to himself, and liked to hit the attendant from behind. In the following years he was generally calm, though unresponsive and unclean. In 1924 he at times wandered around restlessly and tore his clothes. In 1925 he smiled and grimaced, but said nothing. In 1926 he was mute, smiled inanely and sometimes sang to himself. In the following years he was described only as dull. In 1931 he stood around loosely and smiled sometimes. In 1932 he occasionally spoke to himself. After that he remained dull, unclean, and untalkative. In 1937 he did not answer questions, and when told to open his mouth to show his tongue opened it only a little. In 1939 he stood bent over with his arms hanging loosely. In 1941 he died of pneumonia.

Subject Stö had periods of excitement probably of hallucinatory origin; in other respects the medical history was empty. This is unfortunately generally the case with speech-prompt catatonics. A good conversation cannot be carried on with these autistic catatonics, the only striking features are dullness and the occasional excitement. In 1924 there was the report of a short dialogue from which the presence of speech-prompt catatonia could be readily deduced; beside all general remarks the patient claimed to be in Bellevue and that the year was 1918. When I examined him he displayed all aspects of the typical picture of speech-prompt catatonia. His **brother** was also described as dull with occasional excitement. However there were traits which pointed into another direction. The patient seemed negativistic and spent of lot of time under his blanket. His excitement did not appear to be simply due to hallucinations, but was also at times accompanied by singing and articulated shouting. In addition there was impulsive aggressiveness. Thus traits of negativistic catatonia were found, as described above. The choppy gait perhaps indicated how the patient's negativism affected his motor activity. There was also the presence of a "silly smile", which probably represents the ambivalent smile of negativistic catatonia. Negativistic catatonia does not, however, explain the whole clinical picture since negativistic patients are not as dull as this patient, nor do they carry on conversations with themselves, which so clearly point to auditory hallucinations. On the other hand these traits are also present in the brother with speech-prompt catatonia. Fortunately some of the patient's remarks were recorded at the time of the early stages. They show marked talking-past-the-point – he claimed to be in the eye clinic and spoke senselessly of circulus vitiosus, probably sounding like another word in the question ("Schluß" in German). We have seen how external associations may determine the content of the talking-past-the-point. The completely inappropriate response to the doctor's greeting is also very characteristic: "We won't play any more. I don't want to remain in a fraternity." Thus I suspect that the brother had a combination of negativistic and speech-prompt catatonia. The relationship to the brother's condition is therefore present in one of the components.

Summary

The essential symptom of speech-prompt catatonia is talking-past-the-point. The patients often give correct answers to simple, neutral questions. The more difficult and emotionally charged the questions become the more certain it is that the patients will talk past the point. Close analysis shows that there is a curious type of **short-circuiting in speech** causing the patients to say whatever thought is momentarily available instead of answering thoughtfully. Words that had recently been pronounced often return inappropriately as if there was **perseveration.** Often chance external associations, even sound of words, determine the content of the reply. At times completely incomprehensible words are spoken. These perhaps arise due to the simultaneous availability of several contents which then combine inappropriately. Such combinations sometimes take the form of **neologisms.** Every question leads only to a very short speech impulse which quickly becomes exhausted, so that most answers are quite brief. Usually no complete sentences are spoken, rather a few words are thrown together **ungrammatically.** Answers are given to every question no matter what the question is. This confirms the presence of speech-prompt catatonia. On the other hand there is no pressure of speech. Left alone these patients hardly speak at all so that they are often incorrectly deemed as mute. Higher **initiative** is largely lost as is **affect.** Movements are somewhat restricted. **Facial expression** is curiously empty and inexpressive, revealing nothing to the observer about the mental processes. The inapproachability is emphasized by pronounced autism in which the patients seem to have no care at all about their surroundings, as long as they are not directly stimulated to make some utterances.

Sluggish Catatonia

Sluggish catatonia contrasts with speech-prompt catatonia because verbalizations occur sluggishly and in later stages may cease altogether. I have on occasion called this form "distracted" since distractibility is also very characteristic. In common with the speech-prompt form sluggish catatonia displays lack of initiative. What Kleist called "motivationless catatonia" included speech-prompt, sluggish, and some of the proskinetic forms of my classification. Kraepelin considers the alternation between taciturnity and verbal excitement with hallucinations (very characteristic for this form of catatonia) as a trait of his "negativistic dementia".

In later stages of sluggish catatonia the patients are generally so unapproachable that nothing more can be learned about their inner life. Often during the early stages they show a picture of general sluggishness and unapproachability. In other cases patients in earlier stages talk a great deal and even present **fantastic ideas.** One patient claimed he was engaged to the daughter of the emperor and was going to receive 50,000 Marks from the parliament for magnetic cars, street cars and telephones. He stated that he had ordered the army to march, that he had young devils in his chest. Another patient claimed that he had taken a spaceship to Jupiter and had thrown meteors onto the earth. A female patient claimed to have been repeatedly kissed electrically, been made pregnant, and mice and a dog had been put into her body; she had gone to Munich to have all of them taken out. She said she had pinched off the antenna, but still had found no peace. These stories have a **confabulatory character;** they do not represent primary confabulations but rather repetitions of hallucinatory experiences. This will become clearer when we see how sluggish catatonia is a marked hallucinatory condition in which confabulations may also include remarks about voices and bodily sensations. The stories are especially reminiscent of dreamlike experiences and are possibly caused by scenic visual hallucinations. Often they are expansively colored. They probably still play a role in the end states of the patients, but are no longer observable due to the general inaccessibility.

Even if there is no evidence of confabulations it is clear from the beginning that the patients are **hallucinating.** Their distracted inwardly directed facial expression may be decisive for an early diagnosis. This is particularly true when the patients speak little. As the disorder progresses patients become unapproachable and, in the end state, reach the characteristic sluggish picture.

If communication with these patients is attempted they look up vacantly, but usually do not answer. Even to repeated questions they give no answers or their replies are meaningless. Sometimes they seem to be about to speak, but they stop before having produced any sound. If questioning is continued with persistence some sort of answer may be given, such as their name or date of birth. The patients refuse to speak not out of negativism; speech simply seems extremely sluggish, arising only after the most intense stimulation. Peculiarly these patients speak excitedly without inhibition during their excited phases, as we will see.

Less strongly than speech but still clearly enough other reactions of the patients seem somewhat **sluggish** as well. When addressed they look up slowly. They follow orders slowly but correctly if the orders are given forcefully. Aside from sluggishness their movements occur normally. Opposite tendencies such as seen in negativistic catatonia are not observed. Psychomotor sluggishness is complemented by empty facial expression. But there is something else that is important for the diagnosis.

Although the face hardly moves it does have a certain expression, particularly in the eyes. When asked something the patients look up with a somewhat perplexed, empty look; before having turned all the way toward the observer their look slides away again. Sluggishly it wanders here and there and in-between stays put without being fixed on any other particular object. This is observed when the patient should be objectively turned toward the examiner. The facial expression reveals that the patient's attention is not directed at the observer nor to the surroundings, but inwardly. This **distracted facial expression,** often present at the onset, becomes particularly instructive if the patient's hallucinations do not result in conversation with the voices. Generally this is the case. While facial expression reveals the internal direction of the attention, the lips are whispering. Sometimes only lip movements are visible, sometimes the whispers can be heard, sometimes loud voices are produced or even in excitement shouted out. With regard to whispering there is no real confusion with the mumbling of proskinetic catatonia if careful attention is paid. It is not the mumbling while turned toward the observer but the whispering with distracted facial expression. The **hallucinations** are apparently **continuous** – the patients always offer this picture of distractibility.

Verbal sluggishness may be seen as an expression of the fact that the patients seem not to realize what is happening, that they do not even hear what they are being asked. This would explain why occasionally they give short answers and then immediately stop reacting and look away distracted. It clearly shows the difference from speech-prompt catatonia. In the latter the patient's attention is immediately directed to the question and leads quickly to a response, while in sluggish catatonia attention is deficient and response is retarded or completely absent.

The distractibility associated with hallucinations becomes impressively clear in periods of **excitement** occurring occasionally in sluggish catatonia. In earlier stages excitement may lead to attacks on the surroundings to which the hallucinations apparently are related. In later stages patients may become excited themselves in response to their hallucinations. They stand around shouting and scolding into the air, evidently against the voices. In addition there are gesticulations and grimaces which are clearly recognizable as reactions to affect-laden voices. Facial expressions are often threatening, scornful, or otherwise hostile. Such periods of excitement may come every few weeks and last from one to several days. In other cases, especially in male patients, alternation between calmness and excitement is less marked; rather answering voices, always present, becomes louder without real excitement. It may also happen from time to time that patients shout a few words and then are quiet again for a long time. In all these forms excitement is characteristic, always discharged in words, although the patients' speech is otherwise sluggish. This confirms that the patients' difficulty of verbal expression is due to their being inwardly distracted and thus unable to concentrate on questions. When stimulated from within their verbal expressions may become more fluid. During their hallucinatory outbreaks they talk quickly. Other reactions during hallucinatory excitement are not inhibited; in response to the voices they hear they stamp on the floor, hit the wall, spit, threaten, and gesticulate, all this with rapidity.

Unfortunately it is not always possible to understand what the patients shout during their excitement, since speech is often indistinct. It is however usually possible to grasp enough of the contents to recognize the connection with hallucinations. I here present the transcript of a female patient's excited verbalizations.

„It will be good, or else I'll hit you. If I to every apprentice... (laughs to herself)... or I'll hit you again... I don't wan't him either. I can't fear no God and no devil, I don't know what this is... I don't know, what I should eat. I do that too. Do you think I'd let my heart for nothing... now I mean, now I want my peace. If I hear anything again... if N. has to come down again to me. I am not going to P. for nothing. Now I would like to mean... I would like my soup and my peace, or else I'll beat him up... you take everything from me, too... if you make me so angry... what do you think, how poor you make somebody, I'm just a poor child... what do you think really?...?

Pauses during which verbal output ceases correspond to the points in time when patients are listening to their voices. Nevertheless this is not clear in every case. Patients may be observed to shout and scold continuously for longer periods of time so that they probably have no time to hallucinate inbetween. But from such continuous expressions may be deduced that they turn against the hallucinations. Whether the patients are reacting to brief hallucinations and no longer let their imaginary enemy say anything or whether they are able to hallucinate and scold simultaneously is unclear. The following transcript clearly reveals the connection with hallucinations, although the patients spoke with hardly any pauses (the gaps represent spots where I could not understand anything):

„I don't like Hans as much... she has his child, it is... at Haselberger we weren't children anymore... he doesn't even has a shirt, the boy... the nurse has devoted eyes... that you know that shoes belong to me, and such a dress I have too... the man said I was condemned to death. Why can't we sleep? You never succeeded as a cobbler... I don't have chest catarrh... would like warm blankets, that is true... that you always report the children..."

It has not to be assumed that hallucinations are the primary cause of excitement; these patients nearly always hallucinate and are only occasionally excited. Undoubtedly psychomotor processes are decisive for the appearance of states of excitement. Those catatonias where there is loss of movements generally tend to show transient excitement, perhaps serving to discharge built-up energy. This is seen most impressively in negativistic catatonia but is also present in the sluggish form. The milder bad tempers and excitements found in speech-prompt and proskinetic forms likewise could be understood as countermovements to the general impoverishment of initiative.

Excitement in negativistic and sluggish catatonia is of comparable strength, but in other respects is different. Periods of excitement in negativistic catatonia are generally short and sudden and are directed against the surroundings, whereas in sluggish catatonia they are mostly verbal and without relationship to the surroundings. If aggressiveness occurs it is interpolated into the verbalizations. We will find a related form of excitement among the paranoid form of incoherent schizophrenia.

More detailed information about the nature of the **hallucinations** cannot be obtained in sluggish catatonics because the patients do not reply to questions. In the early stages they speak of voices, sensations, and optical phenomena, generally only as indefinite hints. As we have seen, at the onset confabulatory stories often point to optical experiences. It is possible that the patients see as well as hear the people they are encountering in their hallucinations. This might explain why the patients when excited for some time stare rigidly at one spot. From some of their scoldings it seems that female patients feel sexually abused. One male patient clearly had hallucinations of taste and smell. He often spat and on one occasion said it was "always these spots of wax". In the end state all forms of hallucinations may play a role, but the voices are in the foreground; one always recognizable feature is the hallucinatory dialogue.

The posture of patients is also clearly abnormal. Patients, especially women, sit with backs bent, leaning slightly forward. The posture is not stiff, and therefore unlike that of Parkinsonism. Facial expressiveness is greatly impoverished though also without rigidity. Mental expressiveness is largely absent as in speech-prompt catatonia. Considering the general inaccessibility of sluggish catatonics the lack of facial expression is less remarkable than in speech-prompt catatonics. Higher **initiative** is completely extinguished. Patients may be occupied with simple tasks but without stimulation they hardly move, unless of course they are experiencing hallucinatory excitement. Movements are slow, external events that might attract attention produce only incomplete reactions. **Affectivity** is shown only in the form of excitement when they become very irritated with scolding and shouting. Otherwise they display no emotions.

It is difficult to say anything definite about the **thought disorder** of sluggish catatonia. When patients still give simple information their statements may be correct. Often however their thought processes are quite tangential. In the end states answers are given only when the patients are excited, i. e., when they are more strongly distracted by the hallucinations. Their utterances then are not to be taken as answers but rather as responses to internal stimuli. Yet the disconnectedness of their verbalizations is remarkable. I carried out the following dialogue with a female patient who was mildly excited:

(What is your name?) "Maria." (And?) "Yes, hm, mother of sorrows." (What is your last name?) "It's not true, no, no." (When were you born?) "Yes, I don't know, what are you really thinking. I greeted first." (Where were you born?) "No." (Where did you go to school?) "I would have been given that as a present. What are you really doing? Quite small." (Where did you go to school?) "She gave me my first language. My mother comes.

Yes, so. If I went there again at quarter after six not yet. Those are sicknesses." (How old are you?) "It opens in the back. That is advertising with the big one. Yes, he laughed recently. I only said yes. That is not really Thursday or so. Sit around for nothing here, go to bed." (How are you?) "Hm, hm, at this occasional to bed. I would not like the whole hand. So take what I have, the lady can have the rest as a present. She can wait for once. If that wasn't so strenuous. How he got away? That is not true? You have to be so careful?"

This example is more revealing about the patient's hallucinations than her thought disorder; she repeatedly responded to hallucinations but not to the questions. Yet it may be assumed that the complete disconnectedness of the answers reflects incoherence of thought. Listening for the patient for a long period of time, one notes not only the lack of connection with the questions but also the lack of connection between one of her answers and the next. The themes appear to change continually. Thought itself is apparently **incoherent.** The objection that hallucinations are also constantly changing is not valid, for their nature and their course are undoubtedly dependent on the thought disorder. Thus I believe that the thought disorder of sluggish catatonia is characterized above all by incoherence. This hypothesis is supported by the fact that incoherent schizophrenia in which this form of thought disorder predominates, as well as the type of hallucinations, has much similarity to sluggish catatonia.

Case 64: Therese Geig, born in 1873, was suspicious, moody, irritable, and rude for years before her first institutionalization. In 1902 she claimed that she heard voices from the stove pipe and the walls. Later she believed that she had a cable in her head. On one occasion she spoke of a world secret and therefore went to the bishop. In addition she expressed hypochondriacal ideas: she no longer moved her bowels, her blood had thickened, and her bones were swollen. In 1904 she was admitted to Institution B. She sat emotionless, brooded, and every word had to be pressed out of her. She claimed not to have moved her bowels in three years and that her pulse had stopped. In 1905 she was flat and claimed that she was being badly treated like a domestic servant. In-between times she became excited and she scolded. Sometimes she threw herself on her knees, screaming and wringing her hands. She admitted hearing voices which called her a "pig". Quite often she carried on quiet conversations with herself. In 1906 she alternated between dull, mute behaviour and excitement with scolding, gesticulations, and grimaces. She claimed people attacked her and pushed her around. In 1907 she often stood in corners speaking and gesticulating. She was taken home, and at discharge declared that she did not know her mother. At home she was often excited and occasionally spent time in an institution which kept no medical records. In 1911 she was admitted to Psychiatric Hospital M. There she did not speak except to mumble occasionally to herself and to offer resistance. At times she played with her fingers and beat her chest. She was transferred to Institution K. There she was dull, smiled when addressed but gave no answers. She then became excited, continually shouting out swear words. In 1912 she often attacked other patients. In 1913 she never answered questions and at times became excited. During the following years she was dull, gave no answers, but often suddenly burst into laughter and made grunting noises. In 1918 she did not react to being addressed, but whispered to herself, sweeping her hands across the bed between separate words. Intermittently she was excited and scolded: "Pig man, pig, pig." In 1919 she whispered and laughed to herself. In 1920 she scolded, sometimes severely, gesticulated, and frequently whispered to herself. At times she attacked other patients. During the following years she was dull, gave no replies, but often talked to herself and shouted strong swear words. After about 1927 excitement became less common. According to the progress notes she whispered to herself, often with gesticulations and laughter. From 1930 on gesticulations, which appeared stereotyped, became more common than whispering. In 1931 she was occasionally described as being very loud, scolding, gesticulating, and grimacing. In 1933 she whispered to herself and snapped her fingers. In 1934 she continued to snap her fingers, she stuck out her tongue and grimaced. In 1937 she spoke confusedly to herself and giggled. – B. Schulz and I examined her in 1939. She gave no answers, only looked straight ahead, smiled, and whispered to herself. If in order to emphasize a question she was given a push she looked up fleetingly, but immediately thereafter became self-absorbed. Her lip movements were accompanied with various facial expressions; she suddenly looked to the side, whispered something rather angrily, or distractedly looked back and forth. Once she stuck out her tongue, looking annoyed. In-between times she kept tapping one hand with the other. With energetic questioning and push she did give her name. Eventually she also gave her first name, but no other answers could be obtained.

A **mother's sister,** Maria Mül, born in 1843, became ill in 1874, after having given birth to a child and having lost her husband due to sudden death. She became increasingly depressed, believed God had deserted her, and that she was guilty of everything. She was admitted to Institution M., but discharged herself the next day and was then secluded at home for a year and a half. She was very excited, wanted to kill everyone, tore clothes, screamed, banged her head against a wall, and was unclean. Eventually she recovered and was no longer abnormal. In 1893 she became ill again. She thought that her grandchildren were devils and became threatening. She was admitted to Institution H., where she claimed that nothing had made sense anymore and that everything was collapsing. She also claimed to have seen a messenger of God, that her prayers could have saved the world, and that she had become second in rank to the Mother of God. However she had not given her signature to God's messenger, so now everything was lost and she was guilty of all misery. The Holy Ghost had already disappeared, God the Father and God the Son were sick, in six or seven years at the latest the world would come to an end, and God would sink back into the sea of eternity. The patient was clearly desperate and she cried and prayed a great deal. In 1894 her ideas continued and she thought of herself as a destroyer of the world. In conversations she was plaintive and full of self-accusations. At the end of 1895 her mood changed: she became happy, festive, and talkative, claiming that the Christ Child had appeared to her and had announced to her that grace has been granted to all, even herself. No one was damned, the world was saved, and there was great joy. In January 1896 she often knelt in ecstatic posture and joyously expressed that she could see a host of guardian angels outside. A few days later her mood changed again. She became anxious, she shouted, cried, and asked to die. She ran back and forth through the ward calling out that she had destroyed heaven. She tried to strangle herself with her kerchief and begged to be treated mercifully. In February she hardly spoke at all, was quiet and with resignation declared that she was already dead. She remained quite silent until May. Then for a few days she was improved but soon after she was again very anxious. The stars were going to fall from the sky and it would rain until everyone had drowned. In August she prayed the whole day long. She then showed considerable improvement. Until January 1897 only pneumonia and pleurisy were mentioned which the patient survived. Apparently her mental condition had become so unremarkable that there was nothing left so say about it. Finally the patient was described as well balanced, talkative, friendly, and thankful. She was discharged home.

Subject Geig was not described precisely during the onset of her disorder. Voices are mentioned early. There are also hypochondriacal complaints, probably arising from bodily sensations. Once institutionalized Geig displayed the characteristic picture of sluggish catatonia. She hallucinated continuously, spoke, gesticulated, grimaced, and occasionally argued excitedly with her voices. For a period of time her excitement led to aggressive behaviour; later the excitement abated. What remained were periods of relative calm interrupted by more active preoccupation with voices. When I examined her she showed the characteristic picture of sluggish catatonia. The stereotyped tapping of hands was conspicuous (also mentioned in the progress notes). Presumably it was only a form of defense against the hallucinations. Since sluggish catatonics grimace and gesticulate a lot during their hallucinatory episodes it is not strange that certain movements in opposition to the voices are introjected. With her aunt we find a very different picture. She first became ill after giving birth, starting with depression, followed by excitement with recovery after a year and a half. There is no precise description of the psychosis. Her aggressiveness outwardly and against herself as well as her uncleanness were most likely due to the fact that she was kept in prolonged seclusion. She became ill for the second time 19 years later and displayed an alternatingly anxious and ecstatic picture with ideas of salvation. During her anxious periods her ideas of inferiority also had religious coloration. This was the picture of anxiety-happiness psychosis. After the disorder had fluctuated over a period of three years the patient recovered. The diagnosis of a benign anxiety-happiness psychosis was confirmed from the course. Hence there is no possibility of relating this psychosis to that of the subject.

Summary

Sluggish catatonics give sluggish answers in the early stages and in later stages generally do not give any. Later they look around distractedly and move their lips in whispers. From time to time they become more excited. They talk more loudly to themselves or become so excited that they scream and gesticulate into space, scolding against the voices they hear. In earlier stages patients' verbalizations confirm that they suffer considerably from **hallucinations.** Moreover they often present fantastic **confabulations** which probably also play a role in the end state, but which can no longer be demonstrated due to the patients' unapproachability. Not only speech but all other reactions are slowed in the end state. However since the patients speak a lot and move actively during their excitement their sluggish reactions presumably are due to their constant **distractibility** causing them to be nearly oblivious to external stimuli. **Initiative** is extinguished, self-motivation is apparent only in their behaviour toward their hallucinations. **Affect,** too, is only recognizable in their excited arguments with the voices. The thinking of sluggish catatonics cannot be easily judged because of their unapproachability, but it appears to be characterized by **incoherence.**

Review of the Family Picture of Systematic Catatonias

In reviewing the total family picture of systematic catatonias I will start with the **Frankfurt** cases. When I could judge with any degree of certainty, psychoses in the patients' families were usually the same form or the same form combined with another. This was particularly true of close family relationships, i. e., siblings. The few psychoses in encountered parents were not adequately described.

Subject Stö (Case 63) was presented as having speech-prompt catatonia; his brother probably had combined speech-prompt negativistic catatonia. Subject Nied (Case 60) was presented as having manneristic catatonia; her sister was similarly affected. Then there were the three siblings of subject Dob (Case 59), all with the same parakinetic catatonia. Furthermore another patient with parakinetic catatonia had a brother whose medical record also shows descriptions of symptoms of parakinetic catatonia. I could not obtain the medical history of a sister of a proskinetic patient. In no case did I have any reason to suspect that a sibling had a form of catatonia different from that of the subject, nor that the sibling had a different form of schizophrenia.

The psychoses of more distant relatives were basically similar to the proband's.

Subject Ile (Case 61) had proskinetic catatonia, his cousin likewise. Subject Au (Case 62) had negativistic catatonia, as did her cousin. Exceptions included subject Geig (Case 64) who had sluggish catatonia, whereas her aunt had anxiety-happiness psychosis; a female subject with manneristic catatonia whose relative showed a fantastic form; and a female subject with parakinetic catatonia whose relative had periodic catatonia.

The following are the psychoses present in the relatives of patients from my **Berlin series.**

- A negativistic catatonic male patient had a brother whom I also found to be a negativistic catatonic.
- A second male negativistic catatonic patient had a sister whom I also found to be a negativistic catatonic.
- A female proskinetic catatonic had a mother described in a medical record as paranoid, therefore with no connection to the proband's psychosis.
- A male speech-prompt catatonic had a grandfather (mother's father) whom I found to have confabulatory paraphrenia. This is the only proband in whose family I found systematic paraphrenia beside systematic catatonia.
- A female speech-prompt catatonic patient had an aunt, who suffered an anxiety psychosis during her menopause and died of cardiac degeneration. There seems to have been no connection to the proband's psychosis.

- A female patient with sluggish catatonia had a mother from whose medical history sluggish catatonia could be diagnosed as well.
- A female sluggish catatonic patient had a mentally ill grandmother (mother's mother) who was not institutionalized.
- A male manneristic catatonic patient had a great-aunt (father's mother's sister), who was repeatedly in psychiatric hospitals. No medical history could be found. According to the relatives she suffered from compulsive symptoms suggesting a relationship to the manneristic catatonia of the proband.
- A male manneristic catatonic patient had a brother who according to my own examination also had manneristic catatonia.
- A female manneristic catatonic patient had a sister who was twice in a psychiatric institution. She had high blood pressure and drank a lot; it is not clear if she had an endogenous psychosis.
- A male parakinetic catatonic patient had two brothers, who were repeatedly institutionalized. One whom we were able to examine also suffered from parakinetic catatonia.
- A female parakinetic catatonic patient had a sister whom I found also to suffer from parakinetic catatonia.
- A male parakinetic catatonic patient had a father whom I found also to suffer from parakinetic catatonia.
- A male parakinetic catatonic patient had a cousin (mother's brother's daughter) who was periodically psychotic with complete recoveries. There was no apparent relationship with the proband's condition.

Here again patients' relatives were generally found to display the same form of schizophrenia. As my earlier investigations had shown, of all the systematic forms of schizophrenia parakinetic catatonia had the highest incidence among relatives.

Parakinetic catatonia is curious in another way. Not infrequently healthy relatives display a conspicuous type of movement reminiscent of parakinetic catatonia. Earlier remarks about "latent schizophrenics" are particularly true for parakinetic catatonia. It is of course questionable if it is more significant here than in the other forms. Perhaps in "latency" parakinesia can be more easily recognized than other schizophrenic phenomena. Otherwise I found only the rigidity of manneristic catatonics in healthy relatives. Below I present the cases in which relatives of parakinetic catatonics displayed parakinetic movements. This should not suggest that these persons were ill, only that a certain disposition was evident.

Three fathers and two mothers showed unharmonic movements in which gestures and mannerisms were exaggerated and jerky. There were also indeterminate twitches in the body. Facial expressions were at times almost grimaces, and laughter was somehow distorted. Similar phenomena were found in two sisters, two nieces, and a great-nephew of parakinetic catatonics. One of the mothers may have had a mild form of parakinetic catatonia. She had had children by four different men and had married two of them. She did not bother about her sick son and did not even know the address of another. When we saw her she was constantly restless with movements in her whole body. Several times for no reason she suddenly rose up from her chair, only to sit down again. Her facial movements seemed to be more grimaces than expressions. However she had never had a more serious disability and was able to take good care of her husband and her two last children.

The results from the study of the patients' families indicate that systematic catatonias are based on specific propensities always leading to the expected pictures, be they pure or mixed with other forms. The potentials however only rarely lead to a manifest disorder. Keeping in mind the relative rarity of psychoses in the families of patients with systematic forms of schizophrenia there must be significant external factors. These will be discussed later on in this book.

Hebephrenic Forms

The concept of hebephrenia is not homogenous. E. BLEULER sees it as "the big pot into which are thrown all forms that cannot be grouped in the other three headings." He comes to this conclusion primarily by having delimited a special form, dementia simplex, comprising what others call hebephrenia. KLEIST and others see affective disintegration as the essential feature of hebephrenia. They established a sharp delineation of the concept to which I subscribe. HECKER came close to the same opinion that hebephrenia was predominantly, if not exclusively, an affective disintegration. The insidious course is usually considered a characteristic of hebephrenia. This seems unimportant when considering only the systematic forms, all of which tend toward insidious progression. The peculiarity of hebephrenia is that no unsystematic form with periodic or remitting course corresponds to it. Hebephrenia and systematic catatonia have early onset in common. BLEULER'S conceptualization was quite different. Polymorphic pictures difficult to group in any class, as stated above, usually represent unsystematic schizophrenia. Hebephrenia would be an inappropriate diagnosis in such cases. The indefiniteness of the term hebephrenia is reflected in the hereditary statistics of different authors. When the diagnosis was made in BLEULER'S terms hebephrenia was associated with a strong hereditary loading, because a great many unsystematic schizophrenics were included. On the other hand I have found the rate of hereditary loading to be very low, since the cases according to my conceptualization all belong to the systematic group.

The clinical description of systematic hebephrenics is much easier than that of catatonics. There are fewer sub-forms, and fewer symptoms. This does not of course simplify the diagnosis. On the contrary, observations have to be sharper to clearly determine the limits. Mild cases are especially hard to judge. In fact the diagnosis of hebephrenia can be difficult since affective changes, which are the key, are often difficult to ascertain, much more so than catatonic or paranoid symptoms. But if the most important distinct symptoms in the end are clearly recognized their milder forms may be identified as well. At the onset the picture may be still obscured by unspecific process symptoms. There are frequent depressive or euphoric moods. There may also be periods of excitement or inhibition which may appear catatonic. Severe excitement with aggressiveness occurring not infrequently have an irritated rather than catatonic character. Irritated moods may extend into the defect state; in some cases, as we will see, they themselves belong to the picture of the defect state. Delusions and hallucinations may accompany the affective changes in the early stages.

Foolish Hebephrenia

The picture of foolish hebephrenia is known to every psychiatrist. It contains traits usually regarded as typical, reminiscent of the silliness of normal puberty. If, as often occurs, hebephrenia develops during puberty differentiation between the disorder and ordinary pubescent behaviour may be difficult. However the affective flattening of hebephrenia usually becomes very clear.

At the onset there is in the foreground of the picture a tendency to **childish pranks.** As affective flattening increases the pranks soon become meaner. In the milieu of the home the patients often misbehave towards parents, siblings, and neighbours. The misbehaviour is at first minor, but becomes unpleasant when multiplied and repeated. In hospital the pranks or tricks usually take the form of pulling blankets away from other patients, spitting on helpless patients, howling or making other noises when others are trying to sleep. One patient interrupted the institution's religious service by singing along with the priest; another upset his surroundings by loud flatus.

Educational attempts are usually in vain. The patients do yield to greater power, behaving well for the doctor, or staying clear of stronger patients, but this is usually just a matter of waiting for better opportunities. Attempts at education may increase the misbehaviour and result in the patients doing precisely what they have been told not to do. This sort of behaviour easily takes on the character of childish obstinacy.

With the progression of the disorder the tendency toward childish pranks gradually gives way to loss of initiative, to be discussed below. Eventually the patients lose the energy for further pranks. In later stages they may show talking-past-the point; the way they do this indicates that they want to fool the interviewer or to play a joke on him. One of my patients, born in 1901, declared with a smile that he was born in 1891 and came to the world in 1893. Hebephrenic talking-past-the-point seems to be a matter of childish jokes and should not be confused with catatonic talking-past-the-point. In foolish hebephrenia even more characteristic than misbehaviour is the **tendency to laugh.** In early stages it can be very marked and may appear as silly giggling. It looks like the giggling, unmotivated laughter of girls during puberty. In later stages laughing is not so noticeable; in men, even in the beginning, it more often takes the form of smiling rather than laughing. It is characteristic as a reaction to any external stimulus. No matter if the patients are told something happy or sad they respond with smiles. It does become clearer with certain themes. Women giggle particularly strongly when they feel that they are being erotically approached. Laughter is never free but rather more a smiling or giggling to oneself, in fact the patients seem somewhat embarrassed even when there is no reason for this. In very severe cases, when the patients no longer like to speak, the smiles remain as reactions to external stimuli. **The smile therefore becomes the most important diagnostic symptom of foolish hebephrenia.**

It might be assumed that underneath the smile there is a cheerful mood. In fact the patients generally are content and cheerful. In acute phases clear euphoria may develop somewhat reminiscent of mania. On the other hand depressive moods may appear along with suicidal tendencies. These accessory symptoms seem to be the expression of primary involvement of the affective field in hebephrenia. But right from the onset affect generally seems to be somewhat flat. In later stages mood fluctuations occur in the form of transient moodiness. Mood changes are rarer and less clear than those of shallow hebephrenics – to be considered later – but they are noteworthy. Despite their usual contentment these patients may become irritated for a short period. They scold, slam doors, or hit other patients who have annoyed them. Moodiness may be associated with the patients' pranks particularly when they contain an aggressive element.

In foolish hebephrenia, **affective flattening** is considerable; all higher interest is lost. The patients' existence is dull, they do not care about their families and seek no stimulation. In contrast, **instinctual emotions** may become quite prominent. The demand for food and sex is often marked. There is a considerable **ethical dulling** often developing early – the patients lie, steal, and cheat when they have the chance. Women offer themselves readily to men. For more active criminality the patients lack motivation. In hospital ethical dulling may take the form of hitting or otherwise mishandling helpless patients.

Motivation also suffers from the extreme dulling of affect. Events in their surroundings no longer interest the patients and do not stimulate them to activity unless it involves something primitive, such as eating. Initiative is no longer present and in severe states the patients simply exist through the day, hardly moving unless stimulated. They give only monosyllabic answers. In such circumstances patients appear to be catatonic; but on a closer look it is seen that the psychomotor system is not affected: nothing in the way they sit or stand or move is catatonic. There is only severe lack of motivation, with lack of interest and lack of higher emotions. Even in the more severe states of dullness the patients maintain their contented nature and continue to smile in response to being addressed.

Thought, too, seems to suffer from the impoverishment of interest. It is not deformed, only insufficient. The patients tend to give taciturn, meaningless answers. KLEIST describes thought disorders marked by insufficiency and reminiscent of imbecility as "alogical". In severe cases correct answers to the questions of an intelligence test are rarely given. Generally they make only imbecilic remarks and do not respond meaningfully. One female patient answered an intelligence test in the following manner: (What is the difference between child and dwarf?) "There is no big difference, can't even tell." (What is the difference between tree and bush?) "That is such a funny plant." (What does this mean: morning hour has gold in the mouth?[9]) "Yes, the morning hour, that is a good time, it is best in the morning." (No rose without thorns?) "The rose has thorns, the violet has no thorns." Another female patient was even more unproductive and answered my questions in the following manner: (What is the difference between child and dwarf?) "There are children." (What does this mean: He who digs a pit for others falls into it himself?) "That too." (What is this? Patient is shown a Binet picture of a broken window.) "Very pretty, the picture."

No other symptoms can be demonstrated for foolish hebephrenia. Paranoid and catatonic phenomena are absent. In acute phases excitement may appear catatonic, but probably is more a discharge of affect than a psychomotor phenomenon. Besides stuporous states may occur at the onset, probably corresponding to the later moodiness. In most cases the picture even in an acute state is clearly hebephrenic. As a rule there is an insidious course with gradual increase of hebephrenic symptoms.

The brother of a subject was institutionalized for schizophrenia; unfortunately his medical record is lost, here I present the medical history of the subject himself.

Case 65: Ludwig Zins, born in 1903, as a child suffered for a few years from sleepwalking and later had lung tuberculosis from which he recovered. Even as a child he was moody and withdrew easily. In 1931 he thought people were ridiculing and persecuting him. He became depressed, turned on the gas, and was admitted to the psychiatric ward of Hospital S. He asked somewhat inappropriate questions, but seemed otherwise normal and was discharged after eight days. At home he stopped working. Again he believed he was persecuted and also heard voices. At times he was very anxious and insisted that someone wanted to cut his throat. He made another suicide attempt, shooting himself through the lungs. He was readmitted one year after the first admission. He spoke of visual and auditory hallucinations, partly taking a critical attitude towards them. At night he became anxious and saw a black man with a knife standing in front of him. He became nearly unapproachable, grimaced, and tapped stereotypically on his bed covers. He was transferred to Institution E. At first he was withdrawn, did not answer, and grimaced. A little later he smiled. In the following weeks he was sometimes excited and sometimes stuporous. In 1933 he was at first unapproachable and resistive. Later he could be addressed but seemed indifferent and had a disapproving look. Sometimes he misbehaved, spitting on other patients for example. In 1934 he was negligent, messy and unappetizing at meals. In 1935 he "languished indifferently day by day". At night he disturbed the quiet by banging on his bed. In 1936–37 he was similarly dull, but misbehaviour continued. On one occasion he tore up a poster on the wall; another time he destroyed a slipper. In 1938 he was without affect and initiative, answered curtly and repeatedly caused minor destruction. – B. SCHULZ and I examined him in 1939. He constantly laughed to himself, a somewhat shy, embarrassed laughter. When asked to stop, he did so for a short time. While dictation

9 Literal translation of German proverb: „Morgenstund hat Gold im Mund" (English equivalent: „The early bird catches the worm")

about him was given he smiled more intensely. Affectively he was quite flat. No emotional reactions were visible, not even when we spoke of his illness and his institutionalization. The patient, who had done very well at school, could only half solve intelligence questions. There never was a meaningful conversation. (Do you have anything to say?) "I don't know." (How are you?) "Yes, that's how it is. I am content." (No complaints?) "No." (How long have you been here?) "About six years." (Where were you before?) "Munich." (Why are you here?) "I was in Schwabing Hospital for treatment for my nerves in the hospital." (What did you have?) "Nerve shock." (What was it like?) "I don't know that any more." (Are you sick?) "I feel very well." (You are among the mentally ill here!) "No, most are normal." (What is the difference between a box and a basket?) "I don't know, I'd have to think about it." (What is the difference between stairs and ladder?) "Stairs aren't movable, a ladder is movable, one can take it away." (What does this mean: need breaks iron?[10]) "When a man is in need he can do more than when he is not in need."

At first Zins was very anxious. Besides that, stuporous states, excitement, grimaces and stereotypies are noted, perhaps associated with his moodiness. Soon the hebephrenic picture developed with pranks and naughtiness. When I examined him he presented the typical picture of foolish hebephrenia.

Summary

In foolish hebephrenia there is a high degree of affective flattening along with a contented to mildly cheerful mood. Particularly characteristic is the **smiling** or giggling which appears in response to every external stimulus. Flat affect is accompanied **by ethical deterioration** – the patients may become criminal in so far as they have a chance to be, but do not display active criminality. In early stages foolish hebephrenics display a tendency to **childish pranks.** They often have a malicious character, thereby illustrating their ethical deterioration. On the other hand maliciousness may be due to the **moodiness** which does occur, transforming the carefree contentment into irritability. In acute phases different kinds of mood are common: euphoric, depressed, or irritated. **Activity** also suffers from the increasingly flat affect. The patients lose their initiative more and more and live through the day without doing anything. In end states with severe lack of affect and motivation foolish hebephrenics may resemble catatonics. Their postures and movements, however, are not at all catatonic, and their characteristic smile clearly marks them as hebephrenic.

Eccentric Hebephrenia

In contrast to patients with foolish hebephrenia patients with eccentric hebephrenia are somewhat moody and joyless. KLEIST considers eccentric hebephrenia as I described it as his depressive form. KRAEPELIN likewise

[10] Literal translation of German proverb: "Not bricht Eisen" (English equivalent: "Necessity is the mother of invention")

described a depressive form of schizophrenia associated with the affectations shown by eccentric hebephrenics. Once the acute symptoms have abated the depression is no longer a factor while listlessness remains. At the onset there may be euphoric states, though they are considerably less common than depressive and anxious states.

During the early stages of eccentric hebephrenia **compulsive features** are more characteristic. We have already seen them in early manneristic catatonia as the precursors of the mannerisms. Since in eccentric hebephrenia mannerisms also appear, compulsive traits may be taken as early manifestation of defect.

Earlier editions of this book emphasized the presence of **mannerisms** in eccentric hebephrenia. Mannerisms are of course present, but they are more common when eccentric hebephrenia is combined with another type of hebephrenia. Mannerisms and stereotypes may then be plentiful, as we shall see. In simple eccentric hebephrenia they are not very evident; rather, they are replaced by uniformity of behaviour which, by stretching the point, may be regarded as eccentric. The patients may approach the doctor at each visit with the same request such as the wish to be discharged, to be given a new task, or with hypochondriacal complaints which will be discussed below. Refusing or ignoring these requests will not prevent the patients from repeating them at each visit. Often the same words are used. At each visit, and with almost the same words, one of my patients asked for her discharge and then added how long she had been in the institution, what the priest had said when she was admitted, and that her mother had never been in an institution. If I stood still she repeated the same remarks several times. On getting into conversation with such patients, or having come for a complete examination, they like to talk more, even developing some pressure of speech. The monotony remains unchanged, for their speech is always on the same track. Long monologues may revolve around a few themes. In earlier stages the patients may have several themes available, so that a single examination may not allow enough time to identify the uniformity. The best tactic is to avoid asking questions that might stimulate new trains of thought; rather, simply to encourage the patient to keep talking. Then, instead of starting a new theme or saying nothing, the patient will talk for a second and third time about what he has already said before.

Speech usually has a slightly **querulous tone** – patients who have arrived from home may complain about their families, their neighbours, or officials, and so on. Institutionalized patients complain about the food, treatment by the attendants, and of being exploited in their occupations. These complaints about others are generally associated with self-praise. The patients claim to be clever and industrious, and their failures are the fault of others. Their complaining is probably a result of their joyless, discontented mood, while the self-praise is an attempt to make others responsible for their own problems.

Furthermore, **hypochondriacal complaints** repeatedly appear in the patients' remarks. In earlier stages the descriptions of the complaints suggest psychosomatic hallucinations. The patients may claim that there was not enough blood in the brain, that the heart had been displaced, that there was a pulling in the spine, that electric currents went through their body, etc. Such descriptions do not appear in the later stages. Actual sensations arise only in the acute stage, perhaps when feelings of depression, to which the patients tend, reach a greater depth. Later only general complaints are made: the body was weak, there must be stomach trouble, there was something wrong with the lungs, and so on. A wish for better treatment or to be excused from work is usually associated with these complaints.

Talking and complaining always occurs in a monotonous, affectless manner. There is lack of energy even when the patients are querulous. The complaining never becomes excited or irritated, therefore confusion with paraphrenia is not possible. Even if the patients are blatantly ignored, dealt with summarily, or even offended, they do not become excited; at the most they present their concerns somewhat more intensely.

Since their verbalizations are largely independent of what was said to them, foregoing every opportunity to respond naturally, the patients appear eccentric. Their complainings seem to be more manneristic than sensible; it is continued even if it produces no results. Modern psychopharmacological treatment diminishes the patients' drive, hence the tendency to continuous querulousness is reduced. However the monotonous speech with no modulation of the voice is unaffected.

Although the depressive trait persists it still does not obscure the patients' severe affective flattening. The uniformity of the complaints, unchanged as they are by external stimuli, shows the absence of affective depth. The complaints are so monotonous that they are unlikely to be taken seriously. The rest of their behaviour also reveals the flattening – the patients lose interest in their families, and they do not want to participate in any activities. In hospital their **ethical dulling** is seen in the rough way they deal with other patients who have somehow bothered them. Unlike foolish hebephrenics they do not play tricks.

The listless mood of eccentric hebephrenics is occasionally heightened so that a certain degree of irritability may arise. This tendency parallels the irritability in other forms of hebephrenia. In the eccentric form the irritation is generally not marked and hardly interrupts the otherwise unchanging state.

Along with the affective dulling **initiative** also suffers, in so far as it is directed to more sensible goals. On the other hand, a more primitive instinct is maintained. The complaints at times seem to contain a certain compulsiveness, as if the patients were repeatedly driven to complain. This is like the compulsive activities found in the onset of the disorder.

The **thoughts** of the patients are impoverished, as can be seen by the monotony of their action and speech. Intellectual achievement, however, is much better preserved than in foolish hebephrenia. If incomplete answers are given a mild degree of mental handicap may have existed even before the onset of the illness.

Case 66: Rosa Leb, born in 1882, a kindergarten teacher, used to have migraine headaches. In 1909 she became excited, then unhappy, and did not want to be among people any more. She moaned for hours about terrible fears, claiming that people were talking about her, then she laughed for no apparent reason. In 1911 she was admitted to Hospital M., claiming to feel anxious and adding that she had obsessive ideas. Some times she could not get certain thoughts out of her head. Her "feeling of responsibility" stopped her from working. She maintained her belief that people were talking about her. Furthermore she complained of shivering in her arms. With eyes closed she sometimes felt her arms were getting bigger and bigger. She was moved to Institution E., where she was often moody and tearful. In 1912 she attempted to escape, remarking that she had to get away as she could not rid herself of certain thoughts. She felt influenced by a lady doctor. At times she was irritable and excited. In 1913 she was frequently excited, moody, and also became aggressive. In 1914 she continued to be irritated and excited, and she complained of headaches like the migraines she used to have. In 1915 she had many wishes and became excited when these were not fulfilled. From 1916–1918 she was generally busy, but carefully avoided the doctor. Occasionally she was still irritated and also became aggressive. During the following years she maintained the habit of avoiding the doctor. When she saw him she quickly ran into the toilet or hid wherever she could. She generally got along with the nurses. After 1928 she was no longer greatly irritated, but continued to grumble and be dissatisfied. Thus she remained during the following years. She had many wishes, demanded a special diet, wanted a vacation, and was angry with the institution and the doctors. She could however perform some functions. In 1938 she was constantly dissatisfied, but continued her work. She argued and had many "hypochondriacal complaints". – B. SCHULZ and I examined her in 1939. She immediately began to speak and kept on talking as long as she was given an encouraging look. She stuck to the same themes, harping particularly upon the lowering of her pension and complaining about the young, inexperienced nurses. She mentioned again and again that she worked well, fulfilled her duties, and obeyed the rules. At first she was friendly, but gradually she talked to herself into a more querulous tone, her voice remaining monotonous and deeper affect being absent. She could not be brought to become excited, and even provocative remarks that she was stupid bothered her only for a short moment.

Her **sister**, Anna Leb, born in 1887, according to a colleague from her office who had known her since 1902, was always very conscientious, never made a mistake. She read letters ten times and felt responsible for everything. Her overconscientiousness gradually became so bad that she could no longer do her work properly, and often sat staring blankly. She refused a raise of 20 Mark because she felt she did not perform well enough. She often worked over time. When her boss of many years died she became very depressed and in 1926 had to be admitted to Psychiatric Hospital M. She reported that during her vocational education she had been depressed for the first time for about six months. Since 1903 she had suffered obsessional doubting, and always been very scrupulous in religious and personal affairs as well as in her job. She always thought she might say something wrong. She even doubted that her parents were really dead. She sometimes felt that she had only imagined her own name. She had to think about certain letters for so long that she became confused. She said that she had "tainted herself" sexually. Objects that she had always had around her suddenly seemed new and strange. She spoke softly, was depressed, cried, but appeared "nearly uninvolved". The physical examination revealed that she had shaved her pubic hair "to be clean". On the ward the patient was depressed and had little contact with other patients. Her movements appeared awkward and unsure and the muscles of her forehead were distorted. She then improved a little, but always remained "somewhat reserved". She hardly spoke with other patients and avoided the day room because there were too many people there. She was discharged 16 days after admission. The diagnosis was: Psychopathy (psychasthenic, sensitive, autistic, eccentric, with obsessive trends)? Schizophrenia (defect state)?

During the onset subject Leb was anxiously depressed with ideas of reference and clear obsessive symptoms. In the years that followed her moodiness was often accompanied by irritability. At the same time the hebephrenia made its appearance. Irritability continued long into the defect state, because of the patient's tendency toward moodiness from her migraine headaches. Eventually the irritability disappeared or assumed a milder character, as is usual in eccentric hebephrenia. The behaviour of the patient was clearly eccentric, especially her habit of always avoiding the doctor. When I examined her she displayed the typical picture with monotonous affectless querulance.

The sister was not considered schizophrenic in the statistical section of the original classification of B. Schulz. Yet she does represent a case of a similar psychosis in the family. The disturbance began during the period of training when she was depressed for six months. This early onset is characteristic of eccentric hebephrenia. The psychosis, however, did not progress to a severe end-state, rather the process seems to have stopped half-way. Obsessions, otherwise only characteristic of the acute state of eccentric hebephrenia, remained the most essential symptom in the entire course. The medical record shows that it was not a simple obsessional neurosis: she was peculiarly reserved, appeared disinterested despite of her complaints, and had no contact with other patients. Her behaviour was clearly described as eccentric. Her attitude seemed to be partly manneristic rather than anankastic, e. g. having shaved her pubic hair. Therefore her disorder definitely was eccentric hebephrenia which was arrested very early. E. Faust (1953) discussed the Leb family in her paper on latent schizophrenia.

Summary

Eccentric hebephrenia often begins with **obsessive-compulsive symptoms,** which may develop into **mannerisms if** treatment is inadequate. Later the principal features are monotonous speech and querulousness, the uniformity of which gives the impression of mannerisms and eccentricities. The patients' mood is somewhat unhappy. But **affective flattening** is more prominent; complaints are not accompanied by more profound feelings. **Ethical dulling** is also present. During the onset increasing flattening of affect is often interrupted by **periods of moodiness,** which may be associated with irritated excitement. In later stages only mild mood fluctuations appear. From the way patients talk, **thought** appears to be impoverished. Intellectual performance however is only mildly impaired.

Shallow Hebephrenia

In my first classification of schizophrenic forms I did not treat shallow hebephrenia as a separate variety, whereas Kleist delimited an unproductive or apathetic hebephrenia which he related to Bleuler's dementia simplex. From characteristic **states of moodiness** I was finally convinced that this was indeed a separate form. At first the moodiness is no different from that in other hebephrenics. In general such states have an anxious coloration and may be accompanied by ideas of reference and hallucinations.

Of course all hebephrenics are affectively flat; that is the primary basis of the diagnosis. Generally, however, **dulling of affect** is built into other symptoms. The term "shallow" should, in contrast, express that psychic functions appear to operate altogether on a more insipid level. On examination the patients show nothing but the change in their emotional life. One can converse with them – they have opinions, and they answer simple questions correctly – but there is no affective participation. They speak of their years of institutionalization, their mental illness, even the deaths of their relatives, everything that would deeply move a normal person, in an uninvolved manner as if they were discussing the most neutral subject. It is not necessary to lead the discussion to such themes in order to display the shallowness, however. Every conversation, even with neutral content, touches the normal person; at different points he is more or less involved. This inner participation is visible in his behaviour, particularly in his facial expression. In shallow hebephrenia it is altogether missing. The affective emptiness despite the capacity for objective understanding shows that the patient is not emotionally involved.

Because of their emotional shallowness the patients have no wishes or plans for their future. Institutionalized patients may occasionally express the wish to be discharged, but more as a formality when confronting their doctor. When they have a problem, for example, when they are physically sick, or an unfriendly fellow patients bothers them, they express the complaint in words only, while the tone of their voice is quite flat. The impression of flatness is heightened by the distortion of the mood somewhat to the euphoric side. Patients are, of course, never really cheerful but they display a carefree contentment. While the laughter of foolish hebephrenics can make them appear cheerful, the contentment of shallow hebephrenics is void of emotionlism, appearing as a constant carefree state.

Lack of initiative derives from the lack of affect. – There is nothing that the patients feel like doing. Truly asocial tendencies are rarely found though patients because of their lack of affect may sometimes be induced to perform asocial acts. They adapt themselves to the daily schedule; they do not have to be urged or pushed. In this way they remain more active than foolish hebephrenics. Nor do they become uniform like eccentric hebephrenics. There are no essential intellectual losses.

Patients with shallow hebephrenia generally appear as described above. However there are other signs. Much more clearly seen than in the other hebephrenic forms is **moodiness,** often accompanied by **hallucinations.** If the patients are asked about their mood they talk about it in their usual unmoved manner. They report that they are periodically anxious or irritable or they may only mention the hallucinations. These moods occur either after intervals of months or a few weeks, generally lasting several days. Occasionally the patients deny the obvious moodiness giving excuses. The nature of the hallucinations varies; usually these are auditory, although visual and somatic ones also occur. Once the clinical picture of shallow hebephrenia is recognized there is no likelihood of confusion with hallucinations of paranoid schizophrenia, as the patients speak of their symptoms with so little interest. This is due not only to the flattening of the affect, but also to the greater distance they keep from the hallucinations. In fact, they generally understand that the voices exist only in their minds and have no objective basis. The impression is gained of **pseudohallucinations.** Perhaps the periods of calm following the moodiness give the patients an opportunity to assume a more critical stand. Paranoid schizophrenics too may hallucinate less frequently, but they do so without gaining insight. It may therefore be assumed that the internal picture of the hallucinations of hebephrenics has a more special character. If the patients during periods of irritation occasionally complain about the hallucinations, as if they presume them to be real, this should be considered as an expression of affective tension and not of a thoughtful attitude.

I present several **examples:** a patient, usually appearing quite shallow, occasionally became irritated, complaining of voices and of electric currents; after a few days he calmed down. In another patient periods of moodiness were deeper. From time to time he blocked his ears for several days and complained severely about the annoyances. A female patient spoke of anxiety that sometimes came over her while she, at the same time, heard voices which during my reexamination she herself described as "imaginations". Another patient had occasional visual hallucinations but without moodiness. She saw religious pictures and complete scenes as well as people speaking to her. These more dreamlike states give the particularly strong impression of pseudohallucinations. One of my cases, apart from irritated moods, also had euphoric phases with cheerfulness and heightened self-consciousness. This, however, is quite rare. Even during the onset of the disorder cheerfulness is hardly ever observed, yet it is not infrequently found during the onset of foolish hebephrenia.

If the moodiness of shallow hebephrenia has an irritated character, it is often accompanied by **ideas of reference.** Persons in the surroundings are then accused of maliciousness and duplicity. Excitement associated with aggressive behaviour is not rare. It may resemble catatonia, but closer observation shows that it is only a matter of irritated excitement with no catatonic element.

In some cases of shallow hebephrenia moodiness may be very marked, even if otherwise the disturbance is not great. Moodiness may be observed while flattening of affect is still unnoticed. A patient in the Frankfurt Hospital, not belonging to my investigative series, was for years considered to be a psychopath. It was observed that her irritated states were always accompanied by peculiar ideas of reference, and occasionally by hallucinations. Otherwise she was friendly, approachable and appeared normal. However, on getting to know her better, a clear lack of affect could be seen. She cooperated with everything: she stayed in the hospital when her husband wanted it and came home if he suggested so. She spoke of their domestic difficulties, which did exist in reality, with remarkable calm as long as she was not in a bad mood.

Of my cases of shallow hebephrenia, Else Wick (Case 67) and her brother have already been described by FAUST (1941). FAUST, following our mutual examination of the siblings, described them as atypical hebephrenics, since we were at that time not yet familiar with the typical picture of shallow hebephrenia. In both cases, however, FAUST'S description corresponds to the characteristic picture. Briefly it was the following.

Case 67: Else Wick, born in 1881, had in her youth been sensitive and friendly. Apparently she began to seem abnormal in 1927. She became sad, tearful, anxious and felt hypnotized and observed. She was admitted to Sanatorium A., where she reported that strange thoughts were placed in her mind and that her whole life was being told to her. She was perplexed, but then repeatedly ran about with silly laughter. She was discharged after six months. At home she remained anxious, heard voices and saw pictures. In September 1928 she was admitted to the Frankfurt Psychiatric Hospital. She was perplexed, displayed little spontaneity, and spoke of her voices. After two months she was taken home, but remained ill. In 1930 she ran away from home in hallucinatory excitement and was taken to Institution E. There she displayed affective flattening, had no interests, and little initiative. Intermittently she developed an instinctual restlessless and performed incomprehensible acts such as throwing things out of the window. In 1931 more severe excitement appeared during which she ripped clothes and became aggressive. She then quietened down, but was silly, foolish and manneristic. Affectively she was quite "disintegrated". In 1932 she was more excited and heard voices. She behaved this way during the following several years. In 1936 she was "affectively severely disintegrated", but worked on the ward. In 1938 she was occasionally angry and heard voices. – In 1939 FAUST and I examined her. She was friendly, approachable, and very talkative. She gave relevant responses, but seemed to like to digress somewhat. Affectively she was disinterested. She spoke of her voices without affect; she claimed that the voices said neutral things, for example about politics or the kitchen. She remarked, "they aren't voices. It was as if I think it myself." Again without affect she said she occasionally had sexual feelings, as if someone was reaching under her bed covers. She confirmed that these phenomena only occurred from time to time and that, when they did, she was easily excited. During the examination our discussion failed to excite her at all. It did not seem to bother her that we described her as mentally ill. Her only comment, "that really doesn't matter when you are 60 years old", was made with a smile.

Her **brother,** Hans Wick, born in 1882, studied in high school. In 1911 he made accusations against his brother-in-law and wrote love letters to a distant woman. A bit later he became anxious, heard voices, and tried to commit suicide jumping from a window. Recovered from his injuries in 1912 he was admitted to the Frankfurt Psychiatric Hospital. There he said he felt he had been robbed, and was described as being somewhat irritated. He performed gymnastic exercises naked. He was transferred to Institution K., where he was occupied as a working patient, but nothing about his mental state is recorded. In 1923 he was moved to Institution E., where he spoke of his condition without affect and admitted having auditory and visual hallucinations. During the following years he was described as quiet, dull, and indolent but able to perform simple office tasks. In 1930 it was recorded that he was periodically moody for a couple of days during which he performed his work incorrectly. Occasionally he stayed in bed in a mildly irritated state during which he grimaced and spoke to himself. – FAUST and I examined him in 1939. He explained that he had once heard voices although no one was there and that he had been scolded by the voices. He said that he no longer heard them. He said he enjoyed being in the institution because he could go outside when the weather was good. He appeared carefree and occasionally smiled. Considerable affective flattening was noticeable, nothing excited him; even a question about his mental illness did not move him. He spoke of the examination and the opinion that he was mentally ill as if it had nothing to do with him.

Subject Wick's psychosis began remarkably late when she was 46 years old. Perhaps an insidious onset had not been observed before. During the early phase hallucinations with anxious mood played an important role. Later only the bad moods were prominent, occasionally leading to aggressivity. Characteristic flattening of emotions occurred. When I examined her she offered the typical picture of shallow hebephrenia. The voices had no affective meaning for the patient and she recognized them as abnormal. Her brother who became mentally ill at age 29 displayed a considerably milder disease process. During the onset he too was anxious, later becoming dull and indolent, but was successfully employed as a working patient. His bad moods which were often present only occasionally assumed severe forms. Generally they seem hardly to have been noticed. Hallucinations apparently occurred only during the severe phases when he lay in bed, irritated and self-talkative. During my examination I found typical emotional blunting yet carefree, contented mood.

Summary

Shallow hebephrenia is characterized by **affective blunting.** In neutral conversations there is lack of emotional participation; in conversations dealing with themes that should move the patient more deeply, the lack of affect is even clearer. Superficially the patients remain approachable so that ordinary conversations can be carried out quite well. Affective flattening and **lack of interest** result in **lack of initiative** even though the patients adapt to daily life without special pressure. The mood is generally characterized by carefree contentment. This basic state, in which lack of affect appears to be the only obvious symptom, is interrupted from time to time by brief **periods of moodiness** in which the patients may become irritated, anxious, or, more rarely, euphoric. In their irritation they tend to express **ideas of reference,** and are sometimes excited and aggressive. **Hallucinations** affecting all senses are characteristic, most often taking the form of voices. Although the patients when agitated may become angry with the voices like paranoid patients, later they always have clear insight into their pathological nature; therefore the symptoms may be regarded as pseudohallucinations.

Autistic Hebephrenia

When I wrote the book about defect schizophrenic clinical pictures (LEONHARD 1936), I still believed autistic hebephrenia to be one of the paranoid schizophrenias and did not include it among the hebephrenias until I wrote my "Mitteilungen". Periods of excitement which I once thought seemed paranoid surely are evidence of an underlying moodiness. Furthermore autistic patients are affectively flat and should be listed among the hebephrenics.

Autism is considered by some to be a symptom of all schizophrenias. I cannot agree with this notion, at least in reference to the defect states. FLECK (1928) has also disagreed in that instance. Of course the intention of every schizophrenic to some degree is directed toward what he experiences in his illness, but this cannot be considered apart and does not represent a separate symptom. Among catatonics only the speech-prompt form drew our attention to features of autism alongside other symptoms. But in autistic hebephrenia it gives the clinical picture its own peculiar form.

If not approached the patients live only for themselves; they do not develop relationships with other patients, do not speak to them, and rarely approach the doctor. If addressed they usually give evasive and incomplete answers in a disinterested and listless manner. To questions which they could certainly answer they often respond with "I don't know." If they are not in a bad mood they can be friendly, but there is never an open friendliness – there is always some reservation. Their facial expressions correspond to their autism. It is somewhat stiff and appears unfathomable. As in speech-prompt catatonia, no conclusions about internal experiences can be drawn from the facial expression. If the patients' answers are friendly the **facial expression** incompletely reflects the friendly mood. If they give negative answers, the refusal is not sufficiently discernible in the face. This strangeness, which the facial expressions of schizophrenics can sometimes display, is especially marked in autistic hebephrenics. If one comes into direct contact with another person and is not, as has been said of schizophrenics, separated by a pane of glass, facial expressiveness above all makes the obvious difference. The facial expressiveness of hebephrenics is particularly unfathomable and one is unable to get closer to the patients' inner life.

If the patients are given a task they carry it out, but speak as little as possible. On meeting other people they usually walk by as if they had not noticed them. Usually they perform their work well; given somewhat more independent and responsible tasks they rarely fail. They seem a little more serious than other hebephrenics. Their **affect** is not quite as flat. Of course they seldom independently show any interest by themselves. They languish on, maintain no contact with their families, do not become excited by emotional themes, and rarely ask to be discharged. If irritated they may show some excitement, but only if they are already in a bad mood. If attempts are made to provoke these patients their unfathomable facial expression does not reveal how they react internally.

Despite the autism which may conceal some reactions the patients may be described as affectively blunt. Their inner life seems quite impoverished, they never express independent opinions. They have no **initiative,** never begin anything spontaneously, but only do the things they are asked to do.

Their responses to intelligence questions are brief and often insufficient. Their **thought processes** are usually better than those of foolish hebephrenics. Their incomplete answers give less the impression of thought weakness than of refusal or of simple lack of interest. Together with the autism the abnormal **mood** produces this refusal. The patients are somehow sullen and unhappy and never show a friendly trait. They resemble eccentric hebephrenics somewhat but their sullenness is stronger and becomes more intense when the patients are asked to speak or to respond.

It is of great interest psychologically that two of the hebephrenias display an unhappy mood, and the other two forms, the foolish and the shallow, are characterized by a carefree mood. Again this speaks for the hypothesis that in the hebephrenias **antagonistic functional areas** are affected, as we have already seen in the catatonias.

The autistic unhappy state which the patients generally display at times is briefly but impressively interrupted. **Periods of sullenness** occur repeatedly and are reminiscent of shallow hebephrenia, but have a rather different character: one of irritation, in which the patients excitedly utter words, expressing resentment against a person in their surroundings, or containing an accusation or threat. Without saying much the patients may become aggressive, not against anybody but rather against a particular person. I considered this earlier as "paranoid" excitement, since the patients seem to express hostility toward a certain person. More likely however, the irritation is directed against whoever is immediately available. Thus a nurse was brutally attacked by a patient when she was going to give another patient his food first. It is always a matter of unexpected, one-time reactions, without prolonged hostility towards anyone in particular. The patients can be very brutal and use whatever weapons happen to be available. One of my patients hit an attendant on the head with a stick he had taken during occupational therapy, causing the attendant to collapse. Another with no apparent motive shot a relative at home.

These excitations are dangerous because the patients' irritations are rarely noticed as they maintain their reserved manner and inexpressive face. In talking with the patient during one of these states the affective tension can be recognized in the irritated answers and the sullenness of the face. But when no attention is paid to the patient during this period the irritation remains unnoticed unless it breaks forth in some form or other. The patient's autism is then again recognizable. In later stages of the disorder periods of excitement are generally milder and may even disappear entirely. Then patients must be closely observed in order to discern the residual sullenness.

As we saw in shallow hebephrenia **pseudohallucinations** may appear during periods of moodiness. One patient claimed that she saw a nurse hearing her say "Wendelstein". But she knew what she had heard was not real. Another complained that she sometimes saw and heard all kinds of things. It is possible that such phenomena occur quite often but remain hidden because of the autism.

The mother of a male patient became ill during her menopause. More precise information was not available but there is a medical history of her sister.

Case 68: Johann Gro, born in 1915, was always considered strange. During his apprenticeship he claimed, although no such thing had occurred, that his superior had threatened him with a revolver. Later he was stubborn but had practically no contact with other people. When his brother reprimanded him during the harvest season he stayed in bed and threatened to kill the brother. In 1937 he was drafted into the army despite the warning of the mayor. He immediately became "the problem child" of his regiment because he refused training. He kept his back turned on the commanding officer, ignoring him. On one occasion, while on a march when another soldier behind him pushed his steel helmet forward to keep it from slipping off, he turned around and hit the soldier brutally in the face. Finally when he attacked a superior with a poker he was examined and discharged from the military with the diagnosis of "schizophrenia". In 1938 he was admitted to the Frankfurt Hospital, where he was quite dull, expressed no wishes, and gave only monosyllabic answers. His inner life was hidden by an impenetrable facial expression. If attempts were made to get information from him he became rather irritated. After observation which always yielded the same result he was taken home.

His **aunt** (mother's sister), Elisabeth Has, born in 1880, became more and more excited in 1906. She had feelings of anxiety, could not sleep, and heard a voice at night saying that she would be killed. After an attempt to drown herself she was admitted to Institution H. in 1907. There she declared that she heard voices day and night. Her anxiety soon disappeared, but her mood varied considerably. At times she was friendly, at other times she was angry, apparently with her voices. Later she complained of hairs she felt in her throat. She became violent with the doctor. In 1908 her mood continued to change; she discussed her voices with some insight and hinted that they might have partially originated from within herself. In 1909 she again claimed that the voices were real. Furthermore she said she saw shadows flying about her like swallows, and the voices came from these shadows. She also complained that the nurses were sexually molesting her. These symptoms were especially clear during her occasional periods of excitement which were sometimes accompanied by aggressive behaviour. In between she sometimes looked at her hallucinations objectively. She remained withdrawn and unapproachable and usually tried to avoid the doctor. She was taken home in this condition in 1909 after nearly three years of institutionalization.

It is not clear when the psychosis of subject Gro began; apparently it developed insidiously during his apprenticeship. When I examined him he showed the characteristic autistic picture, which, complemented by periods of severe excitement at home and in the army, made up the total picture of autistic hebephrenia. His aunt's condition looked like shallow hebephrenia. She hallucinated in her periods of moodiness, but later showed a degree of understanding of the pathological nature of these hallucinations. As well her experiences of talking shadows, for example, resembled the pseudohallucinations of shallow hebephrenia, but the acts of violence which are described correspond more to autistic hebephrenia. Since in the calm periods she was withdrawn and autistic, I suspect that hers was a combined autistic shallow hebephrenia, only one component of which was present in her nephew.

Summary

Autistic hebephrenics live only for themselves, they are close to no one and seem to be untouched by what goes on around them. Their **facial expressions** accentuate the impression of **autism,** since they are always curiously unfathomable and reveal nothing of their inner lives. Yet it can be said that their inner lives are **affectively impoverished,** as interests and independent opinions are usually absent. It is only in the **intermittent periods of sullenness** that affectivity is visible, and then only as irritability which easily spills over into aggressivity. Intelligence questions are usually answered inadequately, but less so if the patients are ready to respond. Answers tend to be short and disinterested. **Initiative** is reduced although the patients may be led to perform tasks demanding a certain degree of independence. Their **mood** is usually unhappy with phases of peevishness, similar to the affect of eccentric hebephrenics but different from the carefree contentment of the silly and shallow forms.

Review of the Family Picture of Systematic Hebephrenias

Our family research on hebephrenia was unfortunately much less profitable than it was on catatonia. In most cases in which a psychosis was present, no precise information was available about the relative's disorder. Whenever a diagnosis was possible, a similar psychosis was usually found. The eccentric subject Leb (Case 66) had a sister who presumably suffered from the same form. Subject Wick (Case 67) had a brother who also had shallow hebephrenia. The aunt of the autistic subject Gro (Case 68) probably had a combined form of autistic-shallow hebephrenia. On the other hand the father of a patient who had eccentric hebephrenia sufferin g from a periodic psychosis which had no apparent relationship to the patient's disorder.

In my Berlin investigative series, the following psychoses among relatives were found:

- A female patient with foolish hebephrenia had a brother who was institutionalized and whom I found also suffered from foolish hebephrenia.
- A female patient with foolish hebephrenia had a mentally ill great-aunt (mother's mother's sister). No medical history was available.
- A female patient with foolish hebephrenia had a psychotic aunt (father's sister). In institutional treatment she was perplexed, anxious,

affectively shallow, and disinterested. Hebephrenia was possible, but there were no silly traits.
- A female patient with shallow hebephrenia had a mother who according to her medical record likewise had shallow hebephrenia.
- A female patient with shallow hebephrenia had a brother who according to his medical record also had shallow hebephrenia.
- A male patient with shallow hebephrenia had a male cousin who was institutionalized and whom I found to suffer from combined shallow-eccentric hebephrenia.
- A male patient with shallow hebephrenia had a psychotic mother. Institution record show only that she had had schizophrenia.
- A female patient with autistic hebephrenia had a psychotic sister who was institutionalized and whom I found suffering from autistic hebephrenia as well.
- A male patient with combined eccentric-autistic hebephrenia had a sister from whose medical record it appeared that she was suffering from simple eccentric hebephrenia.
- The record of the son of a female patient with combined shallow-autistic hebephrenia apparently showed the picture of periodic catatonia but that patient could easily have been described as having autistic hebephrenia because of his sullen moods.

In the Berlin cases the same pathological forms were usually found in the family.

Paranoid Forms

The paranoid schizophrenias, for which I (with Bumke and others) use the term **paraphrenia** are characterized by **delusions** and **hallucinations.** At the same time there is a thought disorder from which the paranoid symptoms arise. In the description of the various forms of systematic paraphrenia patients bewildered by their symptoms, which may resemble cataphasia, need not be classified separately, as with their increasing thinking disturbance paranoid delusions become more pronounced. Incoherent paraphrenia with the severest thought disorder displays the severest hallucinations. In acute states of paraphrenia there are often **accessory symptoms** making a precise diagnosis difficult. Affective disturbances occur frequently, particularly anxiety. Perplexity is not infrequent as well as ideas of reference. These phenomena may be so conspicuous that they obscure the specific symptoms. Generally the latter are clearly demonstrable. Symptoms outside the realm of paranoia are rare even during the early stages. More profound affective fluctuations are unusual; these suggest early affective paraphrenia rather than one of the systematic forms. The diagnosis becomes more questionable with a more acute psychosis; a stormy onset is much more characteristic for unsystematic schizophrenias. Usually the course of systematic paraphrenias is from the beginning so gradual that no process symptoms are found and only defect symptoms are detected even if at first they are minor.

I was able to classify catatonias and hebephrenias so that there were always pairs with antagonistic pictures. This is not possible with the paraphrenias; although the law of antagonistic forms is valid here as well it is not as easily deduced from clinical pictures. Thus I have not attempted to offer such a classification, and have subdivided the paraphrenias more on the basis of external similarity.

Hypochondriacal Paraphrenia

In hypochondriacal neuroses, as well as in hypochondriacal depression and euphoria, we find bodily sensations. The sensations of paraphrenia have a different, hallucinatory character. They are not presented as bodily complaints , but as an annoyances of external origin. The **hallucinatory character** is confirmed by the fact that hypochondriacal paraphrenics always suffer from phonemes[11]. In this way hypochondriacal complaints in benign forms contrast with the hypochondriacal hallucinations of schizophrenics. Hypochondriacal paraphrenia is in general a hallucinatory form. The delusional aspect is relatively less important. Hallucinations usually appear during the onset, even if accessory symptoms, usually in the form of depressive moods, are present.

Hypochondriacal paraphrenics locate their sensations in different parts of the body, most often in the inner organs. Misperceptions are ascribed to organs of which normally there are no conscious sensations. In so far as sensations on the body surface are reported, e. g. stinging, cutting, pressing, burning, they may be compared with normal sensory impressions. On further questioning, however, the sensations are found to be of a peculiar type. Stinging is compared, for example, with a needle running through the head, burning with a shrinking of the whole skin. It is then recognized that these sensations are misperceptions which a normal person could not have. Quite often the sensations are described only in a general manner; the patients speaks of constant "torture", "vexation", or "abuse". Electric currents are often mentioned. Details must be elicited in order to learn the particular nature of the sensation. I present a few examples. Every sentence refers to a different patient.

A doctor had made thin legs for her, had poked around her anus, and had broken off a piece of her womb. – His blood was being drawn to one point and an electric current was aimed at each tooth. – He was pushed backwards and forwards and he felt a pressure on his eyes. – An X-ray machine was pressed on her brain and she was sexually molested during hypnosis. – "When I went into the kitchen, my breasts were huge and they made my body and legs look as small as a wax figure." – "He aimed a ray at my groin and worked on my whole body, the spinal cord, the lower abdomen, the stomach; all the nerves in my head have been torn." – They put pipes in his brain which they heated artificially; also a tubule into his penis. – His eyes, testes, back, chest, and larynx are burnt out by a machine. – She was kicked in the front and the back by a fellow patient hidden behind a curtain. – Organs were torn out of her, the toes were bent and tangled, and the ankle

[11] The word "phoneme" is defined in the section on "phonemic paraphrenia"

joints were twisted as if the feet were being ripped off the legs. – "They are taking out my insides; I am hindered when I walk; my toes are getting smaller." – Threads are being pulled out of her spine, her ears are being stung. – The current sent shocks through her body, especially through the sexual organs, it lifted her lower abdomen up, it changed her body, it tore the flesh off her calf muscles, it burned and smoked her leg.

These examples reveal the grotesque nature of the sensations. The patients describe the most peculiar events. In trying to determine more precisely what they actually experience, no progress is made. At best the patients use different words with equally little meaning. The impression is gained repeatedly that the patients experience sensations not at all comparable to normal experiences. Often the patients realize this themselves and describe their experience with the qualification "as if" this or that were happening. Yet in reality it is different. Often the patients who otherwise do not produce neologisms coin their own words for their sensations, such as "grongations", "flusions", "pavings" etc.

Since the grotesque nature of the sensations is not clear in every case, I present a few examples in which it might be incorrectly assumed that the sensations are comprehensible.

She must suffer bitter pains, is always beaten, mishandled and pulled about. – She is consistently tormented by terrible pains in her chest.- The blood bottles up, the gums hurt and the nose is blocked.- There are pains in the back, in the buttocks, and in the left breast. – There is a terrible pressure in the head and a feeling of heat in the sexual organs. – Dirty things are constantly done to her. – He is prevented from urinating or moving his bowels. – There is a burning feeling in his back and the chest is compressed. – Her stomach is upset and her intestines are inflamed. – Immoral relations produce sexual sensations. – He is being stung over his body, especially on the back. – She is continuously beaten, mistreated, and sexually abused.

Such complaints are heard along with grotesque descriptions. In milder cases patients may not describe their feelings exactly, but explain that their bodies are being **"influenced"**. However ideas of influence are more common in affective paraphrenia where the sensations in general have a milder form.

Hallucinations of smell and taste do occur in hypochondriacal paraphrenia, but the manner in which these are described is not particularly typical. There may for example be a complaint that the food tastes like poison or that it smells like ether. Such hallucinations are not frequent. **Visual hallucinations** are occasionally found during the onset of hypochondriacal paraphrenia. The patients may see "horrible pictures", "masks and animal heads", "male sex organs", "shadowy outlines of an enemy", "black devils", and the like. Later these visual misperceptions disappear completely or are so fleeting that they are no longer recognizable. Extensive visual hallucinations do not support the diagnosis of hypochondriacal paraphrenia.

On the other hand **phonemes** are a necessary part of the picture. Their absence precludes the presence of the systematic form. They are usually short, and not as sensible as in the phonemic form to be discussed later. All kinds of remarks are made, orders are given which however the patients rarely obey. Sometimes, whatever the patient is thinking is talked about, a sort of **"Gedankenlautwerden"** (thoughts-becoming-loud). This connection with momentary thought is particularly close during the early stages. Later it is more a question of isolated expressions, especially swear words. Nothing more precise can be obtained from the patients. They suffer from the voices and complain of the annoyance, but apparently rarely keep this in their memory. Their complaints refer less to the contents of the voices than the hallucinating itself. Just as the sensations are generally experienced as distressing, so are the voices, regardless of their content. We will return to this when we discuss the phonemic form. A few examples of auditory hallucinations of various patients are presented.

She hears many voices telling her about life; "so much is mixed up and I can't understand it; it would be better if only one talked." – The machines converse day and night, it is never quiet, terrible things are said. It is telepathy, perhaps from his daughter, although it could be from a police machine; the voice can be tuned up or down in a range from high E to low E; all kinds of things are said, he is called a tramp, a thief. – "You hear such things! The people are mean, this mob!". She was scolded because of her hair. – She is being discussed, she does not know by whom. – The devil calls sexual things to her and threatens her. – The police speak to him and urge him to do all kinds of things. – Pastor S. called out in the middle of the night: "Get dressed and come down!" – Another man once said at night: "Don't forget to go home tomorrow." – The commands are always brief, for example: "Go here, go there, do this, do that", she is insulted and accused of theft. – The voices irritate him, mock him, and forbid him to eat. – The voices say dirty things, it is all a mix up. – The voice expresses an opinion about whatever is being done: "He should not go so far", "he does not need to put his shoes on, he must kick the bucket."

Generally the patients have to be urged to describe their voices word for word. What the voices say is not so important to the patient even if he is being scolded. The common remark, it is all "a mix up", probably refers to the fact that the patients themselves realize that much of what they hear does not make sense. Perhaps the voices outweigh what is present in the patients' thoughts since the patients, after speaking about their complaints hallucinate very strongly, yet say very little about what they have heard.

The patients generally relate both the sensations and the voices to **machines,** which allegedly operate from a distance. More rarely they believe that the phenomena are mysteriously caused by people in their surroundings. This may arise in particular connection with so-called **"reflex hallucinations",** which occur in hypochondriacal paraphrenics. Here are a few examples.

She feels it across her abdomen when a nurse slices sausage; she feels needles when people sew. – Severe abdominal pains develop when certain persons come in the door. – He has pains in his teeth whenever he comes to a certain position in the room. – Every time he reaches for the newspaper they burn his eyes and insult him.

Although hypochondriacal paraphrenics hallucinate extensively, they rarely confront their misperceptions directly. They repeatedly complain about them, but rarely scold the voices themselves, as we find with other forms of paraphrenia. This is certainly because the nature and the content of the phonemes are very different from those of real voices. The patients assume as they do with the sensations that the phonemes are artificially produced.

Their defense mechanisms include removing wires which they associate with the hallucinations, or avoiding certain places. Their complaints may become querulous and be repeatedly brought forward empathically, reflecting the painful nature of the symptoms, as well as the patients' basic mood.

They are nearly always in a **sullen dissatisfied mood,** which not infrequently turns into irritability. They not only complain about hallucinations but may also repeatedly demand their discharge. When the wish is denied they become easily excited, showing that their affect is well preserved. This is always seen in the rest of their behaviour. Even after long institutionalization they remain interested in their homes and families, keep up a correspondence with their relatives, and express opinions about events at home. Thus they appear to have **relatively well preserved personalities** to be taken seriously if one does not want to offend them. Probably related to the mood anomaly is the fact that hypochondriacal paraphrenia often begins with a depressive phase, and that suicide is common in the patients' families.

Delusions are present in hypochondriacal paraphrenia only as **explanations.** Occasionally the patients suspect certain people of operating machines against them; they then express hostility, but further elaboration does not occur. Much more frequently they suspect unknown people. These people remain as impersonal for the patients as the suspected machines.

The **thought disorder** of hypochondriacal paraphrenics is relatively minor, too. In simple conversation they are hard to fixate; they often do not stick to the subject, but talk about something else only loosely related. A question to which a specific answer is required has to be repeated. In the end the patients do give good answers. This is a kind of **unconcentrated thinking;** occasionally there are neologisms. If the latter are used only to describe hallucinatory experiences they should not be thought of as a separate abnormality, as they also occur in other contexts as the result of the thought disorder. The disorder is not elucidated by intelligence questions. If the patients can be fixated on the questions they can give strikingly good answers which only sometimes contain slight verbal peculiarities. A female patient is presented with her family history.

Case 69: Rosina Be, born in 1894, had been a quiet child and did well in school. During puberty she had sexual obsessions; lewd thoughts forced themselves on her. The more serious illness appeared in 1925. She had states of anxiety and believed that she had done something wrong. She was admitted to Psychiatric Hospital M., where clear obsessive-compulsive symptoms were observed. She had to curse other people to herself and was always afraid she had done something wrong. She was soon discharged, but remained unwell. Increasingly hypochondriacal phenomena appeared. The patient believed that she had been disfigured, pummeled and raped by the doctors, and she claimed that her brain had been removed. In 1930 she was readmitted to Psychiatric Hospital M. There she reported that the doctors, operating at a distance, had removed her astral body and her brain and that her head was soft. Furthermore voices constantly spoke to her, usually vulgarly. She also said that forks and lances stung her through the wall and sexually abused her. She still had some insight, and wanted conformation that all this was only the work of her imagination. After a trial discharge she returned, claiming that threads were being constantly pulled out of her spinal cord. Later she reported that she had had two children, but added, "that is only hypnotic nonsense". In 1931 she was transferred to Institution E., and asked for an "anti-hallucination drug". Voices and sensations continued, and in 1932 she saw black devils and had the feeling that her head was being separated from her body. In 1933 she said that she was bodily abused and heard the Pope speaking. In 1934 she felt her neck being cut off. Her condition did not change during the following years. – B. SCHULZ and I examined her in 1939. She reported that she heard voices saying dirty and mixed up things. She felt her body was being tortured by an external force. She said her spinal cord was still there and her body did not have any holes; she had only felt this way. She spoke complainingly of her experiences. Her affect appeared well preserved. At one point she abruptly grasped her thigh, perhaps in reaction to a sensation. She did however not admit to this. In general she tended to dissimulation and on one occasion called her experiences "imaginations". There was no evidence of delusions. It was difficult to fixate her on questions. She liked to digress, but answered intelligence questions quite well.

Her **brother,** Alfons Be, born in 1888, became ill in 1907. He vomited blood, although no specific cause could be determined, and then complained of many body symptoms. He was admitted to Psychiatric Hospital M., where he was awkward and slow; he often frowned and sometimes smiled without reason. He complained that his siblings were mean to him. He reported that since vomiting blood he had jerkings under his shoulders and cramps. "The stomach rocked and the intestines were lifted up." His heart did not work well any more; his feet twitched, he was dizzy from the nerves in his feet. After talking the brain began to become heavy, all his muscles twitched. He was transferred to Institution E. His medical record was lost.

Her **father's sister,** Marie Be, born in 1865, was first in Institution R. from 1891 – 1893. She was severely anxious, complained monotonously, expressed self-accusations, saw the devil and heard him speak. The complaining and whining lasted two years. The patient then became calmer and was discharged. At home she was quite healthy for many years. In 1906 she became ill again and returned to Institution R., where she displayed the same picture as that of the first admission. She was anxiously excited, complained, and made uniform demands. She claimed that her siblings were being killed, that she was afraid of committing mortal sins, and occasionally she refused all food. Not until 1909 did she became sufficiently calm to be discharged.

Proband Be had an obsessive-compulsive constitution already obvious during puberty. From the obsessions she developed the later hallucinatory experiences which she observed critically so that she partially recognized them as "imagination". In addition the patient, with her grotesque sensations, voices, and unhappy complaining mood, displayed the characteristic picture. Her brother became ill remarkably early, at the age of 19, and from the onset showed grotesque bodily misperceptions. These were evidence of hypochondriacal paraphrenia, although a second component may have been added. The almost confabulatory elaboration of the experiences – the stomach rocking, the intestines lifted up, and the brain becoming heavy – and the listless behaviour described in the reports suggest the possibility of a confabulatory component, as will be seen in the description of this combined form. Unfortunately the medical record of the later course of his disorder was lost. Even if a combined form was present, its hypochondriacal component established the connection to the sister's psychosis. On the contrary I see no connection to the psychosis of the father's sister, who apparently had periodic anxiety psychosis with very long phases. Because other relatives in the family were also psychotic (although I could not obtain precise information) a second hereditary trait may have been present.

Summary

Hypochondriacal paraphrenia is characterized above all by **sensations** which are generally related to the inner organs, and by the grotesque nature which makes them incomprehensible to a normal person. There are also **voices,** generally insulting, the presence of which is more painful than the content. Partly they are isolated verbal expressions with no recognizable meaning and without the patients paying much attention to the content. In the early stage there is often a closer connection with the patients' thoughts; the symptom of thoughts-becoming-loud ("Gedankenlautwerden") may develop. The sensations are even more painful than the voices. These patients complain, often in a querulous manner, and they generally tend toward querulousness. This is most likely derived from the underlying **unhappy dissatisfied mood** which characterizes hypochondriacal paraphrenia. At the onset there are depressive states more often than in other forms. Affect is well preserved and the patients remain interested in their families. Thought is disturbed in that the patients are fixated only with difficulty and they have a tendency to verbal derailments – unconcentrated thought.

Phonemic Paraphrenia

Verbal hallucinatory paraphrenia, or the phonemic form as I call it for short, can be said to be the mildest form of systematic schizophrenia. Several cases were brought to my attention by coincidence, having been unrecognized at home for many years. This often happens because the onset generally is quite insidious. There are some cases with acute onset. Depressive, anxious or, more rarely, euphoric moods may be present at the beginning. Kleist spoke of "progressive hallucinosis", referring more to this and less to the hypochondriacal form. Schröder too called the attention to schizophrenic verbal hallucinosis.

Phonemes are the predominant symptom; hypochondriacal hallucinations are absent, as are hallucinations of smell and taste. If it appears otherwise careful questioning usually reveals the mistaken diagnosis. A female patient, for example, told me that, when she heard a voice, she felt that her heart was torn. On further questioning she described this apparent sensation as a shock from what the voice said. If exceptionally the patient complains of a body sensation the differential diagnosis is still not difficult; such questionable symptoms are nothing like the massive sensations of hypochondriacal paraphrenia. Only in the acute phase more definite sensations may be present, but then they are only accessory and soon disappear.

The **voices** themselves are essentially different from the auditory hallucinations of the hypochondriacal form. In the latter they often have no sensible content. If they do make sense they refer indiscriminately to with whatever the patient is occupied, often just reflecting what the patient is doing. In the hallucinations of phonemic paraphrenia the relationship to the patient's thinking is much closer. The voices express opinions on the thoughts; they often contradict and sometimes confirm. Usually they are expressed not with isolated words but with full sentences. The patients speak to their voices as they would to real people. They often whisper to themselves and admit having answered the voices. Hypochondriacal paraphrenics never speak with but only about their voices. Phonemic paraphrenics are much less frequently observed in conversation with their voices – patients with incoherent paraphrenia more often (to be considered later) – probably because the patients, still quite sensible, have doubts about conversing with people who are not there. Conversing with the voices does not imply that the voices have a physical clarity. In the hypochondriacal form the voices are probably clearer and more annoying. Phonemic paraphrenics regularly remark that their voices are different from real speech; they are softer, as if they came from a distance. Sometimes the patients state directly that the voices are more like thoughts. They may even speak of **"suggested thoughts".** The characteristics described by SCHRÖDER for schizophrenic hallucinations apply precisely to this form. If the patients speak more clearly about the origin of the voices they deduce it not from the sound of the voice but from the content. They often identify the sex of the voice or they accuse certain people from their surroundings, but their responses are drawn from what their voices said. A female patient declared that a nurse had said that a package had come for her but was sent back. When asked why that nurse she replied, it could only have been that one because she was the only one in the ward – in any case, she had not recognized the nurse from the voice.

The phenomenon of **thoughts-becoming-loud** ("Gedankenlautwerden") may appear. If the voices of a hypochondriacal paraphrenic relate to whatever thoughts are occupying the patient, then something like echoing of thoughts may arise. In the phonemic form the hallucinations seem to reflect the thoughts more directly. This is true not only when voices echo the patient's thoughts but also when they contradict the patient, resulting in argumentative dialogues. Alternatively patients may claim that the whole world knows their thoughts, that nothing is secret, and that everything which should be their own knowledge is discussed by the voices. This shows how sensibly the voices fit into the thoughts of the patients.

The voices of the phonemic form have a second characteristic: their content is quite emotional. The voices of the hypochondriacal form also contain an affective element (usually insulting swear words), but they are non-specific and have no relevant content, so that the patients are rarely badly upset. In the phonemic form, however, the voices take up precisely whatever makes the patients uncomfortable, whatever particularly frightens or embarrasses them. A hypochondriacal patient may be called a thief; the phonemic patient is directly accused of having stolen when something on the ward was missing. The voices may also refer to embarrassing memories, for example, memories of sexual experiences. Rarely do the voices have a pleasant emotional tone. A female patient at times hallucinated that a large amount of money had arrived for her. – A few examples of different patients will demonstrate more specifically the character of the voices.

There is a conversation and reciprocal answering, her whole life is repeated and everything is mocked. – The voices accuse her of leading a bad life and speak of the work house. – From the radio and elsewhere he is told that his work is worthless, that he has wasted everything, which is why he is penniless. – The voices accuse him of many different things, for example, that during the war he had been a coward and was therefore not wounded. – From the forest and from people around her comes a constant sneering and mocking that she was a whore. – A man whom she has never seen continuously speaks to her and says indecent things, particularly during meals, so that she loses her appetite. – There is a constant whispering, for example: "You are proud and unapproachable", or "You are a whore"; but sometimes pleasant things are said. – She hears a whispering at night telling her to get up or else she will receive an injection; there are people who read thoughts. – The radio speaks through the mouth of the nurse about her whole life; she is accused of having a Jew's child; whatever she thinks is repeated. – Three characters speak through the radio, often about sexual topics, or during a meal: "He gets no egg, but still has to work for the institution." – "People hear my mind and know what I think"; "That is how it goes through the whole life"; "They are voices to which you can give good answers." – The voices come from far away through the air, "I have told them to stop a hundred times, but it is no use." – Thoughts are picked out and later people know them, sometimes she has to answer. – Priests no longer treat confessions as secret; the voices accuse her afterwards of what she had confessed.

The voices of the phonemic form are seldom associated with machines, except the radio. The patients believe people, near or far, are speaking to them, or that the voices are those of invisible creatures, ghosts, or witches. Sometimes patients **remark that the voices come out of the inside of their own bodies;** "from the stomach", "from the chest", or "from the head", but they never describe precisely just what they mean. I had a long discussion with an intelligent patient who heard voices from her stomach. She explained that it was like ventriloquism and I had the impression that the voices were not only auditorially received, but also kinesthetically as speech movements (CRAMER'S muscular hallucinations). Since the corresponding kinesthetic impression comes from inside of the body, e. g. the larynx, it is understandable that the patients perceive the hallucinations as arising from inside the head, the chest, or even the stomach. It is important that such voices from inside the body do not lead to the diagnosis of hypochondriacal paraphrenia in which nothing similar is observed.

Once the difference between the voices in the hypochondriacal and the phonemic forms has been recognized, it becomes clear that the two paraphrenic forms have nothing in common, nor that the hypochondriacal form can be derived from the phonemic merely by adding sensations to the voices. The types of auditory hallucinations in the two forms are not the same. We will get to know quite different forms of hallucinations and conclude that the various forms of systematic schizophrenia can be differentiated by the fact that each form has its own kind of hallucinations. There are no common symptoms even though it may at times seem as if there were.

Besides the voices phonemic paraphrenia is occasionally associated with **visual hallucinations,** but these do not play an important role; the patient has to be questioned specifically in order to elicit these, for they are rarely mentioned spontaneously.

The patients may see "faces staring through the door", "a ghostly face that gets bigger and smaller", "peculiar shapes", "people like ghosts", "shadows flitting through the window", "faces in the trees", "animals, faces, and larvae", etc.

Affect of phonemic patients is also different from that of hypochondriacal ones. Unhappy dissatisfaction is absent, the patients' mood is balanced. They do become annoyed in reaction to the nasty voices, but in later stages they are generally more resigned and can talk about the voices more calmly. Hypochondriacs rarely get used to the hallucinations and, in later stages, are still excited about them. In phonemic patients affect is well preserved; they remain participatory and interested. They can be persuaded to perform responsible tasks. Many phonemics are never institutionalized because their affect is relatively well preserved. Affective dulling is present only in the indifference with which they let things happen and accept their fate.

Without more careful examination the impression may be gained that the patients' thought process is almost undisturbed, for in ordinary conversation nothing abnormal may be noted. However if they are asked questions which require some thought, the picture changes. They do not speak to the point, instead, they are circumstantial and vague. The disorder is more obvious in intelligence tests. Here are some illustrative examples.

(What is the difference between stinginess and thrift?) "I am always thrifty, and stinginess – who ever is hungry is stingy." (What is the meaning of courage?) "It can be said he has courage, I have courage, always to work." (Morning hour has gold in the mouth?[12]) "Morning is the best time that you can get up and you are healthy and go to work." (What is the difference between barley and corn?) "Corn is corn and barley is barley, barley and hops make beer." (Explain the picture of a broken window in the Binet test) "The German Michel, he holds his head, he has broken the window." (What is the meaning of justice?) "A person who thinks just, just thought, exact consideration." (What does poverty mean?) "That is when someone is poor. Poor is a broad concept. There are people who are poor and those who call themselves poor but who aren't." (No master has yet fallen from heaven?[13]) "Everyone makes mistakes, even the master, even the tailor, no matter how skilled, sometimes does something wrong." (Stinginess and thrift?) "Stinginess is the root of all evil nor is thrift always fitting. Who does not respect the penny is not worthy of the dollar." (Construct a sentence with these three words: education, school, life) "an educated person has school, and education belongs to life." (Explain the picture of a window promenade in the Binet test) "That is an upside down suitor, because the woman is

[12] See footnote on page 159.

[13] Literal translation of German proverb: "Es ist noch kein Meister vom Himmel gefallen" (English equivalent: "You cannot make a silk purse out of a sow's ear").

there too." (Construct a sentence from hunter, rabbit, field) "I just don't know. The hunter chases the rabbit, if he hits it, the question is, will he catch it."

I have of course chosen examples in which the thought disorder is obvious, and have omitted many better and even some correct answers. It is remarkable that usually sensible patients do so badly in intelligence tests. Although they generally stay on the theme and do not grossly digress, their answers are comments on the theme without being logical; sometimes the remarks approach the correct answer, sometimes they are far off. This may be called **blurry thinking,** an expression for the given theme's being taken up with various vague remarks and without logical clarity. There is a characteristic circling of the issue with no discernible goal. If the correct solution is found it is as if in talking about the theme the correct thought emerged by accident. For the patients themselves their answers do not represent a logical conclusion; they may continue to talk and add some other vague remarks. This thought disorder also differentiates phonemic paraphrenia from the hypochondriacal form as in the latter no obvious breakdown of the thought processes was observed.

Perhaps it is this disordered logic that diminishes the phonemic patients' need for an explanation of their abnormal experiences. As already mentioned, they rarely suspect machines as the source of their hallucinations. They do not offer any other explanations, nor do they present delusions as a prominent symptom. Even though they may occasionally scold people around them from whom in their opinion the voices emanate, no paranoid lasting enmity develops. Clear paranoid symptoms, particularly ideas of reference, negate the diagnosis and instead suggest the reference syndrome of affective paraphrenia. Ideas of reference are common in phonemic paraphrenia only during the onset; in case of doubt the underlying affect is the deciding factor. In phonemic paraphrenia there is a certain indifference, while in affective paraphrenia, especially in the "irritated reference syndrome", there is a profound affective anchoring of ideas.

Case 70: Luise Zu, born in 1897, was anxious as a child and later, during her menstrual periods, had attacks of screaming and cramps in her hands. In 1927 she was admitted to the Frankfurt Hospital after having allegedly heard voices for several years. She remarked that everyone was talking about her and she complained about the people from whom she thought the voices came. She also saw "people in spirit". She was transferred to Institution E. At first she stood in the corners and was antagonistic. In 1928 she was occasionally excited or resistant. She must however have been relatively sensible because she was eventually moved into family care. There is no precise information for 1929 to 1930. In 1931 she fainted on one occasion and hurt herself. She was returned to the institution because she began to scream in connection with her hallucinations. She continued to complain of the voices which were annoying her. In 1932 the auditory hallucinations continued and occasionally led to excitement. In 1933 and 1934 she frequently answered her voices, was often sullen and reluctant to answer questions. In 1935 she continued to "react excitedly to her hallucinations". Her behaviour remained the same in the following years. She is described as industrious and responsible in her work. – MEYER and I examined her in 1939. She seemed somewhat epileptoid from her sweet manner and massive physique, otherwise she was quite approachable. She complained of voices that knew everything about her and made her bad. She also claimed that her thoughts were known. She then became excited because of the voices. Her answers to intelligence questions were partially illogical: (What is the difference between a tree and a bush?) "A bush might be the start of a tree, that would soon grow out of the earth. I don't know if it can blossom and if bushes arise from the blossoms which fall on the ground." (What does this mean: Need breaks iron?[14]) "Yes, yes. I understand that fully. It is more expensive in Frankfurt than in here. I live and die for Frankfurt."

A **father's brother,** Martin Zu, born in 1897, dreamed about wartime experiences after the war was over, and in 1925 was depressed for a period of time. In 1928 his condition worsened; he became anxious, screamed, and attacked his wife. He was admitted to Psychiatric Hospital H. There he was very restless, complained that he had sworn a false oath, cried, and attempted to run away. He became more restless, spoke continuously, sometimes incoherently and sometimes with flight of ideas. A basically anxious mood continued. He was transferred to Institution W. At first he was still excited, but soon became quieter and gave proper answers, albeit in an abrupt military manner. He reported that he heard voices, especially those of his siblings at home. After calming down further he was discharged in July 1928, but at home he still felt threatened and became violent with his relatives. In November 1928 he returned to Hospital H. where he was again anxious, reporting that his neighbours had hidden behind his door and threatened him. For example they called, "now we will beat you do death". People wanted to beat him up. To his own surprise he had a "conversation" with his brother-in-law who lived elsewhere. This came from an "inner voice". He again said that there must be "a radio or something" through which he was threatened and scolded. He was again transferred to Institution W., where he was anxious and complained, "it's the voices". He heard his siblings talking. In 1929 he was mildly perplexed and often listened to voices. He heard his wife saying that he should come home; also that there was a fire in his home town. He remained cooperative, was occupied industriously, but hallucinated constantly. He expressed the hope that the voices would stop once he was at home. In December 1928 he was taken home in this condition. In 1940 his wife wrote that her husband had remained ill and showed little interest in his work. The letter had no more precise information.

Subject Zu probably had a somewhat epileptoid constitution. This assumption is made on the basis of her nature, her physique, and her "fainting attack", and may explain why she was occasionally sullen and resistant. Besides she displayed from the onset a phonemic picture. When I examined her the voices corresponded fully to the description above. Her uncle first had an anxious excitement, sometimes associated with confused pressure of speech. These symptoms which were probably only accessory disappeared and the picture of a simple verbal hallucinosis remained, with voices affectively relating to the patient's thoughts. I assume that this was also a case of phonemic paraphrenia, even though the excitement at the beginning was unusually strong for systematic schizophrenia.

[14] See footnote on page 160.

Summary

Phonemic paraphrenia generally begins with voices which are sometimes present at first with other symptoms, especially affective changes and ideas of reference. The **voices** are of a different nature from those of hypochondriacal paraphrenia. They are closer to the thoughts and emotions of the patients. From their predominantly unpleasant content they are perceived as annoying and not because of their mere presence. Sensations are found rarely in the acute state, except as accessory symptoms, and later they disappear completely. The patients often complain angrily about the hallucinations, but they are not as pained as hypochondrial patients. The **mood** is more balanced and **affect** is well preserved. In every day conversations thought appears undisturbed. If however thought processes are examined more thoroughly, abnormalities are regularly found in which the patients do not logically move toward a goal, but rather talk vaguely around the issue. Their **thoughts** can thus be described as **blurry.**

Incoherent Paraphrenia

As I will show, incoherent paraphrenia is a clearly hallucinatory form. Thought is also severely disturbed. At the onset hallucinations are generally in the foreground, despite some accessory symptoms; the thought disorder develops gradually into its full severity. In later stages the patients are constantly occupied by their hallucinations, more so than in any other form of paraphrenia. We did, however, see similar behaviour in sluggish catatonia.

During the onset, while the patients are still able to give more precise information about their **hallucinations,** they often report sensations, radiation to the body, twitching in the legs, currents in the trunk, sexual molestations, and the like. These are probably accessory symptoms of the acute process as are ideas of reference and influence which often appear in this stage. In later stages I found no evidence of sensations and the behaviour of the patients indicated only voices. Nor did I find definite evidence of visual hallucinations. However, confabulatory remarks occasionally appear in the incoherent remarks of patients, such as: "The child Walter was murdered. He received from a policeman a child who had been shot dead", or "I saw the Tzar of Russia in Darmstadt." In such a markedly hallucinatory form as incoherent paraphrenia, I suspect that these isolated confabulatory remarks are based on visual experiences. In earlier stages when the patients still offer information, neither visual hallucinations nor confabulations seem to play an essential role. On the other hand during the early stages, incoherent patients talk a lot about their voices, as various cases that follow, show.

Hears voices of media, swear words, sexual accusations. – At night he heard the girl talking through the radio as well as swear words. – Someone called: "Now he will be tortured, I have not seen such a stubborn man in years." – She is spoken to by wireless or broadcasting telegraphy. – She is constantly persecuted and threatened by the voices. – He is sent wireless messages and scolded. – "There is a woman who always scolds me, but I don't listen." – There are male and female voices speaking loud and clear from the ceiling and guessing her thoughts; for example, a female voice said, "Tonight under anaesthesia." – She remains in constant contact with her family at home via telephone, whatever she thinks is heard there. – She heard someone call that "it will be decided in two hours". – She heard people talking through the walls.

All of these are examples from earlier stages of the disorder and reveal nothing definite about the eventual character of the voices. Some of the examples reveal a strong tie between the voices and the thoughts and fears of the patients. In later stages nothing definite can be learned from the patients about the voices; either they are very vague or in their incoherent way simply do not answer questions.

The behaviour of the patients clearly show that they **hear voices continually.** They hallucinate even when spoken to or asked questions. It seems that the patient has answered a question when he has, in fact, only answered a hallucination and does not appear to have grasped the question. Even when not asked he talks or whispers. Facial expressions show that the patient does not give his attention to the examiner, only to inner experiences. The patient does not look at the doctor, but his eyes wander about; or he watches the doctor with such a blank look that it is clear that the stare is "aimed inwardly". The patients seem to hallucinate almost without interruption. Day in and day out, they are observed distractedly whispering to themselves. At times the whispering becomes a speaking or a scolding accompanied by gesticulations and grimaces which are unmistakably directed at the voices. If hallucinatory excitement occurs the patients scream, shout, and gesticulate in the direction of the hallucinated enemies. Or they run back and forth, at times screaming out a swear word or something similar. Often the words are obvious answers to voices. The patients not infrequently defend themselves against accusations such as: "That is not true, that is a lie!" Or they scold their opponents with words like, "Get out of here, you pig, you dirty dog." Such remarks indicate that, at least during their excitement, the voices have an unpleasant content. If the patient's words are not comprehensible the angry gestures allow the same conclusion.

Excited periods interrupt the otherwise calm behaviour usually for only a few hours. They may, with fluctuations, last several days. They become milder in later stages, taking the form of loud speaking with the voices. At that point excitement may no longer be present as such; the patients at times speak louder, at times more softly, and sometimes only their facial expression reveals the hallucinatory distractibility.

The continuous hallucinating, occurring even when patients are spoken to, differentiates incoherent paraphrenics from all other paraphrenics. Phonemic patients may sometimes speak and argue with their voices, but only when they are not being externally stimulated. On being approached and asked questions they stop speaking with their voices immediately. Incoherent patients hallucinate so continuously that, as stated above, it is often unclear whether they have answered the doctor's question or the voice. Sometimes when they have been whispering they may suddenly respond loudly to a question, although the content of their responses may still refer to their hallucinations and not to the questions. From this it may be deduced that the hallucinations have a considerable forcefulness, otherwise they could not hold the patients' attention so strongly that even the questions of an examiner are ignored.

Considering the fact that the patients are always distracted by their voices, a high degree of **thought disorder** is obvious. It is not so extreme in any other simple systematic paraphrenia. Incoherence is in the foreground and betrays a much more profound disturbance than the incoherence of confusion psychosis or confused mania. In patients with these two disorders connecting thoughts may be absent, but one can usually fill the gaps with ease, i. e. insert a thought which would make the connection. This incoherence is often due only to a flight of ideas, the speed of which prevents the patients from expressing the connecting ideas. In incoherent paraphrenia the lack of connection is more marked. Completely heterogeneous themes may follow one another in such a way that a connection between the two cannot be discerned. The train of thought in incoherent paraphrenia appears much more peculiar than in confusion psychosis. Some examples follow:

(Do you have any complaints?) "That's impossible, no one likes that any more, the surplus." (Where are you from?) "In German or English?" (How long have you been here?) "You are not allowed to eat, you spit so much." (How old are you?) "My mother is a wealthy woman. She keeps on spitting." (How are you?) "With the bishoprics it is the same as down in Regensburg." (How are you now?) "The earth is beneath, is Regensburg." (What does this mean: The apple falls not far from the tree?) "I never used this for myself. Is that pronounced well or badly? That is sometimes pronounced in one tone." (What is the difference between box and basket?) "Personally I have nothing to do with lots of persons. I was conscious with majesty."

These examples show the complete disconnectedness between questions and answers. This cannot be explained solely from the intervention of hallucinations. The trains of thought that develop spontaneously in the patients are hard to demonstrate, as we shall see, since incoherent paraphrenics are very taciturn and rarely speak more than one sentence at a time. However the disconnectedness between answers and questions, some of which were asked very forcefully, reveals the incoherence of thought quite clearly.

The heterogeneous thought contents do not always remain so separated, they may also interweave, producing **contaminations.** But the train of thought of incoherent paraphrenia always seems to break down; either another thought follows after an interruption, or another thought begins immediately and the two mix together. This form of thought disorder is well known (especially since C. SCHNEIDER and KLEIST), but is too often too generally applied to schizophrenia as a whole. The gross contaminations, as seen in the following examples, are found only in incoherent paraphrenia. These examples come from several patients.

(Are you married?) "No longer now. The wife is divorced, that the penitentiary runs the thing for the child." (What is the examiner's profession?) "One can't say profession, it goes as mankind moves." (What are the names of the planets?) "I believe they are the planets, but penny pieces." (How long have you been here?) "Penitentiary sentence. In the hidden jail, I always said." (Where are you here?) "A German state building." (What do you hear?) "If someone pronounces something, I have no definite evidence." (What is the difference between tree and bush?) "Tree is an unperishable man, for example, a shade tree is not parliament tree, it is a fellow traveler." (Make a sentence out of hunter, rabbit and field) "I am no poetess, but I don't have it, I am connected to the rabbit or the hunter. That is doubly good. That is a member of parliament. Field rabbit. That is good. We actually had a hunt for rabbits, kennel rabbits." (What does this mean: morning hour has gold in the mouth?) "No, I have my own teeth in my mouth, no, no, I must not kiss, otherwise all the people will suffocate." (The apple does not fall far from the tree?) "That is an old proverb, people say this because of people's stayings."

The examples show the incoherence, not only in relation to the examiner's questions, but also in the sequence of the patients' own remarks. For example, the terms "unperishable man", "shade tree", "parliament tree", and "fellow traveler", seem to stand beside each other completely incoherently, just as do the terms "member of parliament" and "field rabbit" in the next sample. Furthermore this illustrates how heterogeneous concepts are mixed together. The concept of divorce leads to the concept of penitentiary, probably because law courts are involved in both, a remark on the planets is interrupted by "penny pieces", the question of institutionalization, probably because of locked doors leads to the "hidden jail". The short formulations containing heterogeneous elements confirm the amalgamation, which may be contrasted with the derailment of other paraphrenic forms.

There are also many verbal abnormalities – the sentences are often completely incomprehensible; sometimes only a single word is pronounced. In addition to this agrammatism, there are many paragrammatic constructions. Neologisms also occur, such as "parliament tree" and "stayings". All these verbal absurdities express the thought contamination.

The utterances of incoherent paraphrenics almost always occur only after questioning, and then perfunctorily with little loudness, even after strong encouragement. Sometimes the patients break off in the middle of a sentence. Of course by themselves they speak with their voices a great deal, but they do not spontaneously approach people around them. Since in general they show little reaction to their surroundings, they appear to be lacking in drive. They tend to sit around, without interest or initiative, and must be urged to do anything. They may appear almost catatonic, but their posture and movements display nothing remarkable. Their facial expressions are only slightly alive, but otherwise free. Careful observation will avoid confusion with sluggish catatonics with whom they have continuous hallucinations in common. But the lack of drive here is not a psychomotor symptom. Rather their interest is directed inwardly toward the voices so that external events are ignored. If these patients have been taught specific activities they carry them out freely with natural movements, and do not need to be urged repeatedly. Incoherent paraphrenics show strong affect when arguing with their voices while emotional reactions to external events are lacking. They express no interest in their relatives; they enjoy visits, but are not distracted from their hallucinations by the visitors.

From the continued presence of **affect** may be deduced that in later stages **ideas of grandeur** occur, as can be detected from the patients' utterances. They believe that they are important people and refer to themselves by such titles as emperor or princess. These are probably wish fulfilments as the personalities become severely defective. In schizophrenic end states a tendency to self-elevation is often found because it is easier for a patient than a healthy person to make a wish a reality. – A female patient is presented with the family history.

Case 71: Anna Mös, born in 1881, did well in school and was a good house maid. She began to change in 1904, becoming obstinate and strange. In 1908 her condition worsened rapidly. She claimed she could read thoughts and felt persecuted by a searchlight: "Do you hear that searchlight? Now they have the sound hole open." She was admitted to Psychiatric Hospital M., where she claimed that she heard voices of men and women but would rather not talk about it. They spoke loud and clear, generally from the ceiling; whatever she thought was repeated. People close to her heard it and guessed her thoughts. Among other things she heard: "She will have her larynx cut." She continued to feel the searchlight and she had peculiar twitching in her arms and legs. Sometimes she was anxious, at other times she laughed compulsively. After four weeks she was taken home, but returned two weeks later. She reported that the people had talked about her and scolded her, and the voices had told her terrible things. She was also being bodily influenced, X-rays were directed under her clothes allowing people to see her body. At times she was excited and complained of annoyances. She was transferred to Institution B., where she said that the most terrible things were inflicted on her by hypnosis. Currents were being led up her nose. She often listened tensely and then shouted back, for example, "Do you mean me? I won't allow that! I don't know you." She claimed the doctor called her a dirty pig. In 1909 she scolded the doctor and asked why he had talked about her. She claimed that someone had put a spiral in her nose. She often spoke and grimaced to herself. During periods of excitement she screamed and scolded out of the window for hours. In 1910 these excited periods with the "screaming scolding" occurred repeatedly. In between she was able to perform tasks, but often spoke to herself. The excited hallucinations continued during the following years; she scolded her imagined enemies and screamed, "The bitch is too big for me to deal with you. If the bitch went out on the street she would be taken home by the police for having venereal disease." In 1911 the following remark was recorded: "What do these strange people want here? This dirty pig should get out." In the interim periods she was calm and industrious. On one occasion in 1915 she screamed, "You pig, you dog, he raped me. I want my mother and my rings." She often scolded into thin air, but sometimes at other patients. From 1916 on her quiet periods were described as dull. In 1918 she was transferred to Institution K. When asked why she was in the institution she said, "the mental institution after healthy walking fell down." Her speech was confused, she soon became excited, scolding and hallucinating. For example she railed against the "appetite whore" In 1920 her hallucinatory excitement continued, and it was impossible to communicate with her sensibly. She occupied herself with knitting. During the following years periods of excitement were rarer but continued. Then she scolded and stamped her feet. When calm she was taciturn. In 1924 she heard many voices and was often excited. In 1925 she talked a lot to herself without becoming more excited. She gave confused answers to questions: (When do you have visits?) "Yesterday and tomorrow. Now I have slept for 16 years." (How old are you?) "43 years. The other girls in my class are a few months older." (What do the voices say?) "No one knows." Her condition did not change in the following years. The excited periods were not so severe as before. Even during the calm periods she talked to herself a lot. She performed simple tasks. – B. SCHULZ and I examined her in 1939. She was at first irritated, wanted to leave the room, and spoke to herself. She calmed down and agreed to stay but spoke and gesticulated further to herself. When she did not speak she maintained a distracted facial expression. (When were you born?) "I don't know, have no book. First name is Heather, Gigerl. King Ludwig, she is emperor. I only know my dog naked." (What kind of dog?) "That is my parents' house all around. One day older than Mrs. Bauer." (Who are you talking with actually?) "I am in the empire of domestic servants." (Are you the empress?) "Stay in the apartment. If you have just been given a bath, you can't put in order." (What is your first name?) "Gigerl, also Anna. I have to take a bath first." (When were you born?) "My parents were married younger." (Where do you work?) "In the gardens." (What you do you do?) "Weeding, helping."

The **daughter of a male cousin,** Elise Mös, born in 1887, had been a rather shy and retiring girl, but in 1908 she changed. She believed someone wanted to hurt her or poison her, and she spoke confusedly to herself. She was admitted to Institution R., where she was anxious, excited, defensive, and spoke confusedly. She whined and was afraid of hell. She was taken home in this condition. She gradually calmed down, began to work again, but remained very quiet. In 1911 she relapsed. She was anxious and restless, and she was returned to Institution R. Her anxiety continued, she was at times restless and at other times inhibited. It soon became apparent that she was listening to voices. She improved but remained perplexed and shy. She spoke to a doctor as if he were her bridegroom. After six months her father took her home, but three weeks later she was taken to Institution D. because she had become so restless at home. Her mood and behaviour varied, being at times perplexed and taciturn, or excited. In 1912 her condition fluctuated between severe excitement and calm behaviour. In her excited states she tore clothes and destroyed objects. She often had a silly laugh after which she was anxious. There were hallucinations during the excitement – the patient spoke out of the window (although no one was outside) and called out, for example, "You dog" or "Yes, Ludwig". In 1913 excitement continued to be associated with hallucinations; she called herself princess Ludwig and spoke disconnected phrases, apparently with voices. For example: "Excuse me – you claim, for the ladies of the house, please – in secret – I have sinned – my grave is to the right – I want to die since the corpse law is written number one – gold and silver from Madama, you are my Mama – I don't recognize you." In 1914 the excited periods continued. In 1915 her father took her home. There she chattered, scolded, cried and laughed to herself. After six weeks she was readmitted to Institution R. There she gave not a single comprehensible answer, for example: (What is your name?) "I was a Bavarian." (How did you get here?) "Yes, I have a dirty one four times." Most of the time she lay in bed, but she became excited from time to time and then scolded and talked confusedly to herself. In 1916 the periods of excitement and confused talking continued. In 1917 she generally lay quietly in bed, but repeatedly talked, scolded and cried for a few hours. In 1918 she died of pneumonia.

Subject Mös displayed a hallucinatory picture throughout her illness. There were many abnormal sensations at first but the characteristic hallucinatory excitement soon appeared. The notation of confusion in her verbalizations appears repeatedly in her record. When I examined her she was constantly occupied with her voices. She did answer questions, eliciting the characteristic thought disorder. Her cousin's daughter became ill very early at the age of 21, the subject herself had begun to change mentally at the age of 23. Perhaps the early onset in these two cases is associated with the acute symptoms in the early stages, especially in the cousin's daughter who displayed severe affective fluctuations with much excitement. Later during the excited periods there was not only scolding but destruction as well. All of them had the hallucinatory character of incoherent paraphrenia, as confirmed by the totally confused remarks noted in the case records. In both cases the diagnosis surely was incoherent paraphrenia. The ideas of grandeur, mentioned once, fit into the picture.

Summary

Incoherent schizophrenia is characterized by very marked **hallucinations** which can at first include abnormal sensations but which later are nearly always auditory in nature. Facial expressions and eye movements show that the patients are inwardly distracted. They also continuously whisper to themselves. Many have periods of hallucinatory excitement in which they speak loudly with their voices and scold, others change with less severe fluctuations. One can deduce from the patients' behaviour that they nearly always hallucinate, even while they are being directly spoken to. This differentiates them from other hallucinatory paraphrenics. To the outside world they show neither interest nor initiative, but rather appear dull and lacking in drive. They answer questions taciturnly and with soft voice. Their verbalizations reveal **severe thought disorder,** in which incoherence is combined with contaminations.

Fantastic Paraphrenia

The fantastic schizophrenias have always excited interest among investigators. Referring to similar phenomena KRAEPELIN spoke of fantastic paraphrenia; SCHRÖDER of halluzinosis phantastica; and KLEIST of phantasiophrenia. However their description never quite conformed to the exact picture of the systematic form, since they did not distinguish fantastic phenomena which may develop in the course of affective paraphrenia from those occurring in fantastic paraphrenia. The onset may be uncharacteristic; affective changes and ideas of reference may accompany the picture, but usually the fantastic features are immediately observed; in any case they are visible much earlier than they would be in affective paraphrenia with a fantastic course.

Fantastic paraphrenia is both a hallucinatory and a delusional form. The **hallucinations** will be described first since their presence in this form should be compared with previously discussed forms which are primarily hallucinatory. As in hypochondriacal paraphrenia body sensations clearly are involved in the hallucinations. The peculiar descriptions of grotesque sensations return as well. Hypochondriacal patients are sometimes a bit inhibited in describing the precise nature of their sensations. Patients with fantastic paraphrenia may also speak of hitting, burning, pushing, and so on, sensations to which they feel exposed; if urged on, without hesitating they continue their descriptions in grotesque or fantastic forms. I present a few examples.

Whole pieces were cut out of her buttock; there were holes into which a hand could be put in. – "My bones and my flesh have been changed as if I had been made into the picture of a man." – A machine drives blood and fat out of her body. – Her sexual organs are made bigger and then smaller and the body is changed as if she were very old. – She felt in her head when someone "peed" into her bladder and made her urinate. – He was continuously molested sexually and had had a testicle in his head which disappeared with pain. – Electricity coming through the wall made her brain and tongue stiff. – A 1.5 g pen was stuck in his head. – The bones were being softened, especially those in the right foot.

On comparing these descriptions with those of hypochondriacal paraphrenics the similarity is obvious. If the ones mentioned above may sound somewhat more fantastic, it is probably because there is a tendency to delusional interpretations, but the grotesque character of the experiences is the same in both forms. Thus we have a symptom that takes the same form in two paraphrenias in so far as one can rely on the descriptions. It is the first symptom of this type that we have met. Closer observation, however, usually reveals differences, for disorders of different functional systems do not produce the same specific symptoms. I do not know how to explain this exception with the grotesque features. Possibly the exception is only superficial; there may be differences which I have not yet been able to discern.

As in hypochondriacal paraphrenia, the sensations are accompanied by hallucinations of smell and taste, whereas the **voices** are somewhat less important in this form; they are not completely absent as the following examples show. In acute phases they are often more important and may take on the character of thoughts-becoming-loud ("Gedankenlautwerden").

Someone took her name away and shouted to her that she had given herself a false name. – He hears voices that say bad things about him. – He is molested by "thought telepathy". – She has a "nervous listening", and is called a whore and a prostitute. – She hears acquaintances talking via "telegraph". – Spirits speak to him. – Voices or media say all kinds of lies and malign her. – She hears the voices of man in authority as if it were reality. – She hears insults and vulgar remarks.

When asked specifically patients with fantastic paraphrenia mention voices but the voices do not play an important role in the patients' thoughts and are rarely mentioned spontaneously. As with other paraphrenics content is usually unpleasant. **Visual hallucinations** are impressive, more so than in the other forms described so far.

At night people come into the room and stand behind her. – From the institution he can see Frankfurt and see life-size pictures. – At night she sees a bird on her bed. – He sees spirits like ones in the movies and also God himself. – She sees man forms, particularly in half-sleep; for example, a naked man without legs. – The priest and the mayor appeared to her. – She sees spirits seated at a table. – She sees a hermaphrodite with female appearance and male powers.

Even more characteristic than such isolated visual phenomena are scenic hallucinations in which a series of events is seen and where hallucinations of the other senses are mixed in. It often becomes difficult to differentiate what was seen, was heard, and was felt. The patients simply report a series of interrelated experiences, without remarking on which sensory area the different details appeared in. At times they speak of mass occurrences, hundreds of people being tortured, mass murder, and the like. I used to call this "mass hallucinations". Lately I have seen such cases but rarely. Scenes of smaller dimensions are also characteristic. Elementary auditory phenomena normally play a role, since the patients often offer as proof that they hear people screaming and machines of torture working. Elementary auditory hallucinations are rare in schizophrenics; they may be present in hypochondriacal paraphrenia, but take the form of incomprehensible verbalizations, and not of noises such as are to be found in fantastic paraphrenia. However, according to the descriptions, the scenic hallucinations are essentially of a visual nature; auditory, somatopsychic, and above all sexual hallucinations are only complementary. Some examples follow.

Horrible things happen in the institution; at night men with masks dragged patients into the basement; the night before he himself had been thrown out of bed. – At night the devil visited her, and then a black tiger with a golden crown, which kicked the bed so hard it flew into the air. – At one time a man was in bed with her and she had a hard time getting rid of him; another time, many people came through the locked door which did not open in the lock but at the hinges. – In the basement tortured souls are constantly crying. – The day before they buried someone, but then said he had jumped from the house and four men took him away. – They buried money in a mountain and covered it with dirt. – In the basement people were being horribly tortured; one, Mr. Bill, is particularly involved; every festival people are imprisoned and then beaten and murdered in the cellar; nearly 50 people are there now, one can hear men and women screaming; sometimes people are pushed naked out into the cold. – At night men come to her bed and abuse her.

These scenic hallucinations are reminiscent of dreams and are, in fact, more often experienced at night. Apparently the same process that create normal dreams favors the development of these fantastic hallucinations. If one suggests to the patients that these are only dreams, they do not want to be told that, they insist it is all real. These experiences, composed of various sensations, are hallucinations unique to fantastic paraphrenia. They do not occur in any other systematic form. However they may appear in the end states of affective paraphrenia. The scene is especially set with unusual, horrible, and sensational events.

The patients reveal their hallucinations through their descriptions. Occasionally they react directly to the voices and call out a few words, generally swear words, or they touch their bodies, declaring that someone had caused them unpleasant sensations, or they spit, claiming that they had tasted something bad.

Besides the hallucinations there is the formation of **fantastic ideas. Absurd ideas** are formed as the patients make claims that are totally outside the realm of possibility; the most basic experiences of daily life are ignored and certainties which are usually considered absolute no longer exist. The laws of nature are pushed aside. From all these extremes there are transitions to ideas which are equally unfounded, but which are not totally impossible. These are not genuine delusions because the relationship to the self in the sense of elevation or abasement is usually absent. An appropriate term might be "delusional fancies", since the ideas are often very transitory and soon replaced by others; however some are retained. Even in milder forms the ideas appear absurd because no attempt is made to explain them. For a patient to claim that a doctor is a socialist leader is not itself unusual; it becomes an absurd idea when there is no evidence for it at all. Other claims have a character of absurdity in themselves. Some examples:

The father died in 1910, but later came to visit her; now he is 104 years old. – There are 6 to 12 popes, Mackensen is Pope Pius XII. – In her home town 96 women were "bastardized" (sterilised), and 46 were "post-bastardized" by the priest. – Her husband was Saint Francis; he was also a freemason and the devil. Mother and husband had not died; rather they were "thrown up" out of their graves. – The Lord, "Jesus Nicholas", whom she knew very well, is crucified every year and is resurrected; she herself had been reborn several times. – Dr. Botton turned into Miss Dr. Emmerich, "I imagine that he switched from man to woman due to hormonal changes." – She died already a hundred years ago and perhaps ran around in Dessau as a dog. – People are buried. (Alive?) "No, they rise again." (People can arise from the dead?) "Yes, of course."

The last answer, "Yes, of course", shows very well how the patients can lose all sense of the impossibility of their claims. The last quoted patient was astounded that anyone would question whether people could arise from the dead; for him it was a matter of course. The claim that the dead have come back to life appears frequently in such patients' absurd ideas. They talk about a meeting with someone previously dead; that he is alive again is not mentioned at all because for them it is as obvious as the patient who said "Yes, of course". Another patient said, "My dead sister is now 65 years old and looks like 24."

Misidentification of persons when it occurs is very characteristic of fantastic paraphrenia. Since apparently there is no reason why this or that person should be misidentified this is merely a special form of absurd idea construction. Persons in the surroundings are often mistaken as historical or well-known personalities. Examples:

His deceased father is in the institution and so is Bismarck. – She had known the director when he was a child, perhaps he is also the Minister of Labour. – In 1924 Mrs. Göbbels was her colleague as a lamp girl and was then called Miss Eberlein. – Mayor Blum sometimes calls himself Gondermann or Haag or Derr. – Both Hindenburg and the former crown prince are in the institution. – Her father was a butcher, at other times he was a priest. – Princess of Zerbst is in the institution as Miss Kunz. – "This place is full of emperors, princes and tzars." – The head doctor used to run a farm and used the name Schulz; another doctor is the brother of her mother. – One doctor is simultaneously Kaiser Wilhelm, Hindenburg, her father, and her husband. – The director used to be a carpenter. – The doctor who used to be called Amundsen, had gone to school with him.

The name and title of a given person may change over a time, but generally remains the same. At Gabersee I was was given many false names and backgrounds: a peasant son from Dorfen, an illegitimate child of the maid Anna, a horse groom on an estate, and a lieutenant in the 4th artillery regiment. It is especially bizarre when one person simultaneously represents several people (a doctor was simultaneously Kaiser, Hindenburg, father and husband), it does not bother the patient because he has no sense of the absurd.

Ideas of grandeur also carry the stamp of absurdity; they are always present in progressive fantastic paraphrenia. The strangest claims are made with no need to explain. Formerly the patients simply called themselves Kaiser, King, Tzarina, Princess, etc., even God without saying anything else. Nowadays they give themselves other high sounding titles. They are disturbed by the contrast between their grandiose ideas and reality. They behave like everyone else in the institutional milieu, are occupied with the same simple tasks, and do not try to exploit their high position outwardly. If they are displeased about something, they may refer to their title in order to underscore their wishes, this is never done with emphasis. On the contrary, when they have an urgent wish, they try to adapt to the external situation as much as possible, try to explain their wishes as reasonable ones, and for the moment not to think of their exalted position. This is a very familiar peculiarity of schizophrenics, to express themselves expansively without attempting to pull rank; this is particularly so for fantastic paraphrenia. In "expansive" paraphrenia (to be considered later) it is different, therefore this observation does not apply to all forms of schizophrenia. Here are a few examples illustrating the ideas of grandeur of patients with fantastic paraphrenia.

"His original race was Turk and Mongol", and he descended from a "high potentate of Eastern Central Asia".- Her mother was a countess, and she herself is supposed to be declared a saint. – She is God and emperor of the world; she will soon marry Jehovah. – Her "birth mother" is Countess of Erbach; she herself is pregnant by her father, Kaiser Wilhelm, who came to her while she was asleep. – He is going to be a doctor of medicine and has a fortune of 500 million Mark. – She is the descendant of a French count and her son was elected leader.

Absurd ideas are responsible for the fantasies which characterize this form of schizophrenia. Scenic hallucinations heighten the impression of absurdity since they also have no normal content but are regularly quite unusual and astonishing. The presence of grotesque sensations completes the impression of the fantasy. The totally fantastic world in which these patients live is well illustrated in the following letter from a patient.

"I am receiving suggestions so intense that I can barely write. Out there in T. crimes have been committed by the crazy doctors. I am not mentally ill, but the men are geologists and they had machines and apparatus with them. The little house has no basement, but there are underground passages and paths, Trotzigburg, where socialist riff-raff and others are found: a hide-out for international criminals. The director of this institution, as you may know, is Friedrich Waldemar of P. He also calls himself Prince of Felseneck and is the socialist leader for Germany. All the doctors are experimental criminals. Their cures are gas, suggestion, long distance electrifying and electric waves with which they illegally radiate my sexual organs. Crimes go on under my house where at night there are the director and about three doctors with the mayor and other men from T. who themselves have learned how to make suggestions. The head priest here is also a geologist of criminology, sometimes he participates in the experiments. This den of criminal doctors at Gabersee must be cleaned out! They also perform assassinations via long distance!"

Despite the variety of symptoms the fantastic traits gives this disorder a unified character, and is evidence that here, as is true of the other disorders already described, a unified functional system is involved. Individual symptoms, in contrast to the fantastic symptoms of affective paraphrenia, are usually united from the onset in a closed syndrome. As the disorder progresses the symptoms become more severe, but new ones do not appear. Affective paraphrenia may imitate the fantastic paraphrenic picture to a certain point. But the similarity may be more an external one. What basically distinguishes the two forms is, as already described in connection with affective paraphrenia, the role of affect.

Affect in fantastic paraphrenics is never profound; the strange ideas lack affective intensity. The patients are capable of describing the most horrible experiences with a certain liveliness and even visible contentment. When they complain about hallucinations they may appear to be rather angry, but the anger is more a pose than genuine. Disregard for their complaints never enrages the patients. Even when their complaints are treated with a joking tone and a smile they show nothing more than superficial excitement, and another joke is usually sufficient to restore the good humor. This is a very different situation from that of affective paraphrenia with its deep affective anchoring in the world of ideas. Yet fantastic paraphrenics do not show severe dulling. They retain a certain interest in their surroundings and in their families, and they are always approachable. Facial expressions and gestures remain natural. The body build is pyknic more often than in other paraphrenics. This is a possible explanation for the fact that some of the patients are quite lively. Others are rather quiet and rarely initiate discussion of their experiences. The mood is in no way remarkable.

The absurdity of ideas of fantastic paraphrenics leads to the interesting question of the **thought disorder.** No doubt this is of significant magnitude, but not as high as the type of ideas would suggest. Patients for whom the dead can walk the streets may nevertheless form correct judgements. An abnormality is rarely apparent during simple conversations about objective facts; confusion of thought which I once associated with fantastic paraphrenia occurs only in combined cases. Patients do not always discuss the exact theme and their remarks are often vague. When they do express their fantasies their thoughts usually can no longer be followed. Asking questions in between may however produce clear answers. The thought disorder is more clearly seen in the results of intelligence tests. I present a few examples of various patients.

(Asked to describe the Binet picture "Greeting") "Matter of jealousy. The man acts gallantly and makes a bow to please the ladies. The other lies on his back. Whether he also tried to court, but was inadequate, or if he was knocked down, I cannot decide." (The same picture explained by another patient) "He holds a cane in his hand and greets." (What is below?) "He lies on the ground." (Why?) "A seizure, in any case." (What is the difference between error and lie?) "A lie is a crime, an error anyone can be." (Make a sentence from these three words: soldier, leave, sad) "When a soldier comes home and the family has a misfortune he is sad." (What is the difference between stairs and ladder?) "The stairs fall and the ladder is straight, it rises." (No rose without thorns?) "If you do not participate, for example, there are roses without thorns, love without pain." (What does envy mean?) "Envy is a grudge, if someone has more than another, that he also gives more, and he is also deceitful." (No rose without thorns?) "The rose is a pleasant flower because of its odour, but unpleasant because of its thorns." (Explain the proverb.) "The smell stimulates the human organs, especially the brain."

On comparing these answers with those of incoherent paraphrenics it is obvious how much more severe the disorder is in the latter. Although I chose answers with showed the deficiency in the clearest manner – some answers were correct – they are rarely nonsensical. One usually can understand why the patients say this or that, even if they are not logical. There is no interruption of logical thought or intrusion of heterogeneous elements; the thought process moves into incorrect paths in a much more subtle manner. This is the thought disorder which has been referred to by the term **"derailment"** (in particular by C. SCHNEIDER). When in the first example the patient remarks that the knocked-over child had perhaps also tried to court the lady, it is certainly a gross error, but it is explicable since it does refer to the correct theme. Or when, in later examples, patients say "envy is deceitful", or "the odor stimulates", these are similar derailments. The answer may show up as derailment right away, as in: "If you do not participate, for example, there are roses without thorns." The correct theme is initiated, but right away the train of thought runs off on a false track.

Linguistic errors are often found in fantastic paraphrenia. A female patient explained, for example, "I am debodied", another said that at night she was "listed", a third spoke of "Youthification". Grammatical errors are common, as in the above examples: "An error anyone can be." It is comprehensible that derailments occasionally may take a more verbal form.

One of the cases of fantastic paraphrenia had a family history. The brother was mentally ill for many years, but stayed at home so that there is no medical record. I present the medical history of the subject herself.

Case 72: Margarete Wa, born in 1863, before becoming ill was religious, industrious, and content. She was depressed for a while in 1906, after the death of a child. In 1908 she was jealous for some time for no apparent reason. From 1915 on she was clearly mentally ill. She claimed that important people were following her and that the priest wanted to marry her. She stayed at home, her disorder apparently gradually progressed during the following years, for when in 1926 Wa became more seriously ill, an advanced clinical picture was already present. She went to the police and said that some Englishmen and Frenchmen were locked up on the top floor of her house and wanted to kill her. "Hindenburg, Braun, Strehner, and Pförtner" were also there. She claimed she was beaten. She was admitted to the Frankfurt Psychiatric Hospital, where she again spoke of foreigners, of Briand, Hindenburg and other "Papas". She was in contact with these people via calls and motor cars. The house was full of a "watch". She was being electrified, anaesthetized, and sprayed with acid. Officially she was called a whore and a prostitute. At night she was "incorrectly" touched. She was somewhat long-winded, but otherwise orderly. She never got the main point of intelligence questions. She was transferred to Institution H. There in 1926 she spoke of her delusions only when urged. In 1927 she hallucinated extensively, talked into the air shaft, and believed she had seen Professor X in the institution. During the following years she was said to be withdrawn but remained friendly. She only spoke of her ideas when asked. Sometimes she complained of being abused and beaten. Occasionally she talked to herself. – In 1931 Faust and I examined her. She said that the former crown price was running around the institution and that Hindenburg was there as well. She claimed to be the descendant of a French count. She also declared that many people were being kept prisoner downstairs; they were being mistreated and murdered. People could hear their screaming. At night she was sexually abused and bodily molested. Professors were at work, everything was under anaesthesia. At night men came into her room and abused her. She also heard insults and scoldings. The patient's explanations were given in a lively manner, but in her talk she was somewhat confused. When questioned her answers could be understood. She solved two intelligence questions in the following manner: (What is the difference between box and basket?) "Box is a box and a basket is a basket." (The apple does not fall far from the tree?) "Sometimes it rolls down the mountain." (What does this mean?) "In my family there is nothing that would cause that." Her overall posture, facial expressions, and gestures were natural.

Summary

Fantastic paraphrenia is an **equally hallucinatory and delusional form** of schizophrenia. Sensations, often grotesquely described, are conspicuous. Visual as well as auditory hallucinations belong to the clinical picture. Especially characteristic of fantastic paraphrenia are scenic experiences in which the visual element predominates, although auditory and somatopsychic components may also be present. Horrible things are often hallucinated, for example, the torture or murder of many people. This is in itself a fantastic trait. Peculiar ideas are regularly expressed by the patients. These ideas typically are not at all limited by natural law or normal experience, and thus appear absurd. Similarly illogical is **misidentification of people** in the surroundings who are most frequently confused with famous personalities. Likewise **ideas of grandeur** have an absurd character in that the patients often raise themselves to incredibly high social positions. But they draw no consequences from their megalomania and cooperate in institutional life. If their ideas are ridiculed or if they are otherwise irritated they may flare up, but the affect is never deep. Nevertheless they maintain an interest in the events around them and in their families. In their overall attitude they remain natural. When presenting their fantastic ideas they may seem to be somewhat confused, but questioning does produce clarity in their thought processes. Further examination including intelligence testing reveals that the characteristic feature of their thought disorder is **derailment.**

Confabulatory Paraphrenia

The essential traits of confabulatory paraphrenia were described by KRAEPELIN as well as by WERNICKE.

At the beginning of the disorder, apart from less characteristic symptoms like mood anomalies and ideas of reference, is often found a peculiar forerunner of **confabulations** taking the form of **dream-like experiences.** We will discuss these in more detail later. In other cases confabulations are present at the onset; these may at first be isolated, but gradually they become consolidated into integrated stories. The patients do not talk much about daily occurrences, rather they tell **sensational tales** such as travels to other continents, adventures with men and animals, happenings on stars or in heaven, even conversations with God, etc. In the fantasies of these patients there are no obstacles; it all flows smoothly in a playful sort way. I present characteristic reports of three patients.

Subject Prin: "We drove from Germany through Czechoslovakia to Greece, then to Skutari, over the high Mount Everest to Nanking. There I lived with a Deusser family on Palace Street. They are acquaintances of mine. At that time the rate of exchange was weak and I wanted to change money: the French Franc was worth two pennies. Because of the war we could not go to the coast so we took the car to Siam and then boarded a ship. It hit a mine. We were picked up by an English steamer and taken to Liverpool. From there I wanted to return to Germany. I took a battleship and tried to land on the German or French coast. But we could not land because the war was still going on, so we landed in the harbour of Montevideo. There I met the captain of the German cruiser "Emden". In order to get away from the English steamer and to return home, I sailed with the "Emden", – it was called "Emden" or, perhaps, "Bremen". We were soon fired upon. The ship had a leak and we headed for the Dardanelles. They wanted to put on a few new sheets of armour, but I don't know if they succeeded. I met an English lady who took me with her to Parliament."

Subject Hart: "Mr. von Hindenburg told me that I am Queen and Empress of Germany, but the pharmacist took away my papers. Hindenburg stood under my window so I could see the wart on his face. He sent a package to me and my son containing valuable objects, above all a crown. But my son was beaten down at the door, and the attacker stole the armoured car in which all the gifts were kept. They also took my money, the 30 million I had brought into the marriage. My great-grandfather made me a lady of the crystal palace and the pharmacist moved into the residential palace where I really belonged.

Subject Nöt: On Venus I saw a big man who extended from the slate floor to the roof and who from the floor up covered the roof. That was God, who had also carried up bricks. The roofer wanted to be paid for the whole roof. God and I would not allow this, so the roofer said that if God had covered some of the roof himself so and so many meters would be taken off the price. I moved into the house, it was paradise. They had built a school house before; they wanted to send all the children of Venus to school so that the parents would not lose them. Moreover a milk pump was built for the children. Workers were brought up from Earth to Venus. God simply took them under his arm. I myself had a sewing machine under my arm. The big children did not want to let the little children enjoy the milk, so the pump was closed. About a year later I returned from Venus to Earth. I simply raised my arms and came back down here, the trip took one day."

The reports of these patients are quite fantastic, in no sense less so than those of fantastic paraphrenics. Closer observation however reveals that this is only a superficial similarity; a true relationship between the absurdities there and the confabulations here does not exist. Patients with confabulatory paraphrenia describe, in orderly fashion, experiences that hang together as a whole. In confabulations gaps may occur now and then – a patient still talking about his stop in Egypt when he is already in Japan, having forgotten to tell how he got from one place to the next. It is totally different with fantastic paraphrenics in whom nonsensical claims appear as isolated facts without relationship to other fantasies and which cannot be cleared up by further questioning. A fantastic patient may say for example that he was an Egyptian king many millennia ago; if one wants to know how that happened, no clarification if forthcoming. He relates no experiences that have any connection with his claim. On the other hand a confabulatory patient who made a similar claim would immediately describe in detail how he became King of Egypt, how he used his power, and so on. Whereas absurd ideas are primarily expressions of nonsensical thoughts confabulatory stories are based on genuine **falsification of memory,** or better, since they are new creations, **errors of memory.** With considerable sensual clarity memories, particularly visual ones, appear to develop without real experiences ever having preceded them. Because of the sensual certainty of the experiences the patients are able to describe the details vividly. If they are contradicted they refer to their memories, emphasizing that they have experienced it all just as normal persons do when referring to real experiences.

It is probably an effect of criticism that confabulatory paraphrenics do not weave their immediate environment into their confabulations, but take it as it really is. They also correctly identify people around them. The confabulated experiences always occur somewhere else, in another country, on another continent, or on a distant planet. Occasionally the patients are somewhat more critical and say that they had their experiences in a **dream** or in a **trance.** They still have enough judgement to recognize the unreality; but since the memories are so clear they are reminded of dreams they have had. Apparently however it is not solely due to the critical capability of the patients that they have incomplete insight into their fantasies, but rather that incompletely formed confabulations occur which resemble dreams.

A patient said that the following happened one night: "I was supposed to have been wounded with a weapon in the head and then taken in a coffin to India, there to be crucified with a holy animal. But I could not leave, so the animal came to me in bed on the ward. I felt it creeping over my legs. It sucked out all my sperm and out of it grew a sphinx. Then snakes came, but the animal ate them all up." On another night: "There were two men there, and they suggested to me all kinds of crimes I had to admit. If it was evil I was supposed to have done it; their proof for murder was a dead cat. Then Miss M. came, insisting on having spiritual sperm for the dead child she was carrying in her body. Then she went over to the small man with glasses on Ward I, and I could perceive how she had intercourse with him. An attendant came by and watched them. Then I was supposed to stand up and take a pistol that someone was passing to me through the window, but I did not think it was necessary."

The patient described interrelated experiences, resembling confabulations, but spoke of them as if they were dreams he had had in the night. They are differentiated from normal dreams by their fantastic formulation and their sensual clarity, since the patient seems to have had the details much clearer in his mind than is the case with dreams. Similar descriptions, apparently fluctuating between dream-like experiences and confabulations, are often presented by confabulatory patients at the onset of the disorder.

A female patient did not speak of dreams, but of the gift of "second sight" which she had only at night. She had seen, for example, how her husband was having an affair with the waitress from Cafe Schneider. She saw how he picked her up and went out with her. She was then shown the building in which the two were, and then their room. She had seen much in this manner. "I wake up and then I see it."

This is apparently a kind of precursor of confabulations. There is an underlying tendency to reification, so that thoughts of free fantasy take on the sensory character of actual memories. During the onset – and much later, if the well developed critical facilities are preserved – these new experiences may be repressed by the waking consciousness; but when conscious awareness is lowered, particularly at night, the confabulations become more prominent. These are certainly the same sensual clear visions, here treated as memories, i. e. confabulations, and there considered to be dreams or "second sight". The second patient said openly that the experiences did not appear during sleep like in dreams, but they were real during the night: "I wake up and I see it." Such confabulations probably developed more readily at night when active thought becomes less tense and imagination is freer. This assumption is given support by the fact that the origin of confabulations cannot be observed. A sudden question that the patient does not expect might easily lead him to a new error of memory in order to answer the question. At that point the patient's facial expression, a short pause or slight distraction might reveal the emergence of a new thought. This is never the case. What the patients report is always prepared; it is already complete, having been previously developed, probably when the patient was not actively stimulated and his thoughts ran free, i. e. when the patients was nearly asleep.

Whether they appear as false memories or as dream states, the confabulations involve all the sensory areas. As in reality they reflect total experiences, involving different sensory perceptions side by side. In general however the visual element is more important than it is in real experiences. A healthy person's memories of an experience are based to a large extent on the auditory verbal sphere. In the patients' account, however, little is spoken and most of the details are picture-like; nor do tactile sensations play an important role. It does happen that patients report how this and that was done to their body. Occasionally they even speak of an operation or an amputation, but it seems that they are referring more to the visual components of the events. Perhaps the predominance of the visual here is explicable the same way it is in normal dreams, as I tried to explain elsewhere (Leonhard 1951). Even though the patients' descriptions at times suggest a similarity between confabulations and hallucinations, genuine hallucinations are present in confabulatory paraphrenia at most during the acute stage, later not.

That the disorder in confabulatory paraphrenia occurs primarily in the visual sphere helps to clarify another peculiar visual phenomenon. Often and particularly at the beginning, before the confabulations have been developed and organized in the patients' thought, the patients claim that objects in their surroundings, especially people, **always look different.** The sizes, shapes, and total appearance keep changing, not before the patient's eyes, but from one day to the next. A person may look one way today and another way tomorrow. I present a few examples.

„One day there was a lady sitting beside me and I knew her; after a little while the same lady was there again, with different hair and eyes, but she was the same." – The stairway kept on looking different, the people have different faces. – Her husband's facial expression was not normal, people often had new appearances, then for a time it stayed the same. – Her children were different, they might look bigger or smaller; or they might retain their sizes but have different facial expressions.

These phenomena which I characterize as **perceptive errors** are of great interest. KRAEPELIN in his paraphrenia confabulans mentioned a patient who claimed that her husband was played by different people, here he was shorter, there he was taller. Perhaps in this order imagined ideas become abnormally perceptible to the senses. When such patients see a person they do not see him outwardly as he really is, but inwardly as a clear picture of him yesterday, when he wore other clothes, his hair was different, his face was redder or paler, or he sat in a different light. Thus the patients would see differences that are really there but which the normal person does not notice because he thinks conceptually rather than pictorially. Perceptive errors are not at all related to misidentifications of patients with fantastic paraphrenia; confabulatory patients never misidentify but recognize people for who they are. Fantastic misidentifications are never associated with perceptive changes; rather they remain, as we saw, completely unfounded like the absurd ideas.

It can be directly demonstrated that confabulatory thought has switched from the conceptual to the pictorial. In conversations about ordinary every day topics there is no thought disorder. The patients speak in an orderly manner and report the confabulations in an equally orderly way. But if asked questions requiring abstract conceptualization they fail because they are caught up in their concrete thinking. This becomes very obvious in the following example. A female patient who had done well at school and then had been a good cook responded in the intelligence test as follows:

(What is the difference between child and dwarf?) "That is seen often; there are even dwarfs among children." (Mountain and mountain range?) "A mountain range is really forest, a mountain, pasture, fields." (Error and lie?) "Error is error and lie is lie." (No rose without thorns?) "The rose bush has thorns, that's nothing." (Lies have short legs?[15])

[15] Literal translation of German proverb: "Lügen haben kurze Beine".

"They have no legs at all." (No master has yet fallen from heaven?[16]) "That is obvious, no one falls down from heaven." (All that glitters is not gold?) "Gold glitters in the sun." (What do you see in this Binet picture?) "He hit him with a cane and they look out of the window."

The answers may appear feeble-minded, but it did not look as if the patient was trying to reach a solution and only made it half way, as in mental retardation; rather she replied quickly and with certainty, clearly believing that she was presenting a complete answer. Subjectively she thought her answers were perfectly correct, e. g. when remarking: "that is obvious" or "that's nothing". Here are further examples from different patients.

(Child and dwarf?) "Dwarfs exist only in fairy tales; I have a child." (Envy and thrift?) "No big difference, envy is more than thrift." (Explanation of the word "loyalty"?) "I have to approach others with an open heart and expect the same from them." (Description of Binet picture of "window promenade") "Child fell, tripped, wanted to greet his father. A woman with child and two women. I don't know what it means." (No rose without thorns?) "Those are words, I have not thought of it." (Morning is the best time?) "Simply that the day will be friendly, without rain or thunder." (Necessity breaks iron?) "If you are rich you can't say this: Necessity breaks iron." (What does that mean?) "I don't know." (Binet picture of "window promenade") "He does not see his good fortune; a woman lies in front of him, she seems to love him." (The same picture shown to another patient) "The child calls attention of people, the child lies on the ground. Maybe it got hurt somehow." (Construct a sentence with the words child, field and flowers) "By the Main there are many fields, where the children pick flowers." (The apple does not fall far from the tree?) "That means the same if you say it this way or that."

The failure is clearest in explanations of proverbs and pictures. That is no coincidence, since it is in these questions that the patient must move from a concrete to an abstract meaning which they fail to do. The individual answers always reveal the failure. Occasionally picture-like aspects shine out in the answers, e. g. when the patients mention rain, thunder, or fields by the Main. I believe that in confabulatory paraphrenia we find a particular form of schizophrenic thought disorder different from those described above, which may be called **pictorial thought.**

This kind of thought disorder also illuminates the patients' behaviour with regard to their errors of memory. In dealing with the concrete they retain their critical ability and recognize that their confabulations do not correspond with reality. Therefore they try to remove their experiences from the reality of their concrete environment and place them on another continent, even on a distant planet. The two concrete factors – reality and confabulation – are no longer in conflict. Weakness in abstract thought encourages this kind of solution, which is of course only an apparent solution. Thus the uncritical fantastic nature of the patients' confabulations may be better understood.

Affect also flavours the fantastic quality of the phenomena. The **basic mood** corresponds to the sensational things that the patients claim to have

[16] See footnote on page 182.

experienced, Münchhausen stories with which the free fantasy likes to occupy itself. The mood is usually somewhat elevated, fairly happy, even mildly ecstatic. The effervescence and fluctuations found in the affect of patients with happiness psychosis or affective paraphrenia is absent in this rather steady mood elevation. Despite this the mood anomaly is remarkable since systematic paraphrenics are otherwise basically indifferent. This is also true of fantastic patients as long as they are not pyknic types who have cheerful temperaments. Only among hypochondriacs is found, in contrast to confabulatory patients, a chronically sullen and unhappy mood. This contrast probably reflects a **fundamental contrariness** in the thought disorder as well. Hypochondriacs cannot concentrate in conversation but are good at intelligence tests; on the other hand, confabulatory patients fail in the intelligence test but can conduct a normal conversation. Because of their mood hypochondriacal patients are somewhat querulous; the good mood of confabulatory patients imbues their errors of memory with a happy tone.

The **boundless ideas of grandeur** of confabulatory patients thus become more comprehensible. They are usually built into errors of memory in that the alleged experiences repeatedly point to the patient's high position, or the patients lay claim directly to the high position, title or wealth. Of course as with fantastic patients, the patients reflect their high rank only with their words. They make no real claims; they do not try to behave in a manner appropriate to their exalted titles. Rather they readily conform to the institutional life without further ado. Their ideas clearly represent delusional wish-fulfilment, and because of the elevated mood are not being corrected. Some examples of different patients show how far these ideas of grandeur can go.

During her travels she was married to princes and kings. – He used to call himself "Charles the Saint", then Sultan who travelled from Mars to Venus. – She is a princess and much more, she will bring the salvation that Christ did not bring. – She is the daughter of Kaiser Wilhelm; Zita of Burma tried to force her to give up her birth rights. – She is a daughter of the Prince of Bernsdorf and has friendly relations with Kaiser Wilhelm II. – She was chosen by God, has contact with him, and has also seen Jesus and Mary. – He is Kaiser Charles V and reigned in Berlin. – He also invented the air plane. – She will inherit 50 million Mark since her father in law, August Friedrich Wilhelm von Volkamer is the richest mine owner in Germany. – She is German empress and the sister of Prince Heinrich of Prussia.

If the patients are contradicted they may fall back on their alleged experiences; they tell how they were crowned emperor, how God spoke with them, how a carriage full of treasures was presented to them, etc. They do not become excited when nobody believes them, but some display a certain sensitiveness and hold back with further stories. The depth of feeling is better preserved than in fantastic patients. They remain interested in their surroundings, they are friendly toward the doctor and like to chat.

Case 73: Waltraut Hof, born in 1923, did well in school, then was actively employed on her father's farm. She worked hard but was seduced and had two illegitimate children. At the age of 31 after the second child was born she started to have somatic complaints which she described in a peculiar manner. A blow against her head which she had actually received "displaced my brain inside my head". She felt something "like thunder" in her blood. She spent two months in a psychiatric hospital. Then she was back home, where nothing remarkable happened. Probably she held back some of her strange ideas, for in 1958, at the age of 35 years, she was admitted to the Berlin Psychiatric Hospital, where she immediately described fantastic experiences. She said that on one occasion a horse which she had led went off and pulled her head off with the reins. She was as if dead, but then was able to go home and continue to work. On another occasion there were men who at night hit her on the head with an axe. She added: "It is a miracle that I am still here." Such experiences she said she had while asleep, yet she spoke as if she was giving an account of what had really happened. Then she said: "I could dream all day." Later she no longer mentioned sleep or dreams. There was a time when she was knocked down and buried. She lay under the ground for seven days, then when she was dug up she heard: "I was not dead at all in the coffin, I heard that." Having been exhumed she was raped by a man until she was full of life again. – On another occasion she went to America, the ship sank, everyone including herself swam in the water. Then she went down to the bottom of the ocean. Here she walked around for several hours, looking for a way to rise to the top. Eventually she succeeded, and emerging near the coast of America, went ashore and looked for President Roosevelt, who claimed that she was to blame for the world coming to the end. In talking like this she stopped for a moment, then corrected herself: "Oh, that was in Moscow!" Subsequently she continued to talk of America. When she cleaned the window at the Roosevelts she fell to the ground from the skyscraper. Things like that happened to her when she was piloting a plane. When crashing without parachute she managed to hold on to a tree. Eventually she was fed up with all these crashes and gave up her occupation as a pilot. Here are excerpts from many other stories she told: During the World War a man came towards her and kept on shooting her in the stomach. In Moscow she was supposed to perform as a singer. She piloted a plane around the world, and on the way she was murdered: "I was murdered so many times until I was home again." She once lived with a king, a cook gave her delicious dishes.

This patient stayed in the Berlin Hospital with interruptions until 1970. Her condition did not change in all the years. The confabulations mentioned above coming from different times were retained by her and repeated in the same fashion. New confabulations were constantly being added. Gradually she became less productive and she no longer enjoyed telling her stories as much as before. She became somewhat more idle and obese. In 1973, after she had been transferred to another psychiatric hospital, I reexamined her. Besides the old confabulations some new ones had developed. She had become a princess in America, and was made "Präsidentin". A shark in the ocean had nearly swallowed her. On one occasion she and a man were together on the moon. Her stories always sounded as if she had some insight in alluding to dreams. When she spoke about her stay on the moon she added: "That was like a dream; was it the truth or a dream? I don't know."

In my reexaminations of this patient I tested more precisely her capacity for abstract thinking. Questions on the differences between concrete objects she answered without error, likewise she correctly constructed sentences using three words. The fact that she had learned well in school was confirmed. But with questions of the difference between abstract concepts and the explanation of proverbs she failed completely: (What is the difference between error and lie?) "Error may change and lie may change. Error is error." (What is the difference?) "Error, there was an error, it can be different, lie is completely untrue, lie is lie." (To give or to lend?) "To lend makes worry, to give, that was intended . To lend, you may make enemies." (No rose without thorns?) "The rose has thorns, the rose is not without thorns." (The apple does not fall far from the tree?) "It cannot fall far away, it falls as far as the tree's bough." (The jug goes to the well until it breaks?) "Wher it is broken it no longer goes to the well, it is in pieces."

From the answers given it is clear that the thinking of this patient stands stil in the concrete. Confabulatory paraphrenia is confirmed by this type o thought disorder. The confabulations present with impressive transitions tc dream states. Perhaps the former good intelligence of this patient made i possible for her never to have completely lost her capacity of looking at hei experiences critically.

Summary

The picture of confabulatory paraphrenia is dominated by **errors of memory** which appear in the form of interconnected experiences. They are generally fantastic in form and refer to other countries, continents, or even planets. Apparently freely imagined contents take on the sensuous character otherwise reserved for memories. The immediate surroundings are not drawn into the confabulations, but are correctly judged. Presumably the patients' critical ability allows confabulations to stand only if they occur in other places. To this self-awareness may be ascribed the fact that some patients speak of **dreams, trances,** or other exceptional states during which the fantastic occurrences allegedly happened. The abnormal reification of imagined contents is probably related to peculiar **"perceptive errors".** Built into the confabulations are **ideas of grandeur** which may be boundless; they are carried by a somewhat **elevated mood** which is also characteristic. An elevated mood participates in the genesis of the fantastic coloration of confabulations with a trend toward a sensational character. The peculiar **thought disorder** operates similarly, allowing patients to be aware of concrete elements, but impairing abstract thought and the ability to solve abstract problems. While patients function well in daily situations, when given intelligence tests a **pictorial form of thought** is revealed.

Expansive Paraphrenia

The megalomania of fantastic and confabulatory paraphrenics is not deeply anchored in the personality. The patients mention their ideas and indulge in them, but their true personalities are only slightly influenced by them. Now we will examine a paraphrenic form in which the patients are much more seriously pervaded by **ideas of grandeur,** and behave accordingly. Because this is the only salient symptom of this disorder it seems appropriate to characterize it as expansive, although the ideas of grandeur are not as boundless as in the fantastic and confabulatory forms. My cases are rather similar to KRAEPELIN'S "paraphrenia expansiva". KRAEPELIN does, of course, describe many accessory phenomena such as hallucinations, ideas of persecution, and errors of memories which do not belong to the picture, but for some of his patients he does describe an "affected, bombastic, unctuous manner" characteristic of expansive paraphrenia. Since hallucinations, common in the acute state, are later absent, expansive paraphrenia is, as I describe it, a purely delusional form. There is considerable thought disorder. If confabulations are considered as separate from delusions expansive paraphrenia is the only purely delusional form. It contrasts with three hallucinatory forms – hypochondriacal, phonemic, and incoherent paraphrenia. In the fantastic form delusional and hallucinatory symptoms are about equal in importance.

In the fantastic and confabulatory forms the megalomania could be well described by reporting its contents. There was no question of expansive behaviour. For expansive paraphrenics I have to describe the behaviour of the patients, as in this way the megalomania may be recognized in its characteristic form. I present a series of examples.

He believes he is a director with a large fortune and wants to marry a "lady of the finer class". – He greets people with a deep bow and his facial expression reveals that he wishes to demonstrate how well he has mastered upper class behaviour. – She keeps away from the other patients, as a "fine lady", and she speaks condescendingly of them. – She displays a majestic manner, taking measured steps and speaking with such affectation that her words have a chopped-off sound. – She likes to wear showy dresses and wears a turban-like hat with pride. – From time to time she leaves the "idiots" and studies her "documents", full of meaningless script, in her room. – She talks haughtily about the other patients, refuses to do any work, claims that her only work is music although she had failed as a music teacher. – The director of the hospital is "irrelevant" for her; she is going to send her complaints to Hitler. – Earlier he considered himself to be the owner of the institution, later he no longer mentioned that, but demanded that he be addressed as "Monsieur". He self-righteously yelled at other patients to show off his lordliness. – She thinks she is Hitler's messenger, writes high sounding letters to him and other important personalities, uses quotes and pathetic pleas without saying anything of consequence. She also compares herself with Joan of Arc and wants to "transmit steely thoughts to Germans in Italy". – He believes he has been chosen by God, but gives no details. He "must not reveal the world of God." On one occasion he had seen a halo, but not again. "Our divinity is not so splendid." Earlier he was more open-hearted and called himself "the greatest prophet and the greatest employer of the world." – He calls himself "illuminarius", i.e., "the enlightened", and feels capable of taking part in government because of "his ability to understand quickly". – He used to want to be an actor and singer although he had no qualifications. He also believed that he was very wealthy. Later he demonstrated his upper class position by using affected language and ceremonious manners when dealing with others.

If the patients are simply treated like others they get used to it and cooperate. Occasionally when called upon to do something unusual, they flare up and point to their high position. They then can become irritated and excited. I used to think that their irritation never went deep and they could be easily appeased. But now I have come across cases where with great persistence patients objected to being asked to come into the room next door in order to be examined like other patients. When they were urged to come they started to shout swear words at the examiner. Their affect did not change when we sat down with them in their room in order to meet them half way. Under no circumstances did they agree to be treated like other patients. Their overbearing manner reflected more a feeling of having been insulted than general irritability, although their affect was seen to be genuine. In general they do not seriously resist the regulations of the ward, their behaviour has more the character of a demonstration without serious consequences. This is the origin of the **megalomaniac pose** so characteristic of expansive paranoia. As the examples showed the patients like to dress up, strut around, speak affectedly, and use expressions which they consider to be upper class, for example using foreign words which they do not themselves understand. They often speak superciliously about other patients, describing them as "common people". With the doctor they assume a jovial manner to emphasize that they are at least his social equal. Sometimes they give commands in a haughty manner; one patient ordered the doctor "out of her chamber", "in the name of my authority". Another gave stereotyped commands to which no one paid attention. Here is an order which he wrote down.

„Order for permissible foods, delivery. It is my wish and desire to serve fasting meals on Wednesday, Friday, and all other holidays. On Tuesday will there be fasting meals, service? While it is a duty for Catholics to eat fast meals on the dates above: into which fasting meals the above will be divided will be determined by the institutional administration. White coffee bread, as well as house bread to stay strictly controlled. All rights reserved."

The piece of writing shows that the patient is trying to demonstrate his station and his education with high sounding and complicated expressions. When a patient always presents the same pose, always issues "orders", and always maintains the same form of speech and movements, his behaviour may appear quite manneristic. Yet it is not a question of genuine mannerism, but rather one of demonstrative forms of expressing megalomania, albeit one lacking in fantasy. Poverty of ideas in general is typical of expansive paraphrenia.

Nevertheless **confabulations** do appear. Patients claim to have met important personalities and to have won their friendship or to have participated as leaders in important undertakings. Quite often they claim that they were mixed up as infants and are really the offsprings of highly placed parents. It is clear that such declarations form the concrete foundation for the patients' expansive posture. Thus the patients' claims do not represent an independent pathological symptom, but are rather part of their attempt to appear important to the people in their surroundings. Their confabulations lack fantasy and are not like the tales of confabulatory paraphrenics.

Patients often try to act as if they were important by using secret words, letters, and numbers. The object is to create the impression of having secret information. One patient prepared a secret script "in order that no abuse will be made of anything of scientific importance that I write down". Below, a piece of writing of another patient is presented. The totally disconnected "balance" i. e., simple addition, visibly has the purpose of making an impression by means of the enormous numbers. The rest is supposed to appear a secret.

„A new stick Wag Fahr Waf difficult

A convent Phäyer

Karolvsd

1 2 3 4 5 6 7 8

G D V X Value

9 10 11 12 13 14 15 16 17

With best wishes G.G.D.D.R.R.

5 4 2 3 8 4

New Balance	9 289 295 743
	276 464 259
	9 565 760 002

God knows where. 9 August 1926 Prussia Horse 22914
Bavaria Horse 428224"

At times patients produce mysterious correlations in order to make themselves seem important. A piece of writing is presented below that demonstrates this. From the first sentence it is obvious that megalomania is behind it while thoughtful performance is absent.

„I want to present a theory which can be used in all areas of science. The numerical system is used with translation into words. Numbers have the following meanings: 4 = the norm, 5 = the support, 6 = blue, 7 = the spy, 8 = warning, 9 = troubles, 10 = work, 11 = help, 12 = the time, 13 = misfortune, 14 = the war, 15 = the guilt, 20 = Kapp-Putsch, 21 = rocks, 22 = the services, 23 = weather, 24 = the writing, 25 = jubilee, 26 = goods, 27 = birthday, 28 = ship, 29 = teaching, 30 = bank. The use of these words is visible in lines, sentences and chapters, through signals with the help of the clock. Only the numbers are in question, their position and transposition must be sought."

Since expansive paraphrenia is a purely delusional disorder, one might expect that the ideas of grandeur would be accompanied by delusions of persecution. However these are absent. What does arise are accusations against people who speak up against the patients' expansive ideas. This is important when considering the differential diagnosis of affective paraphrenia. Only at the beginning do ideas of reference occur. A hostile paranoid attitude does not persist; probably it is prevented by the **blunting of affect** clearly present in expansive paraphrenia. Interest in their families diminishes. The patients remain in the institution without offering much resistance; they can be given day leaves without running away, while they object to being treated as patients. Occasionally they try to subsume their external situation which they cannot change into their megalomania by declaring the institution to be their possession and fellow patients their employees.

Patients with expansive paraphrenia also lose **initiative** through their blunted affect. Only when demonstrating their megalomania do they show some energy. The inclination to speak differs from individual to individual, but most become talkative when stimulated. This form of activity also arises from their expansiveness, since they may speak broastfully about their true position, accentuate that they are not treated with enough respect, complain about other patients' improper behaviour, or talk of their plans to make inventions and collect their money. The patients are pleased when they are allowed to talk and they believe they are taken seriously, although again the affect is not deep. The ecstatic coloration sometimes found in confabulatory patients is absent. Despite the delusions of grandeur there is no change in the basic mood.

In speaking with the patients **thought disorder** is evident. When they use incorrect words and do not complete sentences it may appear as if this were merely an expression of pomposity; in the attempt to sound as elegant as possible they make errors. Beyond doubt this element does play a role. In contorted style the patients try to imitate the language of bureaucracy and use expressions they may have read in legal documents or military orders. This is, however, only a partial explanation of the **verbal peculiarities;** actual failures result from the thought disorder which affects speech above all else. Wrong words, mistaken word constructions, and false word combinations are common, as well as grammatical errors. I will present a series of examples of various patients. I must again point out that I have chosen typical examples. Many questions are answered in an orderly manner. Wrong versions may become more pronounced when patients' ambitions to be important come to the fore. The first example may actually look like a somewhat self-important bureaucratic style.

In response to a doctor's remark, that it was his duty to keep the patient in the institution, the patient replied: "The false sense of duty which you arbitrarily exercise is in reality nothing more than the manifestation emanating from a cunningly suspicious and highly integrated control apparatus which practices an irresistible controlling terror from continuous threats by every manner of the apparatus. This subjective comprehension of duty facilitates all arbitrary acts." – A former music teacher said: "Yes, I was ousted from my profession, my music, I suffered repression of my music; the attitude of my music came back."

Answers to intelligence questions included the following: (What is the difference between box and basket?) "That the box does not exist of better material, whereas for a basket a certain better work may be necessary." (Child and dwarf?) "A child is still in the prime of life and a dwarf does not grow **bigger** himself." (What does honesty mean?) "That depends on the job, if he leaves aside everything that is bad." (Morning is the best time?) "That means, that the profit motive starts early." (Envy and thrift?) "Envy is nothing to notice, an instinct that should not be noticed." (Child and dwarf?) "Dwarf is completeness, a person who has matured." (Construct a sentence with the words winter, oven, trees) "The trees **winter** in the oven." (What is an island?) "A piece of land, a piece of land, that stands in contact with each other." (What is a sieve?) "A household object for the filtration of liquids so that any residues which are in the liquid are repelled." (Staircase and ladder?) "It is a possible way of ascending with different interests for all the people who need it." (Stream and pond?) "Streams are in the mountains with the influx of mountain water. Pond is a massive lake originating from the source, or a pond of the waters." (Box and basket?) "Circumference, wood or **basket windings."** (Error and lie?) "Error is a thing, also personal, lie is an exceptional condition." (Thrifty and stingy?) "Thrifty is advancing, stingy is also decreasing." (Mountain and mountain range?) "A mountain is a simple matter, in the sense of the plural, a mountain range is a **matter of more numbers."** – A patient answered three intelligence questions with three neologisms, presented in a pompous manner: Piuresta, Aquarol, Kapsan.

In these examples verbal errors are conspicuous. Similar ones are found only in cataphasia. Both sentence construction and word construction are distorted. Beyond this there is a conceptual disorder. The concepts are not totally wrong but imprecise and not totally relevant. If a basket needs "better work", that is probably a reference to weaving, but it is not a matter of "better material". If the honest man "leaves aside" everything that is bad, no exact explanation is given, since what is not bad is not necessarily honest. When the proverb about the morning is associated with the profit motive, the concept is totally distorted, since it does not refer to the profit motive but to an honest ambition to make progress. Thus it is repeatedly found that conceptually there is no clarity. The disorder seems to be purely one of thought in the explanation of the concept of an island: "A piece of land, a piece of land, that stands in contact with each other." On the contrary, the island is not in contact with the other land.

In examining the thought disorder of expansive paraphrenics further, the verbalizations are always found to contain something which is not completely incorrect but which does not get to the crux of the matter. Thought appears to be coarsened; concepts do appear, but refinements are absent. Concepts are presented in coarse form; in normal people this could happen at first but is soon corrected. In expansive paraphrenia **coarsened thought** is a quite distinct form of thought disorder found nowhere else. As I have said above, we again find evidence of how inappropriate it is to speak simply of a general schizophrenic thought disorder.

I present the case of a patient whose father was also ill.

Case 74: Hugo Ra, born in 1883, did well in school and in his job as a cook. In 1927 he started to believe that the water supply came from the Main river. From 1930 on he was no longer the same. He never wrote a whole line across the page, leaving the middle empty, and saying that was the "centre". He felt persecuted by neighbours and political parties and suspected his wife was in on the plot. When a potted plant fell onto his balcony he considered it the sign of an amorous advance. Since 1931 he liked the number 19. He claimed that a family living in house number 19 on his street were related to him; he put 19 stamps on his table; he rapped up 19 pennies in paper; and he collected all the pages with number 19 from a calendar. On one occasion he mixed water from the Main with sand in a bottle and poured it onto the streetcar tracks. In 1932 he did it again and was picked up by the police. He was taken to the Frankfurt Psychiatric Hospital, where he explained that he knew the water came from the Main because it tasted funny. When the streetcar passed him it made signs, i. e., it rang its bell twice. He had poured the water onto the tracks to show "them" that they should look after themselves. He had been harassed by neighbours, stamping on the floor and dropping flowers; one neighbour called him a dirty pig. He answered intelligence questions in the following manner: (What is the difference between a fence and a wall?) "Fence is a broken-through fencing, a broken-through bordering.

Wall is a bordering by rocks." (Error and lie?) "A lie is an untrue claim, when he has seen that he has made a mistake. A lie is an untrue, purposeful claim." In hospital he was calm, did not feel wronged, and was discharged after eight weeks. At home he worked little and busied himself with his ideas. In his shop he put up a sign that every 19th customer who made a purchase for 19 Mark would receive a gift. He picked 19 flowers daily, poured water on the tracks of the streetcar number 19, and gave his wife 19 roses on her birthday. He also put up a sign in his window: "Landlord, check the drinking water!" Sometimes he was downcast, threatened suicide, and once even jumped into the river, but was rescued. In 1934 he returned to hospital, expressing ideas similar to those he had expressed previously. Now he spoke of holy times, i. e., from 10–11, or from 4–5, or 7 o'clock, "when Christ prayed". The houses on the left, i. e., "holy" side of the street were more liveable than the others. That was the side "God wished". Seven was found in the bible to mean 19. He was transferred to an institution, where he demanded 1,900,000 Mark in damage payments, since in his apartment he had received Main water instead of drinking water. He greeted only with the left hand, saying "Since 20 June 1938, as a protestant world greeting, I only greet with the left hand. A protestant has the duty to bring others from the erring paths to the Protestant way." He was the greatest prophet and was to become the greatest employer in the world. In 1935 he divided his day into religious ceremonies and stopped eating pork. In 1936 he continued to complain and tore a knife away from an attendant in order to slash his wrist. No deep depression was visible. In 1937 he ate only when the big hand of the clock was going down. In 1939 he claimed the food was full of faeces. In 1941 he was brought to the Frankfurt Hospital for a short reexamination: he claimed that God was in contact with him. When he spoke the weather would be clear, and when he was silent it would snow. As early as 1931 he had seen a halo as a sign from God. In a ceremonious and round-about manner he presented all the ideas he had had since the beginning of his condition. He considered them all very important and would not be interrupted. If asked for more proofs for his relationship to God, he did not know what to say: "Our divinity is not so splendid." When expressing his ideas he expressed a certain affect, but it was not deep. If he was contradicted he became somewhat irritated, but again without depth. The proverb about the morning being the best time was explained by him in the following manner: "What one did in the morning is enough for the whole day."

The **father** of the patient, Herman Ra, born in 1860, became strange at the age of 40. He studied the occult and believed himself to be destined for higher things. He carried on all kinds of experiments with his family; they had to bite into raw potatoes and tell him, with the way they looked, that they were biting into meat or sausage. He performed other such strange things. For a while no fire could be lit nor could the house be locked. In 1910 after he knelt in front of the church praying out loud, he was admitted to the hospital. Little was found and he was soon discharged. He continued to express a strong need for recognition. He became more peculiar in 1922. He read the Bible a lot and at night spoke with the Saviour. He became physically more and more fragile and was finally bedridden. He repeatedly threatened his family with a knife and also threatened to commit suicide. He was finally admitted to the Frankfurt Psychiatric Hospital. He spoke with an unctuous tone about his good daughters and complained about his bad son. He disagreed with what went on at home. "Since I did not want to have anything to do with politics, I read the Bible." He had difficulty walking, was paretic and atactic, nystagmus and atonia were present. The reflexes were increased, the Babinski reflex was positive bilaterally. The cerebrospinal fluid was unchanged. His condition deteriorated; there was delirious handling and grasping, he lost his orientation for time and place. He died three months after admission. The autopsy revealed a small brain with softening; the basal vessels were delicate. No histopathological report was available. The clinical impression was arteriosclerosis or a progressive system disorder.

In reference to subject Ra in the hospital at first a predominance of ideas of reference and significance was assumed. Later there was nothing present beside the posturing typical of expansive paraphrenics trying to appear important. Ra made connections which from their lack of fantasy appeared contrived, for example the emphasis of the number 19 and other play with numbers. I have mentioned these word and number games above. Pouring Main river water on the train tracks might have been based on an actual delusion but here too it probably was another attempt to make himself important. The expansive attitude is present throughout the whole medical history; it is repeatedly externalized by means of some ceremony or other. Whether in 1936 he had a depressive fluctuation or if he took the attendant's knife only because of an argument cannot be determined from the medical history. The thought disorder in this patient is characteristic of expansive paraphrenia. Already at the onset and again during the 1941 examination the disturbance was evident in the way the patient talked.

The father became ill at about the same age as the subject and also became expansive. He too carried out peculiar actions which he considered very important. His disorder remained mild so no clinical treatment was necessary except for a short hospitalization. His overbearing manner continued, therefore expansive paraphrenia was very probable. Not until he was very old did he return to hospital suffering from a severe organic disease. Probably his expansive religiosity – he claimed to speak with Christ – and his unctuous self-righteous manner indicated expansive paraphrenia. Otherwise organic phenomena were in the foreground. If a progressive nervous system disease was present, as was probable since there were no signs of cerebral arteriosclerosis, then it was a case with an interesting combination of systematic schizophrenia and a neurological system disease.

Summary

During the onset of expansive paraphrenia hallucinations may appear; after disappearance of the accessory symptoms the picture is of a pure delusional disorder. The direction of the delusion is expansive. **Ideas of grandeur** usually have a moderate content, but the patients always try to live their megalomania outwardly. While fantastic and confabulatory patients only speak about their high position, but do not behave expansively, expansive patients always try to demonstrate their importance by their behaviour. Here the whole personality is affected by the delusions. In posture and movements, in contact with other people, in their dress, and in their secret writings, the patients try to act important. Their expansive activity is often very uniform, demonstrating the **patients' poverty of ideas,** in contrast with the wealth of ideas of fantastic and confabulatory patients. When the patients talk about their megalomania and present their wishes and complaints they often develop a certain loquacity. Their expansive habits are characterized by a certain eagerness. Otherwise they show **little initiative** and **little interest.** The thought disorder is manifested particularly in verbalizations containing grammatical errors and neologisms. Concepts are not completely erroneous but expressed with lack of precision. I describe this as **coarsening of thought.**

Review of the Family Picture of Systematic Paraphrenias

Unfortunately I have to say that in reviewing the families of systematic paraphrenics all too often the description of an affected relative is insufficient for certain diagnosis. First the **Frankfurt cases.**

Among **siblings** were the following: hypochondriacal subject Be had a brother who was either a hypochondriac or a hypochondriac-combined paraphrenic. Phonemic subject Zu had a brother likewise suffering phonemic paraphrenia. Phonemic subject Haus had a schizophrenic sister who may have had a phonemic form. There was insufficient information regarding the brother of hypochondriacal subject Pu, the brother of fantastic subject Wa, the sister of confabulatory subject Bo, and the brother of expansive subject Ba.

Among **parents:** the expansive subject Ra had a schizophrenic father who was probably an expansive paraphrenic as well. There was insufficient information about the mother of the phonemic subject Enk, the mother of the expansive subject Me, and the mother of the expansive subject Mül.

Among **more distant relatives:** the incoherent subject Mö had a great-niece who most likely had incoherent paraphrenia. Phonemic subject Zu had an uncle who probably also had phonemic paraphrenia. Phonemic subject Kol had a schizophrenic great-aunt who probably had phonemic-fantastic paraphrenia. In earlier editions I described these two cases. There was insufficient information with the regard to the uncle of hypochondriacal subject Scheun and the great-aunt of hypochondriacal subject Hos. The aunt of hypochondriacal subject Be suffered from an endogenous psychosis which went through two phases and showed no similarity to the psychosis of the subject.

Now the **Berlin cases:**

- A female hypochondriac patient had a mother who died in an institution; no record could be obtained.
- A female hypochondriac patient had a sister who according to accounts by relatives had been chronically mentally ill for years; we were unable to examine her.
- A male hypochondriac-paraphrenic patient had an emotionally disturbed grandmother (father's mother) about whom no further information was available.
- A female confabulatory paraphrenic patient had a father who was institutionalized and whom I also found to be suffering from confabulatory paraphrenia.

- A male confabulatory paraphrenic patient had a sister who was institutionalized and whom I found to be suffering from combined confabulatory-hypochondriacal paraphrenia.
- A male incoherent paraphrenic patient had an aunt (mother's sister) who was in an institution; no record was available.
- The healthy mother of a female incoherent paraphrenic patient was unusually difficult to fixate on questions during a discussion; since she also appeared rather suspicious and affectively cool, a relationship to the incoherent paraphrenia of her daughter is possible.
- A female expansive paraphrenic patient had a mother who received treatment in an institution; no record could be obtained.
- A female expansive patient had a mother, victim of the "euthanasia"; no record could be obtained.
- A male expansive paraphrenic patient had a grandfather (father's father) who was institutionalized; no further information was available.
- A male expansive paraphrenic patient whose great-aunt (father's father's sister) became ill at 68, with dermatological delusions compatible with the diagnosis of hypochondriacal schizophrenia; there was no recognizable relationship to the psychosis of the proband.
- A male expansive paraphrenic patient had a cousin (mother's brother's daughter) who according to the medical history suffered from periodic confusion psychosis; there was no recognizable relationship to the proband's psychosis. The cousin's psychosis may have been related to mental disorders in her maternal family with whom the proband had no connection.
- A female phonemic paraphrenic patient had an uncle (father's brother) who heard voices and once attempted suicide; since the voices were reported to have been prominent a connection to the proband's psychosis was possible.
- A female phonemic paraphrenic patient had a grandmother (father's mother) who was institutionalized; no further information was available.
- A female phonemic paraphrenic patient had a mentally ill father about whom no further information was available.

From the above can be seen that in the majority of the few cases for whom specific information about relatives' disorders were available the same form was found; a few were combined with a second form. Unfortunately there were many cases with insufficient information.

Final Remarks on the Simple Systematic Forms of Schizophrenia

Now that I have described all the systematic schizophrenias I have seen, their precise character can be better understood. It is not that a large series of schizophrenic symptoms may form various combinations and thus produce different clinical pictures; on the contrary, the symptoms in each form are of a distinct nature. If voices were present in two forms, then the voices were different in each of the forms; if a thought disorder was present in two forms, it was never the same thought disorder. I believe that I have been able to demonstrate this convincingly in regard to the paraphrenias. Furthermore many symptoms are present only in one form. The typical talking-past-the-point occurs only in one form, parakinesia only in one other. I found repeatedly that particular forms demanded description of particular symptoms, which I take as impressive evidence that the individual forms are truly independent. Furthermore several characteristic symptoms are always found together, forming the total syndrome of that particular schizophrenic form. In my opinion the different symptoms and syndromes are manifestations of closed pathological systems.

Combined Systematic Schizophrenias

KLEIST always assumed the existence of schizophrenic combinations parallel to the combined diseases of the neurological system. During the many examinations that I carried out in Bavarian institutions with SCHULZ and in the Frankfurt Hospital, I came to recognize the importance of combined forms of schizophrenia. I observed total syndromes which seemed to contain two basic syndromes. If simple forms exhibited strict total pictures then combined forms had to do likewise. Of course simple addition was not to be expected since the syndromes could influence one another. The only path to a precise knowledge of the picture of combined forms was through practical observation. So far I have found combinations only within the catatonias, the hebephrenias and the paraphrenias but no catatonic-hebephrenic, catatonic-paraphrenic, or hebephrenic-paraphrenic combinations. Rare exceptions will be considered later. Further tests will have to determine the adequacy of my descriptions of combined forms. Since only a few cases of each combination have been available there is room for much error. Especially where in combination new symptoms arise which cannot be deducd by summation, with the few cases available for study no generally valid conclusions can be drawn with certainty. In the Berlin investigative series, however, I was able to gather more information to which pertinent reference is made below. The examination of childhood schizophrenia added to my knowledge, when I also discovered combined forms which I was able to compare with cases beginning in adult age. This will be discussed in detail.

Reference is made solely to defect states which alone present a pure picture. Accessory symptoms are more pronounced in combined than in simple cases. This can be understood as a result of the greater extent of the pathological process.

Combined Systematic Catatonias

The combinations of systematic catatonia will be considered in the following order: speech-prompt catatonia, sluggish catatonia, proskinetic catatonia, negativistic catatonia, parakinetic catatonia, and manneristic catatonia.

In **speech-prompt-parakinetic catatonia** there is parakinetic unrest; this component is not difficult to recognize. However talking-past-the-point is modified; readiness to answer without thought and in a short-circuited manner becomes pressure of speech. This may be understood as additional impulsivity produced by the parakinetic component. The logical order of thought is lost; the content of the pressured speech is totally confused. Many neologisms are mixed in. Verbalizations do still contain elements of talking-past-the-point. A word that had appeared at one time reappears in further verbalizations so that it is possible to speak of perseveration, as is seen in simple speech-prompt catatonia. When the patients exceptionally answer with only one sentence it is often totally unconnected to the question, much as in talking-past-the-point. During their confused speech the patients turn toward the examiner. They show many facial expressions, but only in parakinetic distortion. Gesticulations are similarly parakinetic.

Similarly in **speech-prompt-proskinetic catatonia** a second component transforms talking-past-the-point into confused pressure of speech. Simple questions are still generally answered correctly, but pressure of speech then develops and the patient is nearly incomprehensible. Many words are frequently repeated, like the verbigeration of proskinetic catatonia. On the other hand shorter senseless answers are reminiscent of talking-past-the-point. Many neologisms also underscore the speech-prompt component. Speech is low in volume but is not like the mumbling of simple proskinetic catatonics. When addressed the patients look up expressionlessly. Sometimes there is fingering of the body or the clothing.

In **speech-prompt-manneristic catatonia** talking-past-the-point is slightly different: in simple speech-prompt catatonia it has two faces, so to speak, since it appears to arise from short-circuited thought as well as to be purposeful. In combination with manneristic catatonia the purposeful element becomes more pronounced. This is probably due to the fact that the mannerisms are not simply automatic actions but are based on the patients' volition, impaired as it is. In this combination answers not only miss the point of the question, but become quite inappropriate. Thus a male patient answered the question of his age with "the same age". Sometimes the patients do not stop with a short answer, as is usual in simple speech-prompt catatonia, but follow with further senseless words. The assumption that a manneristic component should diminish the urge to speak is not correct. "Movement mannerisms", in contrast with "omission mannerisms", are increased. Purposeful talking-past-the-point appears in some neologisms. A male patient called himself a "killer of the dead". The manneristic component also explains why repetitions, which in speech-prompt catatonia correspond to perseveration, appear as stereotypes in the combination. Furthermore patients may repeat each question manneristically. This phenomenon approached echolalia in the case of a female patient who repeated, "What does this mean?", i. e., she not only repeated the proverb but the introductory question as well. Mannerisms in the form of movement tend to recede in combination; the manneristic component restricts itself mainly to verbalizations. Some mannerisms are seen on close observation. A female patient was observed when walking back and forth, briefly standing on tiptoe before turning around. The empty facial expression of speech-prompt catatonics is retained; the manneristic component adds a certain rigidity which is however not pronounced and less noticeable than in simple manneristic catatonia.

In **speech-prompt-negativistic catatonia** talking-past-the-point is even more uniform than in the speech-prompt-manneristic form. Patients may respond with arbitrarily false answers. A female patient, for example, born in 1941, claimed that the present year was 1947 (in reality it was 1975), and that she was 17. During examination patients tend to cling to one answer which has a quality of rejection. Questions are frequently met with, "I don't know". A female patient reacted to every intelligence question with "yes" and even began to say "yes" before the question was finished. Then again patients become silent, with simple negativism replacing the talking-past-the-point. Facial expressions reflect both forms of the combination: they are empty and expressionless, with a hint of refusal, or they show the ambivalent smiling of negativistic catatonia. Apart from talking-past-the-point patients display the symptom of impulse automatism, such as found in simple negativistic and, less clearly, in speech-prompt catatonia. Derived from the negativistic component are the never absent periods of aggressive excitement.

In **speech-prompt-sluggish catatonia** talking-past-the-point takes yet another form. As in speech-prompt catatonia it does not necessarily begin immediately, but perhaps only after simple questions have been answered correctly. It does not however end with short correct or incorrect answers; instead, further remarks follow. Answers are at times hesitant; the patient must be given time and not immediately be asked new questions. The original answers will then be complemented, or expanded senselessly. Furthermore the patients constantly repeat themselves. Sometimes they simply say a sentence twice or, after a few words, come back to what they had said before. The repetitions are not stereotypical, but instead depend on whatever happened to have been said previously. The perseverations of speech-prompt catatonics are more prominent. In giving wrong answers the purposeful character is no longer present; the impression is gained of simple incoherence. This must be the result of the thought disorder of sluggish catatonia, with the incoherence inherent in it. Similarly all other behaviour is determined by this sluggish component. The patients just barely look up when addressed and appear not truly to have grasped the questions asked. Instead they are constantly occupied with themselves. They look around with restless eye and head movements, intermittently staying still as if they were listening. At times they move their lips and whisper audibly to themselves. This is the hallucinatory distractibility of sluggish catatonia. The facial expression is restless when the patients are involved with hallucinations, but is empty and somewhat perplexed if their attention is directed to the examiner.

Of the combinations with sluggish catatonia the connection with the speech-prompt form has now been described. Further combinations follow.

In **sluggish-proskinetic catatonia** the appearance of fantastic-confabulatory ideas is noteworthy. Since these ideas also appear during the onset of simple sluggish catatonia, I presume that they are a general symptom of this form, but become apparent in later stages only if the verbal impoverishment is overcome by a second component. This occurs with a proskinetic component. One of the male patients had survived 700 Indian wars, killed 1400 people, and personally decapitated Darwin. Another patient had been to the moon. There he met God as "Supreme Commander". Hallucinations occur and often cannot be distinguished from confabulations. Visual, auditory, and somatic hallucinations are mixed together. A male patient claimed that his lung had been removed and placed before him on the table, and an express train had gone through his body. Another patient claimed that "the Reich Marshal Göring and his wife" had entered his body. Phonemic hallucinations predominate. As with simple sluggish catatonics these patients are repeatedly distracted by the hallucinations even while being examined. They look around, move their lips, and carry on clear conversations with their voices. Verbal impoverishment of sluggish catatonics is absent because the second component of proskinetic catatonia heightens the drive to speak. This component is clear from the mumbling manner of speech. While the volume does not sink to an incomprehensible level, verbalizations are made in a soft voice and with no modulation. Perhaps the hallucinations act as stimuli and limit the decrease of volume. In the Frankfurt cases I believed that I had found a major thought disorder, perhaps because of the patients' hallucinatory excitement. When quieter, patients could answer intelligence questions correctly.

Fantastic-confabulatory phenomena are also found in **sluggish-parakinetic catatonia.** Supposedly a symptom belonging to the sluggish component comes to the fore, since verbal impoverishment is diminished by the second component. Parakinetic catatonia produces increased motivation in acting and speaking. Confabulations are perhaps not quite so varied as in sluggish-proskinetic catatonia, while the expansive element in the confabulations is more pronounced. A male patient wanted, as President of the Reich, to turn the wheel of world history and move the stars. A female patient, who had died and risen again three times, claimed to have accumulated an immense fortune on estates. Hallucinations involving all the senses are found. The body of a female patient had "recently" been invaded by a fellow patient. Once again voices dominate. Hallucinatory distractibility is more pronounced than in sluggish-proskinetic catatonics. The patients not only move their lips in whispers during the examination, but also speak half-loud in asides. Answers may be quite incoherent since patients are frequently inattentive while listening to their voices. However if fixated on questions, simple answers are given in an orderly manner and some intelligence questions are answered correctly. Nevertheless the patients may diverge from the subject. The tendency of parakinetic catatonics to go off at a tangent probably plays a role here. In fact they often make remarks which have little or nothing to do with the situation at hand. Parakinetic restlessness appears as it does in simple parakinetic catatonia, but perhaps slightly more uniformly and with more parakinetic mannerisms.

In the two combinations described above poverty of speech is canceled out; in two other forms it is more marked, as these themselves are characterized by poverty of speech.

Patients with **sluggish-negativistic catatonia** seldom answer questions. They often whisper with their voices and their faces display hallucinatory restlessness. They differ from simple sluggish catatonics in that the latter, if urged, do sometimes respond with short answers. In further contrast sluggish-negativistic catatonics show a certain restlessness in their movements which cannot be understood as a reaction to hallucinations. They fuss with themselves and with their clothes and they reach out for objects in their fields of vision. They repeat movements, sometimes for long periods of time. One female patient twisted buttons until they broke off. In comparison various proskinetic hand movements are carried out with little energy. The uniform activity of all the patients just mentioned surely arises from the negativistic component of the combination. In simple negativistic catatonia we found impulse automatism and presumed that with the general lack of higher motivation there is increased excitability of lower mechanisms. When addressed sluggish negativistic catatonics pay attention only for short periods, if at all. Their facial expressions become a little more animated only in reference to their hallucinations; otherwise their faces are blank. Impulsive excitement with aggressivity generally confirms the negativistic component.

Patients with **sluggish-manneristic catatonia** also give no or a few meaningless responses; they barely face the examiner. They are always occupied with their voices; their gestures and facial expressions are distorted; they often speak quietly to themselves. Posture and facial expressions are rigid, but not so much as in simple manneristic catatonia, perhaps because the hallucinatory restlessness brings about some facial movements. Mannerisms are common, not as cohesive but as stereotyped movements: repeated reaching to the head, retraction of a leg, or rocking of the body. Posture stereotypies also appear; for example, constantly standing in the same place, or holding the hands on the chest. There is no impulse automatism. The patients respond to a grasp only hesitantly, and if a hand is offered repeatedly, do not respond at all. Generally there is immediate resistance if pressure is exerted on their bodies. The absence of this symptom is important for differential diagnosis, since it is usually present in combination of proskinetic, negativistic, speech-prompt, and sluggish catatonias. It is also found in sluggish-negativistic catatonics who otherwise similarly lack motivation and are as unapproachable as sluggish-manneristic catatonics.

The combinations of **proskinetic** catatonia with speech-prompt and sluggish catatonia have now been discussed. The others are as follows.

In **proskinetic-parakinetic catatonia** there is much restlessness. The patients grimace and gesticulate with distorted movements; certain forms of movements are repeated stereotypically. Simultaneously they constantly fuss with themselves or objects, thus eliciting the proskinetic element. In speech they often intensively direct their verbalizations towards the examiner; this is probably the result of a combination of the motor characteristics of the two catatonias. When addressed they tend to reply with pressure of speech. When stimuli are absent the general restlessness recedes; speech ceases gradually as well. The tone of voice is usually low and speech may become incomprehensible mumbling. Modulation is better than in proskinetic catatonia, probably because of the parakinetic component. Patients speak tangentially; when they continue to speak verbigeration appears, demonstrating the presence of proskinetic catatonia. This occurs more frequently in simple conversations with patients than it does when intelligence questions are asked. The latter may be answered correctly, perhaps because the theme is clearly set. When a male patient was asked simply how he was, he answered as follows: "Nothing at all, not very well, not very well, not very well, you cannot be healthy in the hospital. We also look like we are not very healthy, sometimes we look that way, sometimes we look that way... beside gasping for air we still have hands, something to do, a little more to do, a little to do, doesn't take all day, a little to do, a little to do, look for work, work shoes were dirty, work shoes were dirty, were dirty."

In **proskinetic-manneristic catatonia** verbigeration is equally prominent. I counted the repetitions of one of the patients: he said a short sentence seven times and five repetitions were not rare. The verbalizations are much more easily understood than those of simple proskinetic catatonics since they begin at normal volume and become softer only when repeated several times. Eventually there is only incomprehensible mumbling, but which may contain further repetitions. Presumably the manneristic component transforms the verbigeration of proskinetic catatonia into a mannerism. This supposition is supported by the fact that repetitions may be inserted into stereotyped behaviour. A patient verbigerated in the described form, but he attached a set phrase to his answers which he repeated several times; the phrase was "healthy and uninjured". It was unclear how he hit upon this particular expression, just as it remains unclear how patients develop any particular mannerisms. The fussing of proskinetic catatonia is also accentuated in the combination; here too it occasionally looks like a mannerism. In earlier editions of this book I described a female patient who "bleated" stereotypically. In Berlin I saw a female patient who occasionally spent much time exhibiting purposeless hand movements, after which she again remained calm for a long time. Other mannerisms are present, but are more prominent in earlier phases of the disorder. The rigidity of simple manneristic catatonia is reduced. Performance in the intelligence test cannot be evaluated since the patients do not answer the questions, but instead pick out a phrase and repeat it in verbigerations. Often they simply repeat the question.

In **proskinetic-negativistic catatonia** it could be presumed that the negativity of the one catatonia would be canceled out by the affirmativeness of the other, and to a certain extent this is true. The patients do reply to questions; in that sense they are not negativistic. They answer only once, without verbigerations, therefore they are not exactly proskinetic. In this combination negativism appears in the content of the answers which tend to be rejecting (often: "I don't know"), or non-committal, or represent avoidance. Proskinetic traits may be combined with negativistic traits, for instance, when questions of the examiner are simply repeated. The volume of verbalizations is adequate. There is constant restlessness in the form of purposeless grasping and other hand movements. Similar types of movements are often repeated. The restlessness of the body is greater than that of simple proskinetic catatonics because the hand movements are more sweeping and are probably related to both the proskinetic and negativistic components, as the latter also contains an element of primitive motility as evidenced by impulse automatism. A female patient's fussing with her clothes nearly resulted in her taking them off. The facial expression of the combined form is not lively, but may occasionally become slightly hostile. There is great lack of initiative.

The combinations of **negativistic** catatonia with speech-prompt, sluggish, and proskinetic catatonia have now been described. The other combinations follow.

Negativistic-manneristic catatonia displays the most severe inattentiveness of all the systematic schizophrenias. The patients never answer nor even begin to answer questions. Generally they do not even look up. If they do so very briefly, there is no facial movement; the impression is gained of a quick reflex movement rather than attentiveness. The patients do not offer their hands; if one takes their hands, they let it happen passively. They tend not to react to simple requests; they have to be asked several times to get up from the chair. They do not come into the examination room voluntarily but have to be urged and pushed in. All this indicates the continuation of negativistic and manneristic features. The patients' faces are immobile and rather rigid. There may be mild twisting of the trunk, as is seen in simple negativistic catatonia. Despite the severe reduction of drive these patients show their restlessness in the form of stereotypes, thereby indicating the manneristic component. Movements are even more primitive than in sluggish-manneristic catatonia, and are often merely iterations; for example, single arm and finger movements with pauses, then repeated in the same stereotypical manner. The severe anergia is occasionally interrupted by excited phases, milder than those of simple negativistic catatonics. Unless they are looked for, an observer may miss them if the patients are under continuous medication.

In **negativistic-parakinetic catatonia** parakinetic movements are found similar to those of the simple form, but they are less obvious. A tendency to uniformity appears in occasional simple hitting, knocking or wiping movements. Stereotypically repeated movements are not found, however. There is a clear tendency to disjointed speech; in conversation more or less inappropriate remarks are made repeatedly. The picture corresponds to that of parakinetic catatonia. However the patients' answers are different and show the negativistic component; very few questions are answered clearly. Instead the patients talk-past-the-point, or say, "I don't know" even with the simplest questions. The avoidance does not have the specific character it has in speech-prompt catatonia, but unmistakably displays negativism. Even very easy answers ordinarily at hand immediately which speech-prompt catatonics offer regularly, are in this form given incorrectly or replaced by "I don't know". Periods of excitement like in negativistic catatonia do appear, but are milder. The motor restlessness of parakinetic catatonics probably provides a certain reduction of motor tension.

All the combinations of **parakinetic** and **manneristic** catatonia have now been described except for their combination with one another.

In **parakinetic-manneristic catatonia** the whole body is restless. Although it might have been expected that there would be a reduction of movements, because the manneristic components includes a component towards stiffness, this does not seem to be the case. Here care should be taken not to simply add the symptoms of two components, since parakinetic and manneristic catatonia affect antagonistic systems. Thus something new is to be expected when both systems drop out. We saw similar results in speech-prompt-sluggish catatonia and in proskinetic-negativistic catatonia. The restless movements in the parakinetic-manneristic combination are even more obvious and vary less in intensity. In simple parakinetic catatonics the restlessness may generally decline or even cease when the patients are no longer stimulated. In the combined form, however, the movements continue when the patients are no longer being addressed. The whole body is involved with distorted movements of the arms, the trunk and the face. When standing the patients show their restlessness by taking small steps forwards, backwards, and sideways. In earlier editions I emphasized the slowness of the movements as marking the essential difference between this combined form and simple parakinetic catatonia. This has remained true in some cases. Sometimes, however, the movements take place at a normal speed, unlike the parakinesia of the simple form. The movements are stiff, even when quick, and a certain degree of rigidity is mixed in. In this way the manneristic component may be recognized. Rigidity, present at rest, appears to be continued in the movements. Mannerisms are less prominent in the combination and are maintained less tenaciously than in the simple form. They also change spontaneously so that there may be doubt as to whether they are really mannerisms, not simple habits. A female patient basically refused to shake hands; however, if urged she would give in. A male patient looked stubbornly past the examiner; when told this was abnormal, he did look at the examiner. The parakinesia becomes stronger through the mannerisms. The readiness to move seems to be intensified so that it becomes more prominent. This propensity was also suggested by intensification of purposeless movements seen in proskinetic-manneristic catatonia. The thought process is less impaired in the combination than in simple parakinetic catatonia. This is a result of the intactness of thought in manneristic catatonia. While the patients do tend to make disjointed remarks, these are usually less inappropriate than those of simple parakinetic catatonics. Above all the disjointedness is absent during intelligence tests, when patients answer logically and according to their level of education.

Combined Systematic Hebephrenias

The combinations of systematic hebephrenias are easier to review since there are only four simple forms. I will discuss the combinations in the following order: eccentric hebephrenia, foolish hebephrenia, shallow hebephrenia, autistic hebephrenia.

In **eccentric-foolish hebephrenia** stereotypes in conversation are particularly striking; patients repeatedly express the wish to be discharged. This stereotype is similar to simple eccentric hebephrenia but is much more marked in combination. Patients may repeat the same things over and over again. Highly peculiar was a stereotyped request by a female patient to be "slaughtered". Since she also claimed that she no longer had any life in her body, her wish could have been a symptom of hypochondriacal alienation, which may appear at the beginning of eccentric hebephrenia. There was no appropriate affect behind the patient's statements; on the contrary, she sometimes laughed vivaciously in talking about her wish. The content of the stereotype need not always be a wish. A female patient added a "Yes" to all her answers. Another kept on inserting, "Mother dead". In this case the stereotype must have originated far back in the past since the patient herself was 73 years old. The spoken stereotypes may be complemented by stereotyped actions resembling mannerisms. A female patient did not go through a door without first standing before it on tiptoe. The second component of the combination is manifested in the patients' repeated meaningless laughter. It may be vivacious, going beyond the smiling seen in simple foolish hebephrenics. In their general attitude these patients are more mobile than simple foolish or simple eccentric hebephrenics. Their speech is not so lame. Although there is little voice modulation the volume is normal. When the patients are not laughing they sit quietly and look content. They cannot be excited affectively. The depressive element of simple eccentric hebephrenics is missing if the content of their statements is ignored. Intelligence questions are often answered incompletely but not illogically. Occasionally moodiness appears, with the tendency to aggressiveness, thus revealing the connection with foolish hebephrenia.

In **eccentric-shallow hebephrenia** mannerisms are more marked than in eccentric-foolish hebephrenia. These are "stereotypes of movement" (for example, a stereotyped running out of the room, or talking through the teeth) or, more often, "stereotypes of omission" (not eating certain foods, not talking, stopping before doorways). There is a tendency to produce neologisms the artificial nature of which is clearly manneristic: "error-signs", "beginning-of-the-world-people", "chain-fortress-course", "superstition-sentence", "uncustom". From the name of another patient the word "Minna-Stein-Schlägerei" was derived. Sometimes a neologism serves as an expression for a hallucination. Both mannerisms and manneristic neologisms point to eccentric hebephrenia. The shallow component is most evident in hallucinations which are more frequent than in simple shallow hebephrenia. The voices at times may reflect the becoming loud of thoughts. Sensations are also common; patients feel "rays", "currents", "electricity". Grotesque descriptions characteristic of hypochondriacal paraphrenia are not obtained; the sensations appear closer to normal understanding which probably indicates their pseudohallucinatory character. This was found in simple shallow hebephrenia. In the combination the patients maintain the reality of the phenomena. It cannot be determined if the hallucinations are related to ill humour. Little is learned of the contents, perhaps indicating that the patients do not take the hallucinations too seriously. In the affective sphere both hebephrenic components may be discerned. At rest the patients seem somehow moody and displeased, but when they are stimulated they usually smile in a friendly manner and seem content. In any case affective flattening remains obvious. However general mobility is well maintained and verbalizations have normal volume. Periods of moodiness are common, most likely stemming from the shallow hebephrenia; it may be accompanied by aggressivity. Intelligence questions are solved according to the level of education, but these patients tend to express themselves in an unusual manneristic fashion.

In **eccentric-autistic hebephrenia** mannerisms are found as well (eating habits, peculiar gait, aversion for contact with other people). The uniform speech of simple eccentric hebephrenics is recognizable here. Patients say something, then repeat it after having said something else in-between. They complain and accuse simultaneously like simple eccentric patients; in the combination the repetitions are increased to the point of verbigeration. For example, a female patient who had been asked to explain the difference between error and lie and between stinginess and thrift, replied in the following manner: "It is called stingy, thrifty, it is called thrifty, yes, yes, lie is coming back, thriftiness, lie, stingy, lie, comes back again lie, lie, lie." Such talking occurs especially when more difficult questions are asked, so that the patients cannot find the correct answer quickly. In simple eccentric hebephrenia the uniformity never occurs to such an extent. The second component must therefore be involved in some essential way. Presumably the taciturnity of autistic hebephrenics is based on the underlying poverty of thought. The autistic component also explains the regularly found hallucinations. In the simple form autistic hebephrenics, though less so than shallow hebephrenics, have hallucinations or pseudohallucinations. These reappear in the combination. The patients complain about the voices without talking about what the voices say. Experimental testing brings out incompleteness and derailings in the thought processes. Both the eccentric and autistic components explain the patients' general seriousness and their mild uncooperativeness. While the affect is never deep, moodiness is common.

The combination of **foolish** hebephrenia with the **eccentric** form has already been described. The other combinations are as follows.

In **foolish-autistic hebephrenia** both pictures are discernible side by side. Like autistic hebephrenics if momentarily left unoccupied, the patients sit with a disinterested and mildly sullen expression. When addressed they look up like a foolish hebephrenic and smile. The alternation between these two attitudes may last throughout the examination. In the progress notes the autism is described as predominant; the patients live for themselves, get close to no one, and remain isolated in their activities. The smiling is not seen on the ward because usually no one bothers to address these autistically withdrawn people. When engaged in conversation they do enjoy responding and may even begin to tell stories; but then non-committal answers are given hesitantly. Patients generally respond to intelligence questions with incomplete answers and thought derailments. General drive is higher than in foolish hebephrenia; at work the patients are often quite industrious as are simple autistic patients. Shallowness of affect is considerable.

In **foolish-shallow hebephrenia** the stereotyped laughter of foolish hebephrenics is evident. It is more marked than in the simple form and may appear as lively laughter. At the same time patients remain more active, drive is not reduced as much as in simple foolish hebephrenia. Here the second component is a clearly favorable influence; intellectual performance is better as well. The shallow component is recognizable in its characteristic hallucinations, which appear just as they do in shallow-eccentric hebephrenia. The patients talk freely about the voices, but report little about their content. This is probably another indication of the pseudohallucinatory character of these voices; the content is not taken totally seriously. No episodic appearance can be determined, nor do the patients admit that the voices are abnormal. Hallucinations are more marked here than in the simple form, including somatic ones, such as the influence of currents on the body. Probably both components of foolish-shallow hebephrenia contribute to periods of moodiness. The latter are more severe here than those in simple-shallow hebephrenia, and involve more aggressivity than is found in the foolish form. Since aggression may be directed at weaker patients, it may resemble the malevolence of foolish hebephrenics. Both components also explain that apart from periodic moodiness the patients not only laugh a lot but are generally content in a shallow way. Under these circumstances patients may express playful ideas of grandeur.

All combinations of **shallow** and **autistic** hebephrenia have now been discussed, except for their combination with each other.

As in foolish-autistic hebephrenia, **shallow-autistic hebephrenia** is characterized by both disinterested refusal and friendly responsiveness. While smiling is rare, responses are mostly given in a mood of shallow contentment. During harmless conversations patients are friendly and like to give information. Unpleasant questions lead them to look away and show a sullen unwilling trait. This uncooperativeness is not expressed very strongly; the face lacks animation. In the combined form hallucinations are particularly marked since they are probably induced by both the shallow and the autistic hebephrenic forms. The patients hear many voices of men and women about which they complain, but again they say little about the content. Sometimes the thought-becoming-loud phenomenon ("Gedankenlautwerden") is reported. Sensations are frequently present: "rays" and "currents" influence the body. Deeper affects are no longer present in the patients, even when they complain about the hallucinations. They are orderly in simple conversations although they often repeat what they have said. The slowness of their talk does not suggest automatic repetition, but rather their thoughts are somewhat slow and they repeat what has been said while waiting for the next thought. In assuming that the taciturn character of autistic hebephrenics is associated with impoverished thought the phenomenon may be readily explained. Answers to test questions are generally logical. But more difficult questions have a tendency to lead to derailment of the thought process.

Combined Systematic Paraphrenias

In describing combined systematic paraphrenias the following order will be observed: hypochondriacal paraphrenia, confabulatory paraphrenia, expansive paraphrenia, incoherent paraphrenia, fantastic paraphrenia, phonemic paraphrenia.

In **hypochondriacal-fantastic paraphrenia** all the symptoms of the fantastic component are found. Absurd ideas are expressed, for example: a relative has died, "but he is running around", or, "the deceased husband of Mrs. Schnee nodded to me". Persons are misidentified, for example, a female patient had known the head nurse earlier as a shoe sales lady. There may be ideas of grandeur; a female patient said that she was "the highest". Hallucinations are present in all senses, but comments about these have to be sorted out from the confused speech of the patients. If precise information is looked for the conversation must be focused on the theme repeatedly until the patient finally makes a plain statement. Considerable pressure of speech also hinders exploratory conversations. Many neologisms appear which are not fixed but arise from the subject at hand. At times hallucinations are described, for example: "loud wirings"; in this form they may be held on to. The genesis of neologisms in connection with the immediate theme was particularly clear when a female patient, presented with the proverb, "No rose without thorns", spoke of "thorn-taking". By asking test questions and thus setting the theme the degree of confusion may be ascertained.

(What is the difference between mountain and mountain range?) "Those are mirrorings, water mirrorings." (Child and dwarf?) "But dwarfs are dwarfs, I have nothing like this in my papers. When they stand beside one another, you see the difference, that it is old people, and the child is a child." (Stingy and thrifty?) "If you want to save something, it must also be fulfilled; and if there is too much money, you have to get other people to find you an apartment." (Make a sentence using the words field, hunter, rabbit) "The hunter shoots the game, the rabbits do run around like a nature preserve." (What does this mean: "Necessity is the mother of invention?") "They also sometimes get rid of people with it and make part-time work." (What is the difference between brook and pond?) "A brook is something lying still, a pond is the same." (Stairs and ladder?) "Stairs and ladder seen from the stand point are the same." – A female patient kept on talking when asked to explain the proverb, "A sparrow in hand is better than a pigeon on the roof"[17], she

[17] Literal translation of German proverb: "Ein Spatz in der Hand ist besser als eine Taube auf dem Dach" (English equivalent: "A bird in the hand is worth two in the bush").

answered without being interrupted: "There are birds, men get them, as they come, and when they know it, then something will have to happen about it some day. We cannot accept all relatives. I don't want to have burdensome people, I want to think of myself. Such a death cult I don't find right, when it gets life juice from other people; that kind of thing is unhealthy, unhygienic. How much should you weigh?"

The last example is a good demonstration of the patient's confused speech. Grammatical order is lost as well. This degree of confusion excludes the possibility of simple fantastic paraphrenia. The influence of the second component is found in the hallucinations, since grotesque body sensations are quite prominent. These are, of course, characteristic of fantastic paraphrenia as well, but do belong to the hypochondriacal form. When both paraphrenias merge such body sensations are likely to increase; patients like to describe them, for example:

„They have already taken everything apart from my rear." – "In my face I am shrunk all together." – "My 'chin-stripe' and my teeth are tied together." – "My lung was thickened." – "They pulled my brain together like a thread, it built itself up again and again." – My legs were raised up, my spine was broken, my joints were dislocated, my blood has clotted."

A statement is often made that organs are absent or have been "operated out". A female patient said, "My belly is quite hollow." The statement, such as heard more often in confabulatory combinations, that foreign objects and living things are in the body, may also be verbalized. Somatic sensations are always accompanied by voices; sometimes patients receive orders telling them what to do. Since phonemes appear in both fantastic and hypochondriacal paraphrenia, intensification in the combination might have been expected. However this could not be demonstrated in the patients. Perhaps the many sensations distract somewhat from the voices. Patients complain about both the body sensations and the annoying voices with a certain degree of indignation. This is probably a trait of hypochondriacal paraphrenia which is characterized by a tendency to complain and to accuse. But deep affect is absent, the protest remains superficial.

The gross thought disorder in this combined form can be explained by the presence of a certain excitation of thought in both fantastic and hypochondriacal paraphrenia. There is some kind of blending as the derailing of one form is combined with the lack of concentration of the other.

The thought disorder of **hypochondriacal-phonemic paraphrenia** is very different. The patients can be fixated on a particular theme only with great difficulty. They repeatedly dodge the issue by using indefinite and general turns of phrases. They might be expected to try to avoid only unpleasant questions, but they do the same with neutral topics. Apparently the combination of the thought disorder of hypochondriacal with that of phonemic paraphrenia leads to severe hindering of focusing on a given question.

At intelligence questions patients tend to hesitate and often repeat the questions as if they had not understood them correctly. Then they try to satisfy the questioner with general remarks, or they simply switch to themes that suit them better. They may state that they have never had anything to do with such questions and can therefore supply no answers. It is like the behaviour of phonemic patients who talk vaguely around the central issue and never come to the point. But the hesitation and the repetition of the questions is reminiscent of the difficulty in concentration experienced by patients with hypochondriacal paraphrenia. In both simple forms it is hard to keep the patients on one topic. In the combination this difficulty is extreme. For example, patients offer following answers:

(What is the difference between a box and a basket?) "A box is a box and a basket is a basket." (What is the difference?) "Box is a box and basket is a basket." (Please say what is the difference!) "About what the box is and the basket?" (Yes!) "The basket is of wicker, there are other baskets." (And the box?) "Box first of all is of wood." – (Make a sentence with the words meadow, child, flowers.) "Flowers grow in the meadow, meadows are wet, you don't have to go pick flowers there." (Field, hunter, rabbit?) "Hunter and field I have no idea too." (Sun, curtain, room?) "I don't know curtain. I have no idea of sun-curtain." (Rider, horse, pit?) "It is not in my history, rider and horse are not in it." (What does this mean: "No rose without thorns?") "I don't know anything, haven't been able to buy any roses yet." (What is the difference between thrifty and stingy?) "If you are thrifty you are not stingy." (Child and dwarf?) "The child is good and the dwarf doesn't hurt anyone either."

It is even more difficult to fixate the patients on specific questions when the topic is unpleasant, particularly when attempts are made to learn something specific about their hallucinations. This must come from the unwillingness to talk. I asked a patient about the voices which according to the medical history he heard.

(Do other people know your thoughts?) "Yes, sure." (How do you know that?) "By telepathy." (How does that work?) "When I talk." (What happens then?) "Always pressure high up." (How do you prove the telepathy?) "I was always at home with the doctor." (How do you know that people know what you are thinking?) "Because I was in Berlin and lived in Wittenberg." (Did the people of Wittenberg speak to you in Berlin?) "No." (What does your place in Wittenberg have to do with people knowing your thoughts?) "I live together with my sister." (With whom are you talking when you talk to yourself?) "With my mother and relatives." (Can you also talk with your voices?) "Not always, seldom." – I asked another male patient about his complaint referred to in his record, that he was being irradiated by a magnetic apparatus. (Is something interfering with your body?) "When something happens I was in the hospital. I once had a rash and the attendant gave me a cream." (Is there a magnetic influence in your body?) "My roommate has a radio, but I go for walks in the park." (Does a magnetic current go through your body?) "I don't spend much time with radio." (Is an apparatus directed at you?) "I am not sure, I am not that well educated." (But you talked about it yourself!) "My job is cleaning, I can't tell you anything about an apparatus."

While a certain refusal colors the answers, it is the thought disorder which is decisive. Proof of this is that patients occasionally talk about their voices and sensations even when not asked about them. A male patient suddenly began to press different points on his body and to complain angrily that he was being pushed here and there, although I had nearly given up trying to find out about his hallucinations. When talking about the sensations, he said yes to my question if he had heard someone calling; he himself used the word "voices". On the ward patients often talk to themselves and complain of being abused.

The body sensations do not seem to have a grotesque character, but this may be due to the patients' inability to express themselves clearly. In any case the combination of somatic and auditory hallucinations points to hypochondriacal paraphrenia. However it is not a simple form. The presence of the phonemic component is indicated by the predominance of voices over body sensations. This is apparent in the patients' self-talkativeness seldom seen in simple hypochondriacal nor in simple phonemic paraphrenia. Patients begin to hallucinate as soon as they are left alone, even during an examination. They look around restlessly, now and then appearing to listen and whispering to themselves. Since two forms of clearly hallucinatory paraphrenias are combined this intensified hallucinating, which is not even interrupted by the examination situation, is comprehensible.

In **hypochondriacal-incoherent paraphrenia** hallucinations characteristic of the simple incoherent form are prominent. While sitting across from patients they are observed to be continually whispering to themselves and looking around distractedly; sometimes they even speak loudly to themselves. When a question is asked it may be unclear if the answer is directed at the examiner or at the voices. When addressed they generally respond with only a quick, distracted look, indicating continued hallucinating.

They also pay very little attention to intelligence questions, and tend to offer vague answers: "I don't know", or "No idea". Generally they say very little, but they can be induced to talk. If they are not especially distracted and do respond to a question, an extreme thought disorder is revealed characterized by incoherence and contaminations, such as found in simple incoherent paraphrenia. This is confirmed by test questions; for example the following answers were given.

(What is the difference between tree and bush?) "Tree is built up, the bush becomes wide." (Child and dwarf?) "Child is educated as a person must be, the dwarf is built out." (Error and lie?) "The clever one builds a plough, and the plough is hitched to the horse or tractor." [Note: German "lie" equals Lüge, clever equals klug, plough equals Pflug – trans.] (What does this mean: "No rose without thorns?") "Above the roses, the flowers aren't any." (All that glitters is not gold?) "How does one write that?" (Child and dwarf?) "I think that is why you are making such a remark, yes, I have a breast here." (Error and lie?) "You shouldn't give any more than that." (Lend and give?) "Is nothing lent and nothing given." (Make a sentence with the words meadow, child, flowers.) "I don't want to hear anything about it. I am not looking for a job here." (Sun, curtain, room) "Oh yes, curtain, colored curtain." (Rider, horse, ditch) "The ring is being carried, they are borrowing money from everyone." ("The apple does not fall far from the tree"?) " I am concerned with my family name and to uphold it." (Child and dwarf?) "I haven't nourished any child, I have to first work my way up unattached." (Stairs and ladder?) "I don't know what stairs are. I can't feed myself outside." ("The morning hour has gold in the mouth"?) "Here the teeth could have been in the mouth long ago, I am clean of everything." (Stinginess and thriftiness?) "Perhaps the gentleman wants that to be paid, but I really don't get it, the money."

The second component of the disorder is contained in the confused verbalizations when remarks about body harrassments are made repeatedly. Patients confirm these when confronted with statements recorded in their medical histories. Such sensations are described in very curious ways. For example:

A bullet is in the head, it must be under the skull. A piece of wood is in the belly between the hips, sometimes right, sometimes left; but it can go out through the skin. The chest is being contracted, the body "screwed together", spurts of current are going through it. There are frogs in the body. – The statement that organs are missing is not infrequent: "We have no more lungs; when we came in, everything was taken away."

In such utterances not only grotesque sensations are being described but also absurd conclusions are drawn, demonstrating the severe thought disorder.

The phonemes of hypochondriacal paraphrenia are like those of the incoherent form. In the patients' hallucinatory distraction they confirm that they hear voices. A female patient responded to a question about voices by pointing to the ceiling and saying, "It's the neigbours." Another female patient stated during the examination: "The voice just said something." (What?) "Something dirty." The female patient with "the bullet in the head" claimed that there was a second bullet "with voices" lower down between her mouth and throat, telling her, for example, which way she should go. New words sometimes appear to describe the hallucinations; neologisms may also appear in the otherwise confused speech.

It is not rare for patients to hear their relatives speaking, therefore believing that they too are in the hospital. Here it is not a matter of misidentification. The patients may also hear dead people talking, but this is not necessarily like the "absurd ideas" of patients with fantastic paraphrenia. Ideas of grandeur are sometimes hinted at.

With regard to **hypochondriacal-expansive paraphrenia** I emphasized in earlier editions that the patients are unwilling to admit their hypochondriacal and auditory hallucinations. This was confirmed by the Berlin investigations.

According to the records a male patient said, for example, that he was exposed to electrical current, that radioactive rays were directed at him and artificially activated his heart; there was a knotlike thickening in his head like a magnet; waves went through his head; rays had strengthened his inner organs; small balls and threads were in his head; Lord Zebaoth was sitting under his rib cage and spoke to him. His auditory hallucinations were clearly described in the record. When he spoke it was as if a voice in his throat interrupted him; it reverberated through his head, coming from people who wanted to influence him; his thoughts were picked up by other people; it was as if his thoughts were already noticed by others and he no longer needed to think; the voices said he was knocked foolish; they scolded him and gave him bad advice. The patient also spoke to himself and once explained, when looking toward the corner of the room, that he had just been called a "whore buck" and a "criminal".

Despite this extensive hallucinating it was difficult to learn anything about it from this patient. He talked about it in his everyday surroundings but denied it in the examination situation when he suspected that I wanted to find something wrong with him. This was genuine dissimulating, not a consequence of thought disorder as in hypochondriacal-phonemic paraphrenics.

The cause of the dissimulation may be found in the expansive component, since these patients consider it important to be respected, not to be declared sick and not to be put on the same level as other patients. In the combination expansive paraphrenia is recognizable in the supercilious attitude. The thought disorder is increased by the connection with hypochondriacal paraphrenia. Many of the patients' answers are totally confused, with accentuation of the speech disorder. They attempt to use careful language which is supposed to sound fine and to demonstrate good education and high status. Neologisms are heard, such as "right-speak-evidence", "silision-disease", "wangalistic reality", "sacrusion" (meaning hypnosis), "lieswiles" (meaning: "lying appearance of seeing through love"). As is true in simple expansive paraphrenia neologisms appear in written notes as well. In both written and spoken words grammatical order is generally lost; this is even more so in the combined form. The following answers were given to test questions.

(What is the difference between box and basket?) "Basket can be for shopping employed, a box can be for tool packing and is as a carry-hold for the receiver sending." (Stairs and ladder?) "You can a ladder for the purpose that you can transport it and pick apples on the tree." (Brook and pond?) "Brook, that is an expression, pond that is a talent." (Error and lie?) "Error, that is where knowledge is not sufficient, lie is the salon life, a noris-life appears in disease appearance." (What does envy mean?) "Envy is striving for the goals of other synthetic callings by another occupation." (Make a sentence with the words meadow, child, flowers) "The meadow is an existence for living children for greater joy." (Rider, horse, ditch) "The rider led the horse to a ditch, for the leading eroticism an existence of recognition."

The thought disorder, intensified by the patients' attempts to express themselves more selectively, is so great that talking with them hardly makes any sense at all; yet in conversation their speech is elegant and reserved. The ideas themselves are similar to those in the simple form – moderately expansive. If the patients do not feel themselves sufficiently respected they become upset and irritated, but deeper affect appears only slightly. The tendency to speech is great and is probably enhanced by the hypochondriacal component.

In **hypochondriacal-confabulatory paraphrenia** confabulations about journeys to other countries or to other stars rarely appear in later stages of the condition. On the other hand body sensations, arising from the hypochondriacal component, are imbued with confabulations. According to the patients both inanimate objects and living beings are inside their bodies, for example: wires, handbags, wallets, papers, bees, "a rabbit and two black dogs", people, "a child crying Mama". Besides organs are allegedly absent or out of the ordinary, such as "the heart is of glass". The experiences may resemble the misperceptions of simple hypochondriacal paraphrenia: ribs are broken, nerves extracted, or the body has been sliced. The hypochondriacal component also explains the phonemes which are found regularly in connection with confabulatory processes. For example, they say that the patients' hands should be tied together in order to look inside their bodies. Sometimes patients report that voices had told them about what was happening inside their bodies. Also people who are allegedly inside the patients speak out of their bodies. They have room because, as a female patient explained, they "roll themselves up". Visual hallucinations are common in the early stages of the disorder. Sometimes the patients claim to see what is going on inside their bodies, in such cases it is unclear whether it is a matter of visual hallucinations or confabulations.

A peculiarity of hypochondriacal-confabulatory paraphrenia is lack of motivation and taciturnity. This cannot be derived from either of the two components. I explain the phenomenon by the fact that in hypochondriacal and confabulatory paraphrenia antagonistic symptoms are neutralized (this is discussed in detail elsewhere), so that a new kind of symptom may develop. Lack of motivation may explain the incomplete thought processes of the patients; it may, however, have to do with weakness in abstract thinking occurring in simple confabulatory paraphrenia. Despite the grotesque fantastic picture more extreme logical aberrations are absent.

Of the combinations of **confabulatory** paraphrenia the connection to a hypochondriacal form has been presented. Other combinations are as follows.

Confabulatory-fantastic paraphrenia retains all symptoms of fantastic paraphrenia including absurd ideas, ideas of grandeur, misidentification and hallucinations involving all the senses. The ideas may assume a confabulatory character in that they do not appear only as assertions but are amplified by concrete descriptions. Yet reports are never lengthy; the confabulations end quickly, and nothing is heard of trips to distant countries.

A female patient said: "But right now I am playing all the films of GDR; I was Marika Rökk. I was previously married to Dr. Detzke. Then I had a couple of million children who are now with their father. You can inquire of him if he still wants to mention my marriage." She then spoke of a doctor who "dissects corpses, whatever is in the hospital", and who was supposed to be her husband. – A male patient recounted: "Once we had such a case. He was already buried. Then they had to dig him up. He then called out, made a noise. Then he lived another nine months." – Another male patient had fought in the battle of Tannenberg; his actions led to the German victory. It did not bother him that he was only six months old at that time.

The somatic experiences are not only grotesque as in simple fantastic paraphrenia, they also include a confabulatory element. As in hypochondriacal-confabulatory paraphrenia the reports are very peculiar.

In the body were supposed to be a goat, a pig, a rat (the latter in the head), gnomes, and a doctor. Even an elephant was mentioned. (How does it fit in?) "I don't know either how it got inside." – A female patient once had the rib of Adam and Eve in her body, but her father took it out.

Without confabulatory components the somatic experiences of fantastic paraphrenics would not attain this degree of absurdity. Often there is a picture-like quality in the patients' descriptions. In the combination form, however, the voices of fantastic paraphrenia are not essentially changed; they are described soberly. Visual hallucinations are perhaps included in the fantastic stories; they cannot, however, be delineated. Misidentification of persons occurs more frequently than in simple fantastic paraphrenia. Ideas of grandeur are unlimited as is usually the case in both simple forms.

Patients like to talk; they quickly become confused if allowed to continue talking. Their remarks soon become so peculiar that they cannot be followed; and neologisms further hinder comprehensibility. It is surprising that many questions in the intelligence test are answered correctly. The confusion is, to a certain extent, only a matter of content, i. e., it is related to the peculiar idea construction. Thought derailments are present during the test as the following examples show:

(The apple does not fall far from the tree?) "That is just the same thing, if you want to compare a mother and a child, for example, then the child often has similar traits and can be judged." (Necessity is the mother of invention?) "When in need, you take difficult things, go over to difficult ways, difficult possibilities." (No rose without thorns?) "Roses bloom on good days, and the thorns take the possibility from people to do difficult things, come with them if pulled in." (All that glitters is not gold?) "What is uncovered, what is covered up, that is the same as the cover itself, ways of acting." (What is the difference between brook and pond?) "A brook is a 'flow', a little flowing stream; a pond is a round or perhaps longish, oval, such a hole, water is in it."

Confabulatory-phonemic paraphrenia is the only form among all the schizophrenias, simple and combined, in which visual hallucinations dominate the clinical picture. Simple confabulatory paraphrenics talk about their experiences in a plastic manner, making it appear that they are reporting the content of visual misperceptions. But the events had actually taken place somewhere other than the patients' current location. On the other hand patients with the combined form are telling what they have just experienced in their present surroundings; it may have been yesterday. What they see is generally astonishing.

A cow flew in the window; a man and three women were thrown in. A female patient repeatedly saw "fire crackers and comic figures". A female patient by means of a film transmitter saw angels in the clouds, men and a horse as well. Out of her body came swords, purses, a testament and a hymn book. Another female patient spoke of shapes she had seen, adding: "I see that when I close my eyes." Still another patient had to watch her brothers being tortured and murdered, and yet remained alive. "They cut out their sex organs, then they give them injections in the body and take out their guts. The sex organs of my oldest brother were thrown in the river Spree." The patient was not surprised that her brothers were still alive despite these disfigurements. Ghosts often came to her bed and once a cat came as well.

This patient reported these horrible things without affective participation, rather as a sensation. It is clear that in the patients' consciousness a distinction remains between objective reality and hallucinations. This was impressively clear in two other patients who spoke of their double lives. One experienced – he said "under hypnosis" – being in quite a different place from reality. At night he woke up and was no longer in his bedroom – once he was on the moon. The other patient, a female, claimed that all people existed as two persons: the real ones and the "invisible" ones. She herself saw the "invisible" ones. On one occasion she reported that she had her experiences in her "subconscious". This is much like simple confabulatory paraphrenics who have their experiences sometimes "in trances", or "in dreams", as they say. In the combined form much that is experienced at night is dream-like. A male patient said, "Then the doctor came; I was gone all of a sudden; then I was operated on under hypnosis; then I was awake again; I had a scar on my left cheek." Since phonemic paraphrenia is an obvious hallucinatory disorder it is clear that the confabulations of the combined form turn mainly into hallucinations.

Besides visual hallucinations voices are very important. The voices know the patients' thought and often repeat them. Thoughts-becoming-loud ("Gedankenlautwerden") is an indication of phonemic paraphrenia. In two cases voices came from the body. From one female patient the voices came from near her collar-bone; from another they came from the stomach. Sometimes the people who appear in hallucinations speak; in general, however, the voices appear independently.

Usually these patients like to talk about their experiences; their mood is one of shallow contentment. Ideas of grandeur, regularly present in simple confabulatory paraphrenia, are hinted at in the combined form. Probably the phonemic component, lacking ideas of grandeur, has a dampening effect.

The thought disorder appears to confirm confabulatory-phonemic paraphrenia since the patients during the intelligence tests tend toward "fuzzy" thought and a proclivity for the obvious. For example, I received the following answers:

(What is the difference between box and basket?) "Basket is a little airier, box is when no air gets in, it is probably stronger than a basket. Baskets don't last long either. If you pack lots of heavy things in a basket, it soon breaks. Once I had a little basket that broke." (Brook and pond?) "The pond stands still and the brook flows. Ponds are still water and the brooks, they flow. Brook is long and thin, and pond is a big pond. At home we too had a pond and a brook." (Thrifty and stingy?) "Stingy, if you don't let yourself buy anything; then you are stingy. We were very thrifty at home. There were thrifty people in our village." (Stairs and ladder?) "The stairs are for going; the ladder is for going, too. The ladder is of wood and lengthy, only the ladder has two supports and may be leaned on the wall." (Error and lie?) "A lie is when I fool someone, in court or maybe with other people, if I tell a white lie perhaps and say, there is nothing and maybe something different."

In **confabulatory-incoherent paraphrenia** visual hallucinations are also found, but these tend to become confabulations. A female patient saw an "iced sugar spleen" in a laboratory; she was going to pick it up, but another patient immediately threw it in the toilet. Still another female patient on one occasion saw "in the evening at six o'clock" the last supper of Jesus and the disciples in heaven. Another time her deceased mother sat at the table "with her madonna-hairdo". Generally the confabulatory aspect is even clearer, as the following examples show:

A man was "chased through the fire place" by burning. – A dead woman was lying there after she had been strangled. – The daughter of a female patient had had her body opened by a strange boy; it could not be repaired; however, no scar could be seen. – Her husband's head had been smashed. Her own tonsils had been taken out because a sewing needle belonging to her brother Alfred was stuck in them. – A female patient was shot at so that pieces of the wall fell out; she herself was not hit because the head nurse had rescued her. – A female patient stood before her grave; but then the priest came and said, "My child, you can go home again."

Like confabulations visual hallucinations may have a somewhat dream-like quality. Only single scenes are reported; combined tales of a confabulatory nature are not heard. This is certainly due to incoherent paraphrenia, the second component of the disorder. One case again revealed the "second consciousness" of confabulatory paraphrenia, for the patient, as she herself declared, was at times "benighted". The incoherent component is apparent in the characteristically continuous hallucinating. During conversation the patients often utter aside whispers; when alone they speak loudly with their voices. These include, according to the patients, every day themes as well as confabulations. More can be learned about these patients' utterings than from those of simple incoherent paraphrenics. The extent of thought disorder appears to be reduced by the confabulatory component. This is perhaps due to the confabulatory component's tendency to more concrete thought, thus keeping incoherence in check. Nevertheless it is difficult in the combined form to carry on an orderly conversation since incoherent remarks intervene or thought mix-ups occur. Neologisms appear as well. Test questions produce the following answers:

(What is the difference between box and basket?) "I should become thinner? I was not in the country, I was in the big city; there you find no box and no basket." (Mountain and mountain range?) "You can take the fire-spewing mountain on Sicily, I come from Silesia, I have learned well." (Child and dwarf?) "From dwarfs I don't want a child, for heaven's sake." (Make a sentence from the words sun, curtain, room) "We never have a curtain, I have no curtain in front of the door." (Explain "The apple does not fall far from the tree") "You can still transplant a young tree, but I would not like to transplant my boy, because he can't marry yet." (Tree and bush?) "A tree is brushwood." (Lend and give?) "You should not lend. This happened to me once and was a warning for me. She got herself discharged from the hospital and did not come back." (Need breaks iron?[18]) "When you are desperate, you can in need try to become iron in spirit; thus the spirit becomes stronger through need." (A sparrow in hand is better than a pigeon on the roof?) "You can't get a sparrow, unless you are a real tomcat which can grab and eat it."

These patients are not only more talkative than simple incoherent paraphrenics, they are also more attentive and can momentarily free themselves from the hallucinatory distraction. Their affect is flat. Their mood is one of indifference – the mildly cheerful manner of simple confabulatory paraphrenics is absent. Ideas of grandeur are less exorbitant than in simple con-

[18] The German proverb: "Not bricht Eisen" was translated literally here because in replying the patient used the words "need" and "iron".

fabulatory paraphrenia, even though even in simple incoherent paraphrenia ideas of grandeur occur. Perhaps patients are so much taken up with their confabulations and hallucinations in their consciousness that there is little room for wishful ideas to come to the fore.

In **confabulatory-expansive paraphrenia** patients' remarks are as fantastic as they are in the simple confabulatory form; they are only slightly less plastic, with more unrelated thoughts being mixed in with them. A female patient listed many countries which she had allegedly visited by plane. Only after she had been asked for details she said that she had ridden a camel in the desert and had met Adenauer in the White House. Delusions of grandeur are often more prominent than confabulations. A female patient reported the following:

Four years ago she had received 50,000 Dollars in gold from the American government, and last year 70,000 Mark from the English government. The money is deposited in Wolgast (her home town). Unfortunately she lost the receipt for the deposit. With 145,000 Mark she bought a home at the Baltic Sea, but stayed only one night there. She then drove to Wolgast where her car was taken away. She was previously married to Lord Hamilton who is, however, unfortunately deceased now. Recently she married Virchow. Long ago she lived in Australia and kept leopards and jaguars as pets. Last year she discovered large copper deposits in England which will last for 100 or 200 years. Recently she invented a 40 million watt turbine which provides current for 13 cities. Furthermore she is awaiting a large inheritance since the philosopher Schopenhauer declared her universal heiress. His estate is valued at 21 million Mark. Her grandmother, née Nettelbeck, visited the Schopenhauer homestead with her.

The fantasy of the tales may extend even further. Apparently some patients are inclined to use absurdities for self-aggrandizement. The character of expansive paraphrenics explains that. A male patient reported:

„Among the Amazons I tied a tire around the neck of a bull. Then it swam across the ocean and in 1951 was in Wittenberge. Now it is in Görlitz. There I built a castle on the Elbe, once the capital of Germany. We also drilled for oil. Then Stalin came and made me a Soviet citizen. I also got the patent for atoms and rockets. When I was 10 years old I succeeded in splitting the atom. I am the leader of omnipotence." – The patient flew to the moon with a "flying saucer", and explained to the inhabitants there that they may not wage war with the people on the earth.

Sometimes patients attempt to relate their confabulatory experiences to a second state of consciousness. A male patient said:

„I was suddenly unconscious and re-emerged in another world." He did not know which world it was since there are many worlds. On one of them he had already been hanged, but "accidentally" was let down again. – Another patient reported that he always got out of bed at night and stayed away until he was "wished back".

The expansive component of the disorder cannot be derived from ideas of grandeur, as these could come from the confabulatory element, but rather from the patients' overbearing manner. They like to pose as teachers; they condescendingly and proudly emphasize all they have experienced and the high position they deserve. Outwardly, they attempt to look very elegant. A female patient used so much make-up and powder that her face looked like a mask and she contracted eczema. If contradicted the patients act hurt and are likely to refuse to answer further questions. On the ward they are presumptuous and shun the company of other patients. A male patient presented written statements with curious stylistic formulations characteristic of expansive paraphrenias; one item reads as follows:

„At night time, of the disturbance of quiet, four times to the lavatory, to use. Coming with groans up the stairs, entering the bedroom in front of my bed. The darkness 'rattling' snake noise so much that you are driven out of sleep – upon which he switches on my half-dead quiet – thereby bringing others to madness – killing the nerve. Shame on the attendant who lets this Rühlemann (a fellow patient) run around in the dayroom – such witching products as criminal elements belong in prison, but not in the dayroom of a sanatorium."

Patients often use neologisms in their stories but less so in experimental tests. This apparently is the influence of the confabulatory component which favors concrete thought. The use of neologisms increases spontaneously when patients try to appear important. A male patient used neologisms "Rombas", "Utiliens", "Pedang", and "Choledang", and was proud that he was asked to spell them. Peculiar verbalizations with grammatical mistakes also occurred during experimental testing, as do lapses of thinking. Some examples follow:

(What is the difference between error and lie?) "An error is the unconscious misrepresentation of a representation, a lie is a conscious misrepresentation." (Tree and bush?) "A tree is a trunk growth, a bush a root growth." (Stairs and ladder?) "Stairs are a nailed container of wood, a nailed object with handrails; a ladder is a rung contraption." (What does this mean: "What little Hans does not learn big Hans will never learn."[19]) "That is probably only in behaviour, to place behaviour." (Make a sentence with the words field, hunter, rabbit) "The hunter informs the proprietor of the rabbit that it will not run over the field." (Car, curve, tree) "The car does not drive any curve to the sidewalk, if a tree is growing there." (Need breaks iron?) "Most probably that a truck drives across iron in dire need, perhaps the weight of the truck breaks through the iron." (The apple does not fall far from the tree?) "If the parents have genetic defects the genetically defective children are not to be quartered in these houses." (Stairs and ladder) "That depends on the language, the ladder can play a ladder itself. Stairs can even be self-employed. But those are rung ladders. The real 'listening languages' tell us nothing."

If the patients are told that their replies are not good, they stop answering questions.

Combinations of expansive paraphrenia with hypochondriacal and confabulatory components have been presented. The other possibilities follow.

As with expansive-hypochondriacal paraphrenia investigation of **expansive-fantastic paraphrenia** is hindered by the patients' unwillingness to talk about the hallucinations reported in their medical histories. Expansive-fan-

[19] Literal translation of German proverb: "Was Hänschen nicht lernt, lernt Hans nimmermehr." The nearest English equivalent is: "You can't put new wine in old bottles." Another is: "You can't teach an old dog new tricks."

tastic patients differ from expansive-hypochondriacal paraphrenics in the absurdity of their ideas and in their pressure of speech.

The patients whom I saw took charge of the conversation immediately after having been greeted. In the Berlin cases this occurred in the following fashion.

Before I could even pose a question, a male patient said: "First I am Pseudo, no name, Kaiser Wilhelm, Friedrich Barbarossa, Rothschild, John Glason, Roman I, the Kaiser' s number, judge of the world, prosecutor of the world, Scotland Yard, London, England I, I want to be transferred to Buskau."

Another male patient said: "The whole institution belongs to me, it is no longer called hospital, it is called 'Hydryaffilelibarotkü', reminiscent of villa. What does this talk mean? Here I am in charge, have been called president against my will. Earlier I was 'star-state-president', now I am President of the Organization of the United Nations. There is the order of the materialistic group of peoples, which is a part of the 'immortality people', for example it is incomprehensible that heaven dies, nor the forest, heaven as people and a folk group cannot die either; that fire dies is incomprehensible, likewise water. The organisation has proven to be inadequate. We have permitted them to communicate to us their wishes. We were not willing to maintain the faulty organisation."

When asked for her name a female patient replied: "I am called Wodrig, else I am called Andreani, I am the Soviet star. Dr. Heber (unit physician) is my young son. My mother had two sons first, then she had a child, Gregor Macheina, Andreani, that is my husband. I am here and am called Kaiser Elias Augustus. My husband is alive, that is my son. I finished off Stalin. They call me Maja Kmakurs. I was born in Italy. My father is Kaiser Friedrich the First, the Great, he had been dead since 1938."

Another female patient said at the beginning of the interview: "My fate lies in the Pope's hands. I am reborn Goddess and the Pope wanted me kept in institutions. There are Gods and there are Popes. Pope Pius XII is the Pope who is working on my rebirth together with the good Lord. By praying I must overcome the world, to regain my divine qualities. I can change leprosy up to 6–7 milligram. Leprosy is a 'first-on-disease', it can be set off by 'eye-touch'."

In addition to severe confusion and many neologisms ideas of grandeur and ideas of the most absurd kind are very obvious here. Also appearing are misidentifications, eventually found in all cases. From remarks made by patients visual, auditory, and somatic hallucinations are evident as well. On inquiring further about the misperceptions the patients generally ignore the questions or offer confused verbalizations. Now and then however answers are more relevant. When asked about "pictures" a male patient replied: "With my spiritual eye I see flower shapes, landscapes, and heads coming towards me." Affirmative answers to questions about "voices" were more common. Pejorative, humiliating words may be heard. There may be continuous "talking-through", men and women talk; there is a "telephone line". Another male patient on being asked about voices replied, "When I have radio communications it is probably called that in medical language." Furthermore somatic hallucinations are described, occurring both spontaneously and in response to questions. A female patient said her arms and legs had been taken off and replaced by artificial flesh. A male patient claimed his head was "cut small". Another patient said "quantities of current" were influencing his body to "keep-it-alive". An injection allegedly "deformed" a female patient "to child-like forms". Dirt was pushed into her mouth and throat.

Beside the fantastic the expansive component is also very clear, as manifested by many often quite artificial neologisms. The patients' general behaviour confirms the presence of this component. Patients all display the characteristic haughty, supercilious, and overbearing manner which sometimes becomes an affected strutting. A male patient produced characteristic written items, agrammatical and unimportant in terms of content: "Valued colleague! Please you the Central Radiation Institute, Berlin. Bills are paid." Somewhat surprisingly the patient's confusion is much less evident during experimental tests. Some intelligence questions are answered appropriately. Presumably the confusion of verbalizations arises largely from the abnormal contents and not simply from the thought disorder. Yet in the experimental test there are still many thought and verbal errors. Some examples follow:

(What is the difference between tree and bush?) "No difference at all, the bush may have picked itself from the tree." (Child and dwarf?) "If you bring a child up it does not remain a dwarf." (Stairs and ladder?) "Stairs have handrails; a ladder cannot always been carried." (Brook and pond?) "A brook is the eclipse and in the moveable. That is the swamp in the undergrowth." (Tree and bush?) "It is a question of size and of wealth of fruit of the protection possibility for the environment. The tree has a rocket trunk, a technical trunk and an artificial crown. The bush is predominantly artificial and has conduction pipes." (What does this mean: "Necessity is the mother of invention?") "That you go beyond existing, self-recognized forms." (Child and dwarf?) "Child is forbidden, dwarf is forbidden, because they are malformations. I have been born a child several times." (Tree and bush?) "In its scientific girth the tree is schizophrenic."

In **expansive-phonemic paraphrenia** the supercilious posture of simple expansive paraphrenics is also present. For example a male patient remarked: "This is no hospital, there are people who should have been hanged long ago. I am here as professor, general and major. I don't have to work for the loafers." As in the simple form ideas of grandeur are moderate, although at times elaborated in a somewhat confabulatory manner. A male patient said he had important relatives and was a descendant of the "Maharaja-line". The ideas lack fantasy without suggestion of a confabulatory component. A male patient produced characteristic written items: he wrote endless lines allegedly containing some secret.

Furthermore the patients constantly hear voices discussing everyday things, sometimes confirming ideas of grandeur or contradicting the patients. While expansive paraphrenics in other combinations tend to be silent about the hallucinations, expansive-phonemic patients sometimes report their voices proudly because they believe them to reveal a new capability. A male patient said that by means of voices he could make himself understood all over the world. Probably these patients admit to the voices more readily because here the voices fit in sensibly, while in hypochondriacal and fantastic paraphrenia they appear as senseless insertions into the thinking process. Sometimes the patients talk to themselves and appear to be conversing with their voices. To the second component of the combined form may also be ascribed visual hallucinations. One male patient saw "with his eyes closed", another saw forms that were "dead and white".

While expansive-phonemic paraphrenics are orderly in simple conversation, verbal peculiarities may be mixed in; apparently they are not prominent because of the phonemic element. In experimental tests this type of "fuzzy" thinking is found. Patients talk around the point and produce peculiar verbalizations. Some examples follow:

(What is the difference between brook and pond?) "The brook is liquid" – corrected: "flowing". (Error and lie?) "Lie is something that you say, the untruth, that is lie. Error, you can make a mistake." (Make a sentence with the words car, curve, tree) "The car must observe the curve, so it will not 'smash' a tree." (What does this mean: "The apple does not fall far from the tree?") "The father is supposed to be a crook. The child would be brought up just that way, or in heredity." ("Who sets a trap for others will fall in it himself?") "That is a criminal, he will swear or write something or make a false statement. In the end he will be caught red-handed." (Mountain and mountain range?) "A mountain range can be of greater extension, there are many high chains with different 'elevation data'. A mountain is often just alone, like the Zapfenberg in Silesia, it has a small foothill; a mountain is 700 meters high." (Child and dwarf?) "A child's height can be estimated right away. A dwarf does not grow any further, stays that way; a child attains its proper man-size, woman-size, it is always growing until it is 20 years old." (Error and lie?) "There is the purposeful error, the everyday error, that may be uncovered and rejected. A lie is a shameful remark, an untruth. First a cigarette is taken, then money, then a lie is told, as if it was someone else."

In **expansive-incoherent paraphrenia** the most severe confusion of any schizophrenia is found. In earlier editions of this book I underestimated the severity of the disorder. Here is an example of a patient's response to the question when he entered the hospital.

„With the sending in of the column, which has come down the Tropau. With world-evaluation here and end-evaluation. What can that be, what is in it? The original family castle of Brandenburg is ours, of our origin going back a thousand years. Set up to these things since childhood. For the longest time man lives here in the world, a worn out world course. Gastrological generations. With world-finished lecture, castle and palace, ancient cathedral, are new-built castles, have taken in world view 'seliarisch' world gastrologists with assimilation, that the Altenburg up there, the Eulenburg, now, not that, we are getting done now, my body would be done now, released of world burdens."

Since pressure of speech is present the disorder becomes particularly pertinacious. Confusion is increased by verbal peculiarities with many neologisms, derived from the expansive component. Simple incoherent paraphrenia, too, produces neologisms; however in the combination, the far-fetched quality of the new words points to their origin in the expansive component. Other examples of neologisms are: "Einverleibungswert" (assimilation-value), "Weltrechnungsakten" (world-calculation-acts), "Weltlaufmaterial" (world-course-material), "Abdienstverhältnis" (off-service-relationship), "Zivilamtsfragen" (civilian-office-questions), "Einstammung" (in-stemming), "pflichtgesellschaftlich" (duty-socially), "Armbanduhrbefindebesitz" (wrist-watch-find-possession). A neologism nearly one line long was found: "Eigenwilligfertigkeitslebensaufrechtungserhaltungswillen" (wilful-completeness-life-maintenance-will). Such constructions clearly indicate that the expansive paraphrenic's desire is to appear important. There are neologisms that just appear in conversation and point out the thought disorder. Test questions rarely receive sensible answers. Severe incoherence is evident, but there is a constant mixture of heterogeneous concepts such as is also found in simple incoherent paraphrenia. Grammatical order disintegrates even more than seen in the following examples since when writing them down one attempts to correct the severe errors somewhat.

(What is the difference between stairs and ladder?) "We are totally valued world imaginations longer than we live, than we need." (Child and dwarf?) "That is monopoly, when they are with the conditions, then people are there." (Box and basket?) "An object suffices like the original stem." (Box and basket?) "Earlier the box had an application in civilian, the basket is an application." (Stairs and ladder?) "Stairs are for everyone who was in building activity." (Give and lend?) "Give is a relationship, what we can understand; whoever lends must know why." (Make a sentence with the words car, curve, tree) "The car remained very sick with me." (What does this mean: "The apple does not fall far from the tree?) "The instemming is still to be found, the apple is not in the house." ("A sparrow in hand is better than a pigeon on the roof.) "The sparrow stops the self-service, the pigeon not, service positions, that has international service." (Box and basket?) "Basket is necessary for it, bulkhead, 'young-nickel'." (Stairs and ladder?) "Because no appropriate catalogue, always catalogue written first."

After each question the patients would have gone on talking in their confused manner had they not been interrupted. As in simple incoherent paraphrenia it may appear that the patients, so severely incoherent, are not responding to the questions, but rather to their hallucinations. However they pay much better attention to the examiner than do patients with the simple form. The hallucinatory distraction begins only when the examiner ceases to talk with the patient. The expansive component, by which in the demonstrative form patients seek a value judgment from others, is responsible for a better outward connection. Pressure of speech is to be ascribed to the expansive component.

Ideas of grandeur emerge from the patients' confused speech; they are even more extreme than in simple expansive paraphrenia. Majestic behaviour is also manifest. Confused speech is presented in a lecturing tone as if it were great wisdom. A male patient raised his finger here and there to further emphasize his teachings.

Hallucinatory distraction occurs when the examiner ceases to converse with the patients and when they have broken off their torrent of speech. They then whisper to themselves and appear to have forgotten the examiner. This confirms the incoherent paraphrenic component.

The combinations of **incoherent** paraphrenia with hypochondriacal, confabulatory, and expansive paraphrenia have now been presented.

In **incoherent-fantastic paraphrenia** there is considerable thought disorder but not as severe as in the combination described above, and less obvious than in simple incoherent paraphrenia. Now and then sensible answers are given. This is perhaps because of the improved attentiveness of fantastic paraphrenics without noticeably increased thought disorder. Nevertheless the experimental test evokes the incoherence, derailments, and contaminations of incoherent paraphrenia. While grammatical order is lost, neologisms are seldom found.

(What is the difference between box and basket?) "You hold the box tight, and the basket lets the water in, if you put it down." (Brook and pond?) "That the water is deeper than a plate." (Mountain and mountain range?) "Because I really can't run the mountain, nor the mountain range." (Child and dwarf?) "The dwarf is a small man, and the child is an offspring. I knew a dwarf; as a big man he became much smaller." (Make a sentence with the words meadow, child, flowers) "The child waters the light flower, but the meadow does not have the man, and the man betrays the woman." (Field, hunter, rabbit) "The rabbit is shot, if it is an animal that has young ones, and the hunter takes the wool and sells it." (What does this mean: "The apple does not fall far from the tree?") "When a father and a mother let themselves in, and leave the children, then the child will be enlightened, what the next man wants of her."

Sometimes a remark is not an answer to a question of the examiner, but rather to a hallucination. A female patient interrupted the test question with the remark: "The little Christian just asked if I was able to save the man." Later she said: "The man was called, looked like a monkey fur, quite odd like a monkey, ran back and forth, served cups."

Although patients hallucinate constantly they are more attentive than simple incoherent paraphrenics. The fantastic component in which there is good attention is recognizable here. In the combination it is also noticeable from the delusions and the hallucinations. The patients misidentify persons in their surroundings and connect them to absurd ideas. For instance someone is named who has already died, but was seen by the patient in the hospital. Another who had supposedly been shot is serving as attendant. A man was changed into a woman. A female patient had 18 children from 18 men. One of the fathers is a doctor in the hospital. In general, however, the formation of ideas is less obvious than the hallucinations. This is understandable since from the incoherent element a markedly hallucinatory form is combined with the expansive form. Ideas of grandeur, though less prominent, are not lacking. The patients like to talk about the voices they hear, for example, the "hear-voices". They also hallucinate in the presence of a doctor, whispering aside during the interview. Somatic hallucinations have the grotesque character of simple fantastic paraphrenia, at times with absurd elaborations. For example:

An apparatus is placed on the body and has taken everything from the abdomen. Artificial glass tubes have been put in the body. The back of the knee was broken and has been 'taken out'. A female patient said: "I don't have a belly-child in now." Earlier, a child had been placed inside and taken out again. A piece of wood was "formed in" in her leg. A female patient had to walk all crooked because her whole back was artificially stuffed. The connection of a grotesque somatic hallucination with an absurd idea is evident in the remark of a female patient: "They have taken my heart out and made it into a chicken."

In **incoherent-phonemic paraphrenia** likewise there is confusion intensified by pressure of speech. A rational answer is given only now and then. Test questions confirm the severe thought disorder along with incoherence and contaminations; grammatical order is usually lost. Neologisms also appear but are not demonstrative and do not suggest an expansive component; they certainly arise from the incoherent paraphrenia. I received among others the following answers:

(What is the difference between box and basket?) "Both can be packed, but the 'carry-leading' is easier with the basket than the box." (Mountain and mountain range?) "How is that to be understood? You can go up more easily on a mountain, a mountain range is rather more difficult with a lack of air." (Tree and bush?) "The tree has a crown, the bush is a shrub, with 'shadow-leading' the tree." (Make a sentence with the words meadow, child, flowers) "You make the meadow, the grass grows, the grass, the child gets itself a new life, picks flowers, and here is the new air." (Peasant, harvest, rain) "The climate here brings to the peasant a rich harvest, when humidity is just opposite to the state of the sun." (Mountain and mountain range?) "Mountain is a higher landscape, a mountain range is a mountain range, an extra thing." (Give and borrow?) "If one gives something for a birthday, can also be so, and to borrow is, you have to ask the other."

These examples do not indicate that the patients always wanted to continue talking. Having recognized the uniformity of their talk, I let them continue and received among others the following answers:

(What is the difference between box and basket?) "Basket has a handle and box you need a string, that you tie the box, it must be this way." (Brook and pond?) "A brook is something small, narrow, but the pond is very wide, more important than the brook, that is so small, the brook is so small compared to the pond." (Child and dwarf?) "A child is something born, isn't it, has to go to school and always has to go back to its parents. Well, and dwarf, we have seen, little of comprehensibilities of dwarfs, so comes a little, like our children. The children are in care, they can know it, too, yes." (Make a sentence with the words sun, curtain, room) "The sun makes a sunshade with the curtain, then the sunshade is the curtain, right? That is the way it is. Yes, are we finished now? Yes, yes." (What does this mean: "Morning has gold in the mouth") "In the morning man is fresher at work and in thought, fresher, and is worth a lot of money for some. When a man is fresher it is worth the gold." ("Constant drops hollow the rock?") "It depends if it is a sharp substance that hollows the rock. There is no such substance, you can have essence, such a rock does not break, it does not wear out either, you have to have essence."

When the patients stopped talking they did not lower their voices, as if they wanted to continue with their line of thought. This peculiarity indicates the "fuzzy" thinking of phonemic paraphrenics who talk around the point without coming to a conclusion.

In conversation many repetitions are found. In response to the question of whether he still had much to do, a patient replied:

„We don't have much to do, no. It depends, how big the job is, how big the job is. It goes to 100, to 200. It is really all good, is alright, is no rubbish, they don't even know rubbish, you don't get anywhere with rubbish. You want to go on. You still have to go on, the normal, 100 to 200, you want to go forwards, they still plan 100, we have to keep and care it all. If you have a business, everything must be in order, look like it. About forwards, they say, you can expand and improve the business, it can be improved."

Communication with these patients may be quite good despite their confusion, because they have few delusions. The formal thought disorder is not aggravated by disturbance of the content. On the other hand there are gross auditory hallucinations. A male patient said: "They talk day and night." The voices speak of everyday themes, but often say bad things about patients and insult them, so that the patients scold back. "Invisible" people are talking, men and women. Sometimes the voices come from the patients' bodies, which is not unusual in simple phonemic paraphrenics, particularly from the throat. On the ward the patients often argue with their voices. Visual hallucinations may also occur; these are derived from the phonemic component. A female patient saw "weather certificates on the wall". Somatic hallucinations do not occur.

The patients continue to be preoccupied with their voices while being interviewed. They speak and whisper to the side and pay no attention to the examiner. As soon as they are addressed, however, they look up, and their hallucinations seem to stop. We have already seen how attentiveness is better maintained in fantastic-incoherent paraphrenia than in the simple incoherent form. The same holds true in the phonemic form, which – like the fantastic form – is characterised by relatively good attentiveness.

The combinations of **phonemic** and **fantastic** paraphrenia have now been presented, except for the discussion of their combination with one another.

Phonemic-fantastic paraphrenia is marked by confusion which is not, however, as great as it was shown in the earlier editions of this book. Patients do give some sensible answers; a test question may be answered correctly. The confusion of this combination is different from that of other combinations, yet similar to that of incoherent-phonemic paraphrenia, in the tendency towards repetition. After introduction to the examiner a female patient poured out the following:

„Robbery and thief they played, now I have the air-work, now I have taken the men prisoner, those are black eyes, those are thieves. Do I work in Wuhlgarten, brain hospital, Berlin, Germany, yes, and there I have Catholics, Protestants, and godless, Berlin, Wuhlgarten, Germany, I build up and is catholic Germany, I build up. And then we have the Protestants and the godless, I build Germany up, I build up German states, yes, yes. Then there are the three men whom I have taken prisoner now, yes. There I work all day, day and night, and that is air-work, Catholic, German, Protestant, I have everything, build up Catholic states, yes."

These repetitions by phonemic paraphrenics clearly indicate the "fuzzy" thinking which never comes to a conclusion. Derailments of thought, as expected from the fantastic component, are easily demonstrated by test questions. Repetitions are also present.

(What is the difference between child and dwarf?) " What a dwarf is I don't know, a child is a child, are girls and boys, girls and boys in our German society." (What does 'error' mean?) "You have made a mistake in pronunciation or the person." (Lend and give?) "What is lent, you always get back, giving is giving. What you have given, was it a gift? And it wasn't an enemy of the State?" (Make a sentence with the words field, hunter, rabbit) "How should I do that? That is potato earth, grain land, and meadow and hay for the animals, for the milk animals." (Car, curve, tree) "The car is car, curve is to fly round the corner, and I have Catholic corners in Germany." (Brook and pond?) "Pond can be round, a pond can be normal, and a brook is also that way, is also like a pond, brook and pond is also similar, is the same thing, another designation." (Give and lend? "Giving is nice, somebody can give someone something; if you borrow[20] something, you are in want, you are in distress; that is right too." (All that glitters is not gold?) "Yes, that is right too, even if they aren't seeing right, he sometimes sees gold when it isn't gold, it is bitter for unseeing eyes."

All fantastic symptoms are present in the combination form, but sometimes they are somewhat mitigated. Persons are misidentified; a female patient gave me a false name and title. Absurd ideas are expressed, but fantasy is not very great. A female patient captured the French, who had "recently" bombarded Berlin, and brought them to the morgue. The brother of a female patient had no "birth mother" and therefore could not drive a tractor. The star of a female patient stood in the sky for an hour each day. Ideas of grandeur are also moderate; patients speak of an imminent entrance into the government or of famous ancestors, such as Goethe. A female patient claimed to be an "important personality". Such moderate self-aggrandizement is sufficient for simple fantastic paraphrenics. Phonemic paraphrenia, as a purely hallucinatory form, seems to limit the formation of delusions. On the other hand, hallucinating is increased considerably. Simple fantastic or simple phonemic patients are rarely observed during their hallucinations. In the combination, hallucinations are nearly continuous and can be interrupted only briefly if a question is posed. While the female patient, whose confused speech is quoted above, was speaking, she was constantly gesturing defensively at her hallucinations; she wiped her head in front of her face and threw her hands upward. During the rest of the examination she made blowing noises, sometimes accompanying her defensive movements with mumbled words clearly directed at the voices. She wanted to get rid of "sight-disturbance-threads" with the wiping motions. Movements of the arms in front of her body were probably a defense against somatic hallucinations. She elicited different forms of hallucinations; she carried on "air-conversations" with her voices. In the "air-space", she saw not only the "sight-disturbance-threads", but also "men with black eyes" and other faces. Her body was damaged by an electric current causing pain and burning in her back and knee. In addition there was "extracting from the body". If the patient was not addressed, conversation with the voices and defensive movements were ceaseless. Another female patient spoke and gesticulated continuously so that no questions could be asked. It was unclear whether she was talking to the examiner or to the voices. On one occasion she said: "She always collapses, then she goes to Friedrichshain, and is chased by the plane. One of them sits full of lice, another is nasty, one with the pie-injection, the other with the drill." This patient also experienced somatic hallucinations. She felt knife stabs in her head, her hands and breasts were "turned the wrong way", her lungs, heart, gall-bladder and kidneys were poked. On one occasion her head was switched with the head of another patient, on another occasion her

[20] In German "borgen" stands for both "lend" and "borrow".

hands with the hands of yet another patient. These utterances probably reflect the presence of visual hallucinations as could be expected to arise from both the fantastic and phonemic components.

Family Picture of Combined Sytematic Schizophrenias

Since we are concerned with a more precise comparison of psychoses within families, information on disorders derived from patients' records is not included; even extensive medical histories of patients with combined forms are inadequate. Therefore I present only the cases which I saw myself, in order to make a truly differentiated diagnosis. Only combined catatonias and paraphrenias are included since I was unable personally to interview any of the relatives of combined hebephrenics. In no cases was an endogenous psychosis fundamentally different from the proband's disorder found in the relative.

- A male speech-prompt manneristic catatonic had a brother with sluggish-manneristic catatonia. Therefore one component corresponds.
- A male proskinetic-negativistic catatonic had a male cousin with proskinetic-manneristic catatonia. Again one component corresponds.
- A male parakinetic-negativistic catatonic had a female cousin (father's sister's daughter) with simple parakinetic catatonia.
- A male parakinetic-manneristic catatonic had a sister with simple parakinetic catatonia.
- A female confabulatory-fantastic paraphrenic had a son with confabulatory-incoherent paraphrenia. One component corresponds.
- A male confabulatory-phonemic paraphrenic had a brother with simple phonemic paraphrenia.
- A female hypochondriacal-confabulatory paraphrenic had a brother with simple confabulatory paraphrenia.
- A male confabulatory-fantastic paraphrenic had a female cousin with simple fantastic paraphrenia.
- A female confabulatory-phonemic paraphrenic had a mother with phonemic paraphrenia.
- A female fantastic-phonemic paraphrenic had a brother with simple fantastic paraphrenia.

The family of a male proskinetic-parakinetic catatonic is particularly interesting. This patient had a sister and a brother both of whom had simple parakinetic catatonia. If I had not been able to examine them, the family would have had to be listed as an exception, since their mother had affective paraphrenia. One of her daughters, i. e. a sister of the proband, also had affective paraphrenia. The proband's father became depressed in his old age and died soon after our examination from arteriosclerosis. This family will be described in more detail in the chapter on "endogenous mixed psychoses".

Comments

Karl Leonhard investigated systematically large numbers of patients over a long period of time, starting with the by now historic investigations at the district psychiatric hospital Gabersee in Upper Bavaria; he continued in Frankfurt at the Kleist University Clinic, and then in Berlin as Director of the Department of Psychiatry of the Charité. He also carried out extensive investigations in the district psychiatric hospitals of the former German Democratic Republic. In later years he paid special attention to the etiology of chronic schizophrenic disorders as well as the psychopathology and etiology of schizophrenias of childhood and early childhood. Leonhard never deliberately mixed together the findings of the various studies from the different groups of patients; instead he continued to try to replicate and extend his findings and his insights. Because of the political and economic situation in the former German Democratic Republic he was forced – in order to include more recent findings – to assume that the reader of the last edition of his book had a knowledge of his earlier results. For the sake of scientific discussion the latter are now being made available. In order to enable rapid orientation a synopsis of the investigative series with statistical findings follows:

Regarding **"Statistical Findings from Investigations before 1968":** The patients were seen in the university hospitals of Frankfurt and Berlin in the years 1938 to 1968. In the evaluation of one group from Frankfurt were included 117 manic-depressives, 187 patients with cycloid psychoses, 222 with pure depression or euphoria, as well as 337 with unsystematic or systematic schizophrenia (as well as 99 patients from the Research Institute for Psychiatry in Munich). The second group from Berlin included 176 patients with manic-depressive psychosis, 219 with monopolar phasic psychoses, 470 with cycloid psychoses, as well as 499 with schizophrenic psychoses. The findings in this section regarding age of onset, sex incidence, and course were derived from examinations of more than 2000 patients.

Regarding **"Statistical Results from Investigations after 1968"** and **"Etiology of Endogenous Psychoses":** In order to evaluate his hypotheses regarding the significance of psychosocial circumstances in the genesis of endogenous psychoses and the role of heredity in unsystematic schizophrenias Leonhard after 1968 produced a new collection, separate from his earlier series, of 1465 patients. These noteworthy investigations Leonhard undertook in various psychiatric district hospitals, primarily on units for chronic patients.

In investigations of twins during the course of years LEONHARD was able to recruit 45 monozygotic and 47 dizygotic twins; he investigated the twins separately and in pairs according to his differentiating diagnosis.

Regarding **"Early Childhood Catatonia":** In this section LEONHARD discusses primarily 117 patients whom he examined in departments supposedly for "mentally retarded" children. In order to verify his findings he added to the retrospective investigations 109 children whom he then investigated prospectively. Independently from the above he examined patients in the Departments for Child and Adolescent Psychiatry" in the Berlin hospital; he studied the distribution of age and diagnosis of endogenous psychoses of early and later childhood, and he also studied adult patients whose illness began in childhood.

Age of Onset, Sex Incidence, Course

Statistical Findings from Investigations before 1968

Having consolidated my method of differentiated diagnosis, I have undertaken, with the help of my co-workers, a review of the earlier statistical findings. There were no fundamental changes but it was necessary to make some important modifications. To a certain extent, this statistical section has been shortened since, in the mean time, ANGST (1966), and, independently, PERRIS (1966) and later WINOKUR et al. (1969), have published extensive statistics on unipolar and bipolar phasic psychoses, confirming the different natures of these disorders. These authors have, like myself, found that both the **number** and **duration** of the phases as well as the **extent among relatives** were different, although they did not base their reasoning on these differences. PERRIS demonstrated that, in the relatives of unipolar patients, primarily unipolar forms reappeared and that the relatives of bipolar patients suffered primarily from bipolar forms. ANGST examined sex incidence among psychoses in the relatives and found very significant differences. The number of women greatly exceeded that of men among the affected relatives of the unipolar group, but this was not the case for the bipolar group. Frau VON TROSTORFF (1968) confirmed ANGST'S results, even though she did not find the differences quite as great.

Age of Onset, Sex Incidence, and Number of Phases in the Phasic Psychoses (Including the Cycloid)

My Frankfurt studies of the polymorphic forms were based on 117 manic-depressive patients, 77 patients with anxiety-happiness psychosis, 56 patients with confusion psychosis, and 54 patients with motility psychosis. My co-workers in Berlin produced statistical results in 167 manic-depressive patients, 101 patients with anxiety-happiness psychosis, 86 patients with confusion psychosis, and 44 patients with motility psychosis. These numbers give an approximate indication of the proportion of disorders seen clinically. It is unclear whether motility psychoses are actually more rare in Berlin than in Frankfurt or if there was a diagnostic difficulty in differentiating between confusion psychosis and motility psychosis. The number of cycloid psychoses examined in Berlin was increased by a further series so that their number on Table 1 cannot be compared with that of the manic-depressive disorders.

Table 1. Average Age of Onset in Polymorphic Phasic Psychoses

	Number of Patients			Age of Onset		
	Male	Female	Both	Male	Female	Both
Manic-Depressive Illness	55	112	167	38.2	33.3	35.8
Anxiety-Happiness Psychosis	99	95	194	30.4	41.1	35.5
Confusion Psychosis	69	99	168	27.4	28.5	27.9
Motility Psychosis	24	84	108	22.9	29.7	27.7

Table 1 presents the average age of onset based on the Berlin cases. Manic-depressive illness and anxiety-happiness psychosis have the latest age of onset, while confusion psychosis and motility psychosis begin much earlier. Manic-depressive illness displays the widest range of onset; a manic-depressive girl had her first episode at age of 9 years, while another female patient had her first episode aged 70. Of the patients with anxiety-happiness psychosis the two youngest – a boy and a girl – were 14 and the two oldest, both women, were 69 years old at onset. In confusion psychosis, a boy had his first episode at age of 13 years while the latest onset was at 52 years for a female patient. For motility psychosis the earliest onset was at 14 years of age for a boy and a girl while the latest onset was for a woman aged 44 years.

No new statistical examinations were undertaken in Berlin **for pure phasic psychoses.** Table 2 presents average age of onset in the Frankfurt series. As may be seen, many of these pure phasic psychoses are very rare disease forms.

Since the number of single pure forms is very small the results have limited significance. This is particularly true for pure euphorias. It is perhaps noteworthy that the age of onset for the three cases for **confabulatory euphoria** is so young. It might be thought that confabulating was the normal tendency of youth to fantasize. I can, however, confirm that the most severe of

Table 2. Average Age of Onset in Pure Phasic Psychoses

	Number of Patients		Age of Onset		
	Male	Female	Male	Female	Both
Pure Melancholia	12	50	40.5	37.1	37.8
Pure Mania	9	11	39.5	30.0	33.7
Agitated Depression	5	22	42.7	39.0	39.5
Hypochondriacal Depression	11	16	38.4	38.8	38.6
Self-Tortured Depression	6	24	35.6	40.2	39.3
Suspicious Depression	12	16	38.1	36.7	37.3
Apathetic Depression	3	7	38.0	34.4	35.5
Unproductive Euphoria	2	1	28.5	27.0	28.0
Hypochondriacal Euphoria	3	5	39.0	45.8	43.3
Exalted Euphoria	3	0	43.7	–	43.7
Confabulatory Euphoria	2	1	16.5	17.0	16.7
Indifferent Euphoria	0	1	–	35.0	35.0
Total	68	154			

the three cases confabulated as much after the age of 40 years as he had during the onset at 15.

Table 3 presents sex incidence in the polymorphic forms based on the results of my Berlin colleagues.

Note that **anxiety-happiness psychosis** is as common in men as it is in women; the number of men is, in fact, somewhat greater. In the other polymorphic forms the predominance of women, particularly in motility psychosis is considerable.

The sex incidence in the pure forms from the Frankfurt series is given in Table 4.

The predominance of women is greatest in **melancholia, agitated depression,** and **self-tortured depression.** In the **euphorias** on the other hand the difference is slight. Since the numbers for each separate form are too small to be evaluated, Table 5 compares all of the depressive forms with all of the euphoric forms. Pure melancholia and pure mania are included.

The euphoric forms are as common in men as in women; the predominance of women does not hold here. Since euphorias are much rarer than depressions, men also tend more to depressive disorders, though not as much as women. As can be seen in Table 5, women are affected two to three times as often. An interesting question is whether this ratio **within manic-**

Table 3. Sex Incidence in Polymorphic Psychoses

	Male	Female	Both
Manic-Depressive Illness	55	112	167
Anxiety-Happiness Psychosis	99	95	194
Excited-Inhibited Confusion Psychosis	69	99	168
Hyperkinetic-Akinetic Motility Psychosis	24	84	108

Table 4. Sex Incidence in Pure Phasic Psychoses

	Male	Female	Both
Pure Melancholia	12 = 19.4%	50 = 80.6%	62
Pure Mania	9 = 45.0%	11 = 55.0%	20
Agitated Depression	5 = 18.5%	22 = 81.5%	27
Hypochondriacal Depression	11 = 40.7%	16 = 59.3%	27
Self-Tortured Depression	6 = 20.0%	24 = 80.0%	30
Suspicious Depression	12 = 42.9%	16 = 57.1%	28
Apathetic Depression	3 = 30.0%	7 = 70.0%	10
Unproductive Euphoria	2 = 66.7%	1 = 33.3%	3
Hypochondriacal Euphoria	3 = 37.5%	5 = 62.5%	8
Exalted Euphoria	3 = 100.0%	0 = 0.0%	3
Confabulatory Euphoria	2 = 66.7%	1 = 33.3%	3
Indifferent Euphoria	0 = 0.0%	1 = 100.0%	1
Total	68 = 30.6%	154 = 69.4%	222

Table 5. Sex Incidence in Pure Forms Summarized

	Male	Female
Pure depressed forms	49 = 26.6%	135 = 73.4%
Pure euphoric forms	19 = 50.0%	19 = 50.0%

Table 6. Sex Incidence in Relation to More Manic or More Depressive Symptoms in Manic-depressive Illness of Frankfurt Patients

	Male	Female	Both
Predominantly manic	5 = 19.2%	16 = 17.6%	21 = 17.9%
Predominantly depressive	8 = 30.8%	22 = 24.2%	30 = 25.6%
More or less equal emphasis on manic and depressive poles	13 = 50.3%	53 = 58.2%	66 = 56.4%
Total	26 = 100.0%	91 = 100.0%	117 = 100.0%

Table 7. Same as Table 6 for Berlin Patients

	Male	Female	Both
Predominantly manic	2 = 5.9%	7 = 9.1%	9 = 8.1%
Predominantly depressive	20 = 58.8%	50 = 64.9%	70 = 63.1%
More or less equal emphasis on manic and depressive poles	12 = 35.3%	20 = 26.0%	32 = 28.8%
Total	34 = 100.0%	77 = 100.0%	111 = 100.0%

depressive disease is the same. Here however I must use both my earlier and later results because differences were found between the populations of Frankfurt and Berlin, as I have discussed elsewhere (LEONHARD 1963b). The tendency to depression was much greater in Berlin than in Frankfurt. Table 6 shows the results for Frankfurt. Table 7 shows the results of the Berlin investigations of UNGER and VON TROSTORFF.

The difference between the **Frankfurt** and the **Berlin** results (men and women together) is so great that it must be considered to be of significance. The primarily manic cases were twice as common in Frankfurt as in Berlin. For the depressive forms the reverse was true and the differences even greater – in Berlin depression was nearly three times as common as in Frankfurt. The cases with equal distribution at both poles were much more common in Frankfurt. Since I directed the examination in both cases the diagnostic methodology was the same. Thus I must conclude that the population of Berlin and its suburbs tended much more to the depressive pole of manic-depressive illness than did the population of Frankfurt and its suburbs. It might be tempting to relate this phenomenon to the calmer and more serious tendencies of **Northern Germans,** but the comparison would be invalid since temperament and disease are not at all the same, and many periodic depressive patients who have never had a manic phase have a hypomanic temperament. Whether a similar difference in temperament existed between Berlin and Frankfurt I cannot say, since I made no precise determinations in my Frankfurt series. As seen in my work with KORFF and SCHULZ (LEONHARD et al. 1962), there were many hypomanic temperaments in the families of the Berlin manic-depressive patients. I think it is possible that an investigation of the Frankfurt cases would not produce a significantly higher percentage.

Men did not show a greater tendency to manic formations in either **Frankfurt** or **Berlin.** Thus the relationship in the pure forms did not hold here, as seen in Tables 6 and 7. Manic forms in both men and women were always less common than depressive ones.

The Berlin population's greater tendency toward the depressive pole reappears in **anxiety-happiness psychosis.** Table 8 presents the results from Frankfurt and Table 9 those from Berlin. It can be seen that the tendency toward the anxious pole was greater among the Berlin patients than among those in Frankfurt – 68% as compared with 53%. Again there was a greater tendency toward the depressive pole among women, just as in the pure forms (but not in manic-depressive illness). In both Frankfurt and Berlin, women were relatively much more often anxious than men.

There was no difference in the behaviour between the Frankfurt and Berlin groups in **confusion psychosis** or in **motility psychosis.** In both, the excited form outweighed the inhibited, particularly in motility psychosis. Table 10 gives the results obtained by LÀSZLÓ in Berlin. The table also shows that a bipolar course was frequent in both forms.

Table 8. Sex Incidence in Relation to More Ecstatic or More Anxious Symptoms in Anxiety-happiness Psychosis of Frankfurt Patients

	Male	Female	Both
Predominantly ecstatic	1 = 4.5%	2 = 3.6%	3 = 3.9%
Predominantly anxious	9 = 40.9%	32 = 58.2%	41 = 53.2%
More or less equal emphasis on ecstatic and anxious poles	12 = 54.6%	21 = 38.2%	33 = 42.9%
Total	22 = 100.0%	55 = 100.0%	77 = 100.0%

Table 9. Same as Table 8 for Berlin Patients

	Male	Female	Both
Predominantly ecstatic	2 = 4.0%	0 = 0.0%	2 = 2.0%
Predominantly anxious	30 = 60.0%	39 = 76.5%	69 = 68.3%
More or less equal emphasis on ecstatic and anxious poles	18 = 36.0%	12 = 23.5%	30 = 29.7%
Total	50 = 100.0%	51 = 100.0%	101 = 100.0%

It should not be thought that there are only excited and inhibited phases, since both poles may be evident in the same phase. These are not **mixed states** but **alternating states.** Mixed states, in the narrow sense, are absent in cycloid psychoses, since in the typical forms only one psychic area is affected. A mixed state does not exist when ecstasy is combined with akinesia. Rather, psychomotor phenomena in each pole indicate that happiness psychosis extends beyond its own picture and contains traits of motility psychosis. Similarly anxiety psychosis may co-exist with excitement or inhibition when traits of motility psychosis or confusion psychosis are added. Such combinations indicate the polymorphic picture of the cycloid psychoses. In the clinical section I included such accompanying symptoms under the concept of mixed psychoses. In the statistics this would have meant the addition of many more phases of cycloid psychoses. Thus many anxiety psychoses with excited thought or action would have had to be counted as mixed states. Table 11 shows separately anxiety-happiness psychoses in which affect alternates between ecstasy and anxiety, confusion psychoses in which thought shifts between excitement and inhibition, and motility psychoses where there is a shift between hyperkinesia and akinesia.

Furthermore the **number of phases** in the different forms are presented in Table 11, as determined by HEESE and LÀSZLÓ. Manic-depressive illness

Table 10. Sex Incidence in Relation to More Excited or More Inhibited Symptoms in Confusion Psychosis and Motility Psychosis

	Male	Female	Both
Confusion Psychosis			
Predominantly excited	12	16	28 = 32%
Predominantly inhibited	10	13	23 = 27%
Equal emphasis on both poles	15	20	35 = 41%
Total	37	49	86 = 100%
Motility Psychosis			
Predominantly hyperkinetic	4	13	17 = 39%
Predominantly akinetic	2	6	8 = 18%
Equal emphasis on both poles	2	17	19 = 43%
Total	8	36	44 = 100%

Table 11. Average Number of Phases in Polymorphic Phasic Psychoses at the End of Observation Period (in Parentheses Number of Patients)

	Male	Female	Both
Manic-Depressive Illness	3.8 (55)	4.4 (112)	4.2 (167)
Anxiety-Happiness Psychosis	2.4 (50)	2.5 (51)	2.4 (101)
Confusion Psychosis	3.5 (37)	3.0 (49)	3.2 (86)
Motility Psychosis	5.3 (8)	3.3 (36)	3.6 (44)

displays the greatest number of phases. Anxiety-happiness psychosis is close to manic-depressive illness in terms of age of onset, but has the smallest number of phases. Motility psychosis is closest to manic-depressive illness in terms of phases. In the Frankfurt investigations the differences were greater. There I had found an average of 5 phases in manic-depressive illness and 3.3 in motility psychosis.

Whether the inference of the table is valid, that women have more phases in manic-depressive illness and men in motility psychosis, remains undecided. The number of men with motility psychosis was too small to ensure significance. In **Frankfurt,** on the other hand, women with motility psychosis had more phases than men. There was no fundamental difference between men and women with manic-depressive illness.

In the **pure forms** counting phases is easier because there is never more than one kind of phase. There are no phases of the opposite pole, no alternating phases and no mixed phases. Table 12 shows the figures in depressive psychoses.

As the table shows, women have a slightly greater number of phases than men. It is noteworthy that women display a greater number of phases in every single form, except in apathetic depression where the figures were equal. The significance of this result is underscored when we look at the euphoric forms presented in Table 13.

Table 12. Average Number of Phases in Pure Depressions at the End of Observation Period (in Parentheses Number of Patients)

	Male	Female	Both
Pure Melancholia	1.8 (12)	2.6 (50)	2.4 (62)
Agitated Depression	1.6 (5)	1.8 (22)	1.8 (27)
Self-Tortured Depression	1.9 (6)	2.1 (24)	2.0 (30)
Hypochondriacal Depression	1.3 (11)	1.8 (16)	1.7 (27)
Suspicious Depression	1.8 (12)	1.9 (16)	1.9 (28)
Apathetic Depression	3.0 (3)	3.0 (7)	3.0 (10)
Total	1.8 (68)	2.2 (154)	2.1 (222)

Table 13. Average Number of Phases in Pure Euphorias at the End of Observation Period (in Parentheses Number of Patients)

	Male	Female	Both
Pure Mania	3.6 (9)	3.5 (11)	3.5 (20)
Unproductive Euphoria Hypochondriacal Euphoria Exalted Euphoria Confabulatory Euphoria Indifferent Euphoria	4.4 (10)	2.1 (8)	3.4 (18)
Total	4.0 (19)	2.9 (19)	3.4 (38)

The number of phases in the euphorias is markedly higher for men. This is less true for pure mania than it is for the five other euphorias which, because of their rarity, are presented together. This result supports the observation that **men tend more to euphoric disorders than women.** This tendency seems to reappear in the pure forms as an increase in the number of phases.

The number of phases in manic-depressive illness in both Frankfurt and Berlin was considerably higher than in the pure forms. Table 14 presents the Berlin results. The table shows that manic-depressive illness has many more phases than the pure forms do. This is true for both men and women. The difference essentially confirms the separate position of the unipolar forms. This was also found by ANGST and PERRIS.

It might be thought that the greater incidence of family psychoses found in manic-depressive illness would lead to a greater penetrability and therefore to an increase in the number of phases. If this were so, the cases of manic-depressive illness with family psychoses would have more phases than cases without family history. That is, however, not the case. In Frankfurt, cases with family history had 5.1 phases, and cases without family history 5.0.

While the number of phases of manic-depressive illness is greater, **the duration of the phases in the pure forms is much longer.** Earlier I deduced this from clinical observation alone and without statistical proof. Frau HEESE corrected this omission in the Berlin group. As Table 15 shows there are significant differences.

The astonishingly long durations of pure euphorias are due to the many chronic courses which began long before my examinations of the patients. This is particularly true of hypochondriacal euphorias which make up 7 of the 15 cases; some had already been present for many years. Pure depressions also tend toward a chronic course, but not as much as pure euphorias.

Table 14. Average Number of Phases in Manic-Depressive Illness and Pure Phasic Psychoses of Berlin Patients (in Parentheses Number of Patients)

	Male	Female	Both
Manic-Depressive Illness	3.8 (55)	4.4 (112)	4.2 (167)
Pure Phasic Psychoses	2.1 (61)	2.7 (159)	2.5 (220)

Table 15. Average Duration in Months of Phases in Manic-Depressive Illness and Pure Phasic Psychoses (in Parentheses Number of Patients Included in the Calculations)

	Male	Female	Both
Manic-Depressive Illness	6.5 (55)	5.1 (112)	5.5 (167)
Pure Melancholia	7.1 (22)	8.5 (57)	8.1 (79)
Pure Depressions	16.2 (34)	9.3 (92)	11.2 (126)
Pure Manias and Pure Euphorias	46.6 (5)	31.5 (10)	36.5 (15)

I found no chronic course in self-tortured depression. This was true in both the earlier and later investigations, carried out by Frau HEESE. She found a phase length of 5.8 months for self-tortured depression, i. e. hardly longer than for manic-depressive illness.

Table 15 suggests that euphoric phases generally last longer than depressive ones. However, this is true only of pure forms. In manic-depressive illness manic phases are much shorter than depressive, as shown in Table 16. Here it is seen that manic phases of manic-depressive illness are on the average only about half as long as depressive phases. This is another important difference from the pure forms.

The differences in duration of phases may be considered as confirmed by the research of ANGST and PERRIS, even though they both differentiate only between cases with **unipolar** and **bipolar** courses and did not find the differences to be as great as those in the above presentations.

The average duration of phases in cycloid psychoses, as determined by LÀSZLÓ, is presented in Table 17.

The phases in cycloid psychoses are, on the average, shorter than those in manic-depressive disease. This is particularly true of motility psychosis. Since the number of phases is lower in cycloid psychoses than in manic-depressive illness they can be considered **particularly benign psychoses.** This should be emphasized precisely because cycloid psychoses are often confused with schizophrenias.

It is difficult to determine to what extent varying responses to **therapeutic measures** influence different phase lengths. In estimating phase length of untreated patients no clear answer is available. Our observations fell in the period when electroshock therapy still predominated and psychopharmacological treatment was just beginning. On the basis of clinical experience I believe that manic-depressive illness and cycloid psychoses respond well to

Table 16. Average Duration in Months of Various Phases in Manic-Depressive Illness (55 Men and 112 Women)

	Male	Female	Both
Manic phases	2.7	3.3	3.1
Depressive phases	7.6	5.5	6.1
Mixed phases	4.8	6.6	5.8
Total	6.5	5.1	5.5

Table 17. Average Duration in Months of Phases in Cycloid Psychoses (in Parentheses Number of Patients Included in the Calculations)

Anxiety-Happiness Psychosis	3.9 (101)
Confusion Psychosis	3.1 (86)
Motility Psychosis	2.8 (44)

therapy, while pure phasic forms are more difficult to influence. It is particularly difficult to interrupt a chronic state of pure depression or pure euphoria.

Incidence of Psychoses in the Families of Phasic Psychoses (Including the Cycloids)

The higher family history of psychoses in bipolar than in unipolar forms has been confirmed by ANGST, PERRIS, and others. In the subdivision of psychoses we go beyond the findings of these authors. In observing manic and depressive phases we can decide not only from the course whether they are unipolar or bipolar, but as early as the first phase we believe we can decide if a basic tendency to bipolarity exists or not. From the clinical picture of the phase we can separate the "polymorphic" manic-depressive illness from the "pure" manias and "pure" depressions. In addition there are the equally "polymorphic" cycloid psychoses with their course in phases of a bipolar nature. In Table 18 are presented the results Frau VON TROSTORFF obtained in Berlin.

Table 18 shows that the family history of manic-depressive illness is **more than twice as high as that of the pure forms.** Furthermore it is seen that pure depressions and pure euphorias have a lower incidence when compared to pure melancholia and pure mania. This had appeared, to a certain extent, in my earlier work. Moreover **cycloid psychoses have a much lower incidence than the equally polymorphic manic-depressive illness.** The incidence is nearly the same as in the pure forms. The correspondence is, however, unimportant since the different clinical pictures disprove the possibility of any connection. In the Frankfurt series this lesser incidence in cycloid psychoses was not yet recognized. Probably at that time I was not able to distinguish them with certainty from periodic catatonia which has a high family history.

Table 19 essentially confirms for parents what we found to be true for siblings. But the difference of incidence between manic-depressive illness and pure forms is not as great as it is for siblings. Pure depressions and euphorias

Table 18. Incidence of Endogenous Psychoses in Siblings of Patients with Phasic Psychoses

	Number of Patients		"Corrected" Number of Siblings		Number of Psychoses		%	
Pure Melancholia and Pure Manias	43	102	64.0	173.5	4	8	6.3	4.6
Pure Depressions and Pure Euphorias	59		109.5		4		3.7	
Manic-Depressive Illness	105		170.5		18		10.6	
Anxiety-Happiness Psychosis	61	135	102.0	170.5	5	8	4.9	4.7
Confusion Psychosis	49		48.0		2		4.2	
Motility Psychosis	25		20.5		1		4.9	

Table 19. Incidence of Endogenous Psychoses in Parents of Patients with Phasic Psychose

	Number of Patients		"Corrected" Number of Parents		Number of Psychoses		%	
Pure Melancholia and Pure Manias	43	102	78.0	187.5	5	10	6.4	5.3
Pure Depressions and Pure Euphorias	59		109.5		5		4.6	
Manic-Depressive Illness	105		190.0		18		9.5	
Anxiety-Happiness Psychosis	61	135	119.0	259.5	8	12	6.7	4.6
Confusion Psychosis	49		94.5		3		3.2	
Motility Psychosis	25		46.0		1		2.2	

also have a considerably lower incidence in the parents than pure melancholia and pure mania. The incidence in cycloid psychoses is nearly as great in the parents as in the siblings. There were clear differences within the cycloid group, since here – as opposed to the situation with the siblings – anxiety-happiness psychosis has a greater incidence than confusion psychosis and an even greater one than motility psychosis. This is certainly due to the difference in time of onset; patients with motility and confusion psychoses start a family less often so that affected parents rarely produce affected children.

The incidence of the pure forms is yet to be determined for **each of them.** Although the numbers are at times too small to be significant I present the results of the Frankfurt investigations. Table 20 shows the findings for siblings.

The variations in Table 20 are not likely due to chance. Melancholia has, for example, an incidence of 6.5% based on 123.5 patients, while agitated

Table 20. Incidence of Endogenous Psychoses in Siblings of the Various Pure Forms

	"Corrected" Number of Siblings		Number of Affected Siblings	
Pure Melancholia	123.5		8 = 6.5%	
Pure Mania	36.5		1 = 2.8%	
Agitated Depression	60.0		4 = 6.7%	
Hypochondriacal Depression	61.0		1 = 1.6%	
Self-Tortured Depression	55.0		2 = 3.6%	
Suspicious Depression	48.0		1 = 2.1%	
Apathetic Depression	20.5		2 = 9.8%	
Unproductive Euphoria	2.5	31.5	0	1 = 3.2%
Hypochondriacal Euphoria	15.5		1	
Exalted Euphoria	7.0		0	
Confabulatory Euphoria	6.0		0	
Indifferent euphoria	0.5		0	

depression, based on 60 cases, has approximately the same incidence. On the other hand hypochondriacal depression, based on 61 cases, has an incidence of only 1.6%. Additional results for the parents are shown in Table 21.

Chance is improbable when the incidence is similar for parents and siblings. The incidence of **pure melancholia** in both cases is relatively high. Also the incidence of apathetic depression is high in siblings and parents; however the absolute numbers are very small here, so that firm conclusions cannot be made. Agitated depression has a high incidence among siblings; suspicious depression has a high incidence among parents. Pure mania and **hypochondriacal depression** have a similarly low incidence in parents and siblings.

Summary: Principal Findings in Phasic (Including Cycloid) Psychoses

In the **polymorphic forms** the average age of onset for manic-depressive illness was 35.8, for anxiety-happiness psychosis 35.5, for confusion psychosis 27.9, and for motility psychosis 27.7 years.

In the **pure forms** the average age of onset varied between 30 and 40 years. Separate figures are found in the earlier editions of this book. In the **euphoric forms** there were not enough cases for definitive evaluation; the average age of onset was somewhat lower than in the depressive forms.

When the Frankfurt and Berlin cases were combined the **sex incidence** in the **polymorphic forms** was as follows: manic-depressive illness: men 75, women 152; anxiety-happiness psychosis: men 137, women 122; confusion psychosis: men 110, women 131; motility psychosis: men 49, women 142.

It is remarkable that there were so many more women than men with motility psychosis; we will later find an explanation for this. That there were more women than men with manic-depressive illness confirms a well known fact. It is noteworthy though that in the related anxiety-happiness psychosis men slightly outnumbered women.

Table 21. Incidence of Endogenous Psychoses in Parents of the Various Pure Forms

	"Corrected" Number of Parents		Number of Affected Siblings	
Pure Melancholia	124		8 = 6.5%	
Pure Mania	40		0 = 0.0%	
Agitated Depression	54		1 = 1.9%	
Hypochondriacal Depression	54		1 = 1.9%	
Self-Torturing Depression	59		3 = 5.1%	
Suspicious Depression	56		3 = 5.3%	
Apathetic Depression	20		1 = 5.0%	
Unproductive Euphoria	6		0	
Hypochondriacal Euphoria	6		0	
Exalted Euphoria	6	26	1	1 = 3.8%
Confabulatory Euphoria	6		0	
Indifferent Euphoria	2		0	

In the **pure forms** pure mania occurs only a little less often in men than in women (9:11), whereas pure melancholia occurs far less often in men than in women (12:50). In the **pure euphorias** (without mania) there were even more men than women (10:8), while in the **pure depressions** (without melancholia) the ratio of men to women was 37:85.

An interesting question is whether in manic-depressive illness the depressive pole is more prominent in women. This was not so in my cases. However there were great differences when I compared the **Frankfurt** patients with the **Berlin** patients – men and women combined. The Berlin patients had far more depressive phases (63.1%), the Frankfurt patients only 25.6%. Of the latter 17.9% had primarily manic phases, the former only 8.1%. Of the Frankfurt patients manic and depressive phases occurred in approximately equal proportion in more than half of all cases (56.4%), whereas in the Berlin patients only a little more than one quarter (28.8%) had this proportion. Such great regional differences are remarkable. This has to be taken into account in separating bipolar from pure phasic psychoses using as sole criterion the course of the disorder. In spite of the differences in Frankfurt and Berlin the depressive pole was generally more prominent, as the figures show.

Similar circumstances, to an even greater degree, apply in **anxiety-happiness psychosis,** considering the manic pole as equivalent to the ecstatic, and the depressive pole as equivalent to the anxious. A predominantly ecstatic form in Frankfurt was seen in only 3.9% of cases, in Berlin even less in 2%. Predominantly anxious forms in Frankfurt were seen in 53.2%, in Berlin in 68% of cases. Approximately equal proportions of both poles in Frankfurt were found in 42.9%, in Berlin in 30% of cases. Between men and women there too was a difference: both in Frankfurt and Berlin the anxious pole was more prominent in women: in Frankfurt the ratio of women to men was 58.2% to 40.9%, in Berlin 57% to 43%. Here the circumstances were not the same as in manic-depressive disease.

With the regard to **confusion psychosis** and **motility psychosis** I could not find any difference between Frankfurt and Berlin. In both places the excited pole was more prominent than the inhibited pole, although in confusion psychosis only slightly more so. For our last comparison the ratios of excited to inhibited phases were 32% to 27% in confusion psychosis, and 39% to 18% in motility psychosis. In confusion psychosis as well as in motility psychosis women more than men tended towards the excited pole.

At the end of the investigation I calculated the average the **number of phases** of the **polymorphic forms.** The number decreased from 4.2 in manic-depressive illness to 3.6 in motility psychosis to 3.2 in confusion psychosis to 2.4 in anxiety-happiness psychosis.

The **average number of phases** in **pure phasic psychoses** is less than in manic-depressive illness. In the Berlin series the ratio was 2.5:4.2; among the former the euphoric forms had 3.4, the depressive forms 2.1 phases.

While the number of phases in manic-depressive illness is greater, **the duration of the phases** is significantly longer in the pure forms. In manic-depressive illness the average duration of manic phases is 3.1 months, of depressive phases 5.5 months. In pure melancholia it is 8.1 months, in pure depressions 11.2 months. The phases of pure mania and of pure euphoria last even longer, some take a chronic course.

Pure phasic psychoses generally tend to have a **chronic course,** the euphoric forms more than the depressive. The rarest form with a chronic course I found to be self-tortured depression. The duration of a phase was only 5.8 months, hardly any longer than in manic-depressive illness. While in the latter manic phases are considerably shorter than depressive phases, in the pure forms the opposite obtains. A chronic course was most often seen in hypochondriacal euphoria. Some time ago I reported this in more detail and at that opportunity described the impressive clinical pictures of pure euphorias (LEONHARD 1981a).

A **chronic course** should not arouse the suspicion that the diagnosis of a pure phasic psychosis is no longer justified. There is no progression whatsoever. As is usually the case in these forms the disorder begins acutely and rapidly attains its characteristic picture. Thereafter there is no further change even when the disorder becomes chronic. This is very much like the course of chronic mania which in fact in most cases does belong to the pure forms. I used to think that this applied to all cases, but then found manic patients with a chronic course among some manic-depressives. In any case a periodic course may precede chronicity. Similarly, pure euphorias may have at first a periodic and then a chronic course.

In chronic mania of the pure form not infrequently there is a transition to a chronic hypomanic state. In reverse it may develop out of a hypomanic temperament – it may not be clear at what point the temperament becomes a disorder. It may even happen that a clear distinction between temperament and disorder cannot be made when there is a chronic course. Transitions between temperament and disease occur quite often in pure forms. In pure melancholia likewise it may not be possible to determine if the disorder continues or abates into a hypomelancholic temperament.

I wanted to know if cases of manic-depressive illness with **family history** perhaps had a more severe course, i. e. had more phases. I checked the Frankfurt cases and found no difference. Patients with family history had an average of 5.1 phases, patients without had an average of 5.0 phases.

The influence of **therapeutic interventions** is difficult to evaluate. Our recent clinical experience shows that on the whole manic-depressive illness and cycloid psychoses respond well to pharmacotherapy, whereas this is not the case with pure phasic psychoses, especially those with a chronic course.

Age of Onset, Sex Incidence, and Course in the Schizophrenias (Investigations before 1968)

In recent years the Berlin research has added much to the statistical findings in studies of the schizophrenias. 499 cases were included in statistical evaluations. We had become particularly interested in the question of heredity and did not repeat all the Frankfurt investigations. Age of onset was not determined in the Berlin cases, but Table 22 gives the results for Frankfurt.

The fact that simple hebephrenia seems to have a lower age of onset than the combined forms is probably a chance finding because of the small size of the sample. In both of the other groups, the simple and combined forms have about the same age of onset. In periodic catatonia early onset is noteworthy. I did see some late onset periodic catatonias which were not included in the series. This does not change the fact that many cases of **periodic catatonia** begin very early, often before the age of 20 years. **Affective paraphrenia** shows a somewhat earlier onset than other paranoid forms. In **cataphasia** the onset is still earlier. The actual age of onset is probably even lower than the table indicates. In the Berlin cases the age at first psychiatric admission was determined. Cataphasics had an average age of 28.7 years at first admission. The Berlin figures are higher because most patients do not enter a psychiatric hospital as soon as the illness begins. For **periodic catatonia** the Berlin results show an age only slightly greater than the Frankfurt results (24.2 to 22.6 years). This may be due to the often abrupt onset requiring immediate hospitalization. In **affective paraphrenia** the age of the patients at the first admission in Berlin was much later than the age of the patient at the onset of symptoms in Frankfurt (42.3 to 31.1 years); this may be due to its at first mild and slowly progressive course.

Sex incidence was determined in Frankfurt and Berlin in the same manner so that the results in both investigations were combined (Table 23). Slightly more women were examined in Frankfurt but this is unimportant. In Berlin 499 statistically evaluated cases were included (249 men and 250 women).

In **hebephrenia** there is no significant difference between men and women. In contrast, systematic catatonia is much more common in men. Even more noteworthy is by how many the female systematic paraphrenics outnumber the male. This confirms the impression gained in institutions with primarily chronic cases. The men's wards are characterized more by dull, autistic patients, whereas the women's wards have more hallucinatory delusional patients. It is also worthy of note that women predominate in **peri-**

Table 22. Average Age of Onset in Subdivisions of Schizophrenia

	Years
Simple Systematic Hebephrenia	22.2
Combined Systematic Hebephrenia	26.2
Simple Systematic Catatonia	24.5
Combined Systematic Catatonia	24.9
Simple Systematic Paraphrenia	35.8
Combined Systematic Paraphrenia	34.3
Affective Paraphrenia	31.1
Cataphasia	29.5
Periodic Catatonia	22.6

Table 23. Sex Incidence in the Various Schizophrenias

	Male	Female	Both
Simple Systematic Hebephrenia	49	46	95
Combined Systematic Hebephrenia	11	16	27
Simple Systematic Catatonia	89	51	140
Combined Systematic Catatonia	40	29	69
Simple Systematic Paraphrenia	71	103	174
Combined Systematic Paraphrenia	29	46	75
Affective Paraphrenia	25	69	94
Cataphasia	57	19	76
Periodic Catatonia	39	48	87
Total	410	427	837

odic catatonia. This fact contributes to the greater unrest in the female wards than in the male wards (at least before the introduction of neuroleptic treatment). The concentration of cycloid psychoses in women also plays a role here. The difference between men and women is particularly great in **affective paraphrenia** and **cataphasia.** These differences should be kept in mind when considering variations in the manifestation of schizophrenic symptoms. We will return to this point later.

Within the systematic forms differences in sex incidence are found as well, even if the samples are at times so small that the results are perhaps accidental. The results for hebephrenia are shown in Table 24. The sample is so small that no clear differences can be recognized. This is not the case for catatonia and paraphrenia, as Tables 25 and 26 show.

In **negativistic catatonia** women predominate; in the other catatonias men are in the majority. Since in the first investigation of Gabersee patients,

Table 24. Sex Incidence in the Simple Systematic Hebephrenias

	Male	Female	Both
Foolish Hebephrenia	9	10	19
Eccentric Hebephrenia	11	15	26
Shallow Hebephrenia	11	8	19
Autistic Hebephrenia	15	11	26

Table 25. Sex Incidence in the Simple Systematic Catatonias

	Male	Female	Both
Manneristic Catatonia	19	14	33
Parakinetic Catatonia	19	5	24
Negativistic Catatonia	7	10	17
Proskinetic Catatonia	17	6	23
Sluggish Catatonia	12	10	22
Speech-Prompt Catatonia	18	7	25

Table 26. Sex Incidence in the Simple Systematic Paraphrenias

	Male	Female	Both
Fantastic Paraphrenia	7	17	24
Confabulatory Paraphrenia	11	20	31
Hypochondriacal Paraphrenia	9	16	25
Phonemic Paraphrenia	14	23	37
Expansive Paraphrenia	21	8	29
Incoherent Paraphrenia	7	20	27

women predominated only in the negativistic form of catatonia and, as this was also true in Frankfurt and eventually in Berlin, the result is certainly valid.

In the paraphrenias men predominate only in the **expansive form,** otherwise women are in the majority. This result corresponds to the Gabersee observations. There women also predominated in all paranoid forms except the expansive forms. This was confirmed in Frankfurt. The differences in some forms are quite significant, particularly in fantastic, hypochondriacal, incoherent, and expansive paraphrenias. It is safe to conclude that men and women, or, as may be said, male and female psyches, are not equally disposed towards the different forms of systematic schizophrenia. In some forms connection to particularly female qualities may be assumed, while the expansive form is connected to a particularly male trait: the striving for personal prestige is certainly more characteristic in men than in women.

Table 27. Endogenous Fluctuations Toward Poles in 499 Schizophrenics (Percentages in Parentheses)

	All			No Endogenous Change			
	Male	Female	Both	Male	Female	Both	
Systematic Schizophrenias							
Simple	116	104	220	85 (73.2)	78 (75.0)	163 (74.0)	(75.7)
Combined	45	44	89	35 (77.7)	36 (81.8)	71 (79.8)	
Unsystematic Schizophrenias							
Affective Paraphrenia	12	54	66	5 (41.7)	24 (44.4)	29 (43.9)	(25.1)
Cataphasia	46	11	57	15 (32.6)	2 (18.2)	17 (29.9)	
Periodic Catatonia	30	37	67	0 (0)	1 (2.7)	1 (1.5)	

The fundamental difference between **systematic** and **unsystematic** schizophrenias is also indicated in the different course of the disorders. The fluctuations of the latter toward one pole or the other were determined in our Berlin cases. Frau VON TROSTORFF has reported this: the fluctuations were only minor; some cases, which she had counted as combined systematic, on repeated examination were found to be simple systematic. Table 27 presents the results. Fluctuations towards the "positive" pole were counted when a patient had a clearly excited or euphoric phase which introduced or interrupted the progressive course of the disorder. Fluctuations towards the "negative" pole were counted when the patient had an inhibited or depressive phase. Patients with fluctuations towards both poles were not counted under either rubric. The figures on the far right of the table show a great difference: systematic schizophrenias had fluctuations toward both poles in only 1.6% of the cases, while the figure for unsystematic schizophrenias was 49.4%. The left-hand column is also interesting, since a simple progressive course without fluctuations was found in 75.7% of the systematic schizophrenias, but in only 25.1% of the unsystematic cases.

A simple chronic course in the unsystematic cases is found as Table 27 shows, particularly in affective paraphrenia. Although it is extremely rare for non-fluctuation to be found in periodic catatonia we did find one patient in whom a residual state of periodic catatonia was diagnosed, there having been no evidence of an acute phase.

Depressive fluctuations are not infrequent in systematic schizophrenias. They occur generally during the onset and may be understood, at least par-

Table 27. (continued)

Fluctuation Towards "Positive Pole"				Fluctuation Towards "Negative Pole"				Fluctuation Towards Both Poles			
Male	Female	Both		Male	Female	Both		Male	Female	Both	
5	1	6		24	25	49		2	0	2	
(4.3)	(0.9)	(2.7)		(20.6)	(24.0)	(22.3)		(1.8)	(0)	(1.0)	
			(2.6)				(20.1)				(1.6)
2	0	2		7	6	13		1	2	3	
(4.4)	(0)	(4.5)		(15.6)	(13.6)	(14.6)		(1.8)	(4.1)	(2.9)	
1	3	4		2	15	17		4	12	16	
(8.3)	(5.6)	(6.1)		(16.7)	(27.8)	(25.8)		(33.8)	(22.2)	(24.2)	
6	1	7	(7.6)	7	3	10	(17.9)	18	5	23	(49.4)
(13.0)	(9.1)	(12.3)		(15.2)	(27.3)	(17.5)		(39.1)	(45.5)	(40.4)	
1	2	3		4	3	7		25	31	56	
(3.3)	(5.4)	(4.5)		(13.3)	(8.1)	(10.5)		(83.3)	(83.8)	(83.6)	

tially, as reactive. The table shows depressive fluctuations to be even somewhat more frequent in systematic than in unsystematic forms (20.1% versus 17.9%).

When investigating the Berlin patients we also determined how often they had been discharged from hospitals. Table 28 shows that systematic schizophrenics had an average of 1.5 and unsystematic cases an average of 2.3 discharges. Table 29 presents the cases which remained institutionalized after the first admission, again the difference is significant. It must be remembered that the first admission for most patients took place many years

Table 28. Average Number of Discharges from Hospital (499 Schizophrenics Compared with 470 Cycloid Psychotics)

	Male		Female		Both	
Systematic Schizophrenias						
Simple	2.0	1.7	1.4	1.3	1.7	1.5
Combined	1.2		1.1		1.2	
Unsystematic Schizophrenias						
Affective Paraphrenia	2.8		1.5		1.8	
Cataphasia	2.3	2.6	4.4	2.1	2.7	2.3
Periodic Catatonia	3.0		2.3		2.6	
Cycloid Psychoses						
Anxiety-Happiness Psychosis	2.9		3.5		3.2	
Confusion Psychosis	5.4	4.0	3.6	4.2	4.3	4.1
Motility Psychosis	4.1		5.1		4.8	

Table 29. Number of Patients Never Discharged After First Admission (Among 499 Schizophrenics and 470 Cycloid Psychotics – Percentages in Parentheses)

	Male	Female	Both
Systematic Schizophrenias			
Simple	28 (26.4)	39 (39.4)	67 (32.7)
Combined	25 (45.4)	24 (49.0)	49 (47.1)
Subtotal	53 (33.0)	63 (42.6)	116 (37.5)
Unsystematic Schizophrenias			
Affective Paraphrenia	5 (38.5)	20 (36.4)	25 (36.8)
Cataphasia	8 (17.0)	1 (8.2)	9 (15.3)
Periodic Catatonia	5 (17.2)	10 (28.6)	15 (23.4)
Subtotal	18 (20.2)	31 (30.4)	49 (25.7)
Cycloid Psychoses			
Anxiety-Happiness Psychosis	2 (11.8)	2 (13.3)	4 (12.5)
Confusion Psychosis	0 (0.0)	2 (10.0)	2 (6.1)
Motility Psychosis	0 (0.0)	4 (13.8)	4 (10.3)
Subtotal	2 (5.0)	8 (12.5)	10 (9.6)

back, because we included only "end-states" in our series. However for the unsystematic cases it was not necessarily a matter of an unchanging residual state; fluctuations of the course may result in changes of the patients' condition decades after onset and thus lead to discharges and readmissions.

In the Frankfurt cases the number of remissions in systematic schizophrenias was also determined. Table 30 shows clearly how strongly systematic schizophrenias are characterized by a progressive course. Cases with only one remission are not common and cases with more than one remission are exceptional.

Number of Psychoses in the Families of Schizophrenics

Our research on the appearance of psychoses in families of schizophrenic patients has in recent years been so thorough that I base my remarks on that work alone. Five hundred cases of schizophrenia, observed over many years, have been statistically evaluated. Frau VON TROSTORFF (1975) has reported this. Minor modifications of her presentation are mentioned here; some cases which she had counted as combined systematic after repeated examination have turned out to have been simple systematic. Furthermore, an affected sister was added to the cases of periodic catatonia.

A great difference between systematic and unsystematic schizophrenias is apparent (17.5% versus 70.2%). Since systematic schizophrenias have predominantly chronic courses and the unsystematic forms often have remissions or periodic courses, it is curious that research in heredity, which sometimes considered different courses, had never revealed this difference. Aside from my colleagues only MITSUDA (1967) found a similar difference between "typical" and "atypical" schizophrenias. I think it can be explained why other authors did not arrive at the same results. Cycloid psychoses have nearly always been included with schizophrenias so that periodic forms with low incidence of family psychoses have been mixed in with periodic and remitting forms with high incidence. Table 18 showed that the incidence among siblings of cycloid psychotics was only 4.7%. In Table 32, on the other hand, we see the incidence of 11.2% among the siblings of unsystematic schizo-

Table 30. Course in Systematic Schizophrenias

	Simply Progressive			One Remission			More Than One Remission		
	Male	Female	Both	Male	Female	Both	Male	Female	Both
Systematic Hebephrenias	20	19	39	7	5	12	3	1	4
Systematic catatonias	38	27	65	11	10	21	1	5	6
Systematic paraphrenias	35	54	89	8	18	26	2	5	7

Table 31. Psychoses in Parents of 499 Schizophrenic Patients

	"Corrected" number of fathers	"Corrected" number of mothers	Both	Affected fathers	Affected mothers	Both		Affected fathers in %	Affected mothers in %	Both in %	
Systematic Schizophrenias											
Simple	190.5	214.5	405.0	3	3	6	13	0.7	0.7	1.5	2.2
Combined	82.5	92.0	174.5	2	5	7		1.1	2.9	4.0	
Unsystematic Schizophrenias											
Affective Paraphrenia	64.5	66.5	131.0	0	2	2		0.0	1.5	1.5	
Cataphasia	55.5	57.5	113.0	6	8	14	42	5.3	7.1	12.4	11.6
Periodic Catatonia	54.0	62.5	116.5	8	18	26		6.9	15.4	22.3	

Table 32. Psychoses in Siblings of 499 Schizophrenic Patients

	"Corrected" number of brothers	"Corrected" number of sisters	Both	Affected brothers	Affected sisters	Both		Affected brothers in %	Affected sisters in %	Both in %	
Systematic Schizophrenias											
Simple	202.0	235.5	437.0	4	4	8	16	0.9	0.9	1.8	2.4
Combined	103.0	122.5	225.0	4	4	8		1.8	1.8	3.5	
Unsystematic Schizophrenias											
Affective Paraphrenia	77.0	92.0	169.0	3	15	18		1.8	8.9	10.7	
Cataphasia	67.0	52.0	119.0	5	5	10	41	4.2	4.2	8.4	11.2
Periodic Catatonia	37.5	40.5	78.0	6	7	13		7.7	9.0	16.7	

phrenics. If the two forms are considered together the high incidence in periodic or remitting cases will not show up in the findings.

Table 31 shows how many parents were affected. Again the significant difference between the systematic forms (2.2%) and the unsystematic forms (11.6%) is clear. The difference between simple systematic schizophrenias and periodic catatonia is particularly great (1.5% versus 22.3%). On the other hand, affected parents are rarely found in affective paraphrenia which is an unsystematic form.

Among siblings of affective paraphrenics (Table 32), however, the rate is high (10.7%). The difference between parents and siblings indicates recessive heredity. This will be discussed below. For cataphasia and periodic catatonia there were significantly fewer affected siblings than parents.

Within the systematic schizophrenias there is a higher incidence of combined forms in parents and siblings. This does not invalidate the findings; each form of systematic schizophrenia should be considered separately. With the exception of **parakinetic catatonia** the numbers based on our cases are too small to evaluate.

In the Frankfurt cases it had already been evident that psychoses among relatives were more common here than in other systematic forms. This was confirmed by the Berlin cases. In Berlin I had 16 simple parakinetic catatonics with 32 parents, and – corrected by the WEINBERG method – 22 siblings. There were 2 affected parents and 3 affected siblings. This is much more than for the total groups of simple systematic schizophrenias. Furthermore in Berlin there were 14 cases in which parakinetic catatonia was combined with another systematic catatonia. The patients had 28 parents and 23.5 "corrected" siblings. There were 3 affected parents and 3 affected siblings, both absolutely and relatively more than the simple forms. The greater incidence for combined cases reappears here. What is more important is the fact that, with parakinetic catatonia in the family, many more psychoses develop, be they in simple or combined forms. In Frankfurt the fact that the concentration of cases was not only found in the simple forms but in the combined forms as well suggested that, despite the small sample, it was not merely accidental. Similar observations in Berlin confirmed this idea.

The particular situation of parakinetic catatonia indicates that the group of systematic schizophrenias cannot be treated as a single genetic unit. To the extent that genetic causes are assumed, a difference in the probability of manifestation will be suspected; to the extent that psychosocial causes are considered, different external constellations leading to one or the other systematic schizophrenia may be assumed.

Statistical Findings from Investigations after 1968

The average ages of onset of **systematic catatonia, hebephrenia,** and **paraphrenia** were 23.0, 23.2, and 35.5 years, respectively. In all forms the onset was somewhat later in women than in men. For women the figures were 25.7,

26.9, and 38.0 years, and for men they were 21.5, 21.3, and 31.8 years, respectively.

In **unsystematic schizophrenias** likewise the average age of onset was somewhat later in women than in men. The figures are as follows: affective paraphrenia – average age for both sexes 36.3, women 37.1, men 33.1 years; cataphasia – average age for both sexes 25.6, women 28.5, men 24.3 years; periodic catatonia – average age for both sexes 24.8, women 26.5, men 23.0 years, respectively. This uniform tendency persists, even though affective paraphrenia is much more common in women and cataphasia in men, showing different dispositions in the two sexes. Perhaps the later onset in women is not an objective fact, because women may enter treatment, especially admission to hospital, later than men. At home a man may not be missed as much as a woman, especially if there are children. Moreover the outward appearance of the illness is more noticeable in men. During the time when our patients' illness began, most women were still at home with domestic tasks; problems in the occupation of men often necessitated earlier medical intervention.

In checking heredity, Table 33 shows the frequency of the various forms of endogenous psychoses in **both sexes,** as well as that of each form occurring in parents and siblings. Under the systematic schizophrenias it is seen that paraphrenia is much more common in women, whereas catatonia is much more common in men. There are two exceptions: the expansive form of paraphrenia is more common in men, and the negativistic form of catatonia is more common in women. This was so for the Frankfurt cases as well as the Berlin cases. Taking both series together, the ratio of men to women in expansive paraphrenia was 48 to 20, that in negativistic catatonia 11 to 20. In unsystematic schizophrenias, cataphasia was much more common in men, whereas affective paraphrenia was much more common in women.

With regard to the **course** of the psychoses, in the **systematic schizophrenias** it is nearly always gradually progressive. If remissions do appear they make such a little impression that there is considerable doubt as to whether the pathological process really recedes, or whether only external circumstances have a calming influence. However with regard to the **unsystematic schizophrenias** remissions belong to the clinical picture. In periodic catatonia, as the name implies, intermittent improvement should never be absent. In very rare cases it may happen that on examination a characteristic residual state is found and it may not be possible to determine if and when an acute severe phase occurred. In affective paraphrenia remissions were found in 57.4%, in cataphasia in 71.2% of cases.

The progress of the disorder in systematic schizophrenias apparently cannot be arrested. Modern measures, psychotherapy and social therapy may improve the adjustment of these patients to the normal environment, but according to my observations these measures do not change the pathological process itself. When during the last 15 years many schizophrenics were re-examined I found that the clinical pictures were just as severe as I had found them 50 years ago in Gabersee. As the result of modern treatment only the

Table 33. Number of Psychoses of Parents and Siblings of Patients with Endogenous Psychoses

		Number of patients	"Corrected" number of parents	Affected parents	%	"Corrected" number of siblings	Affected siblings	%
684 Systematic Schizophrenias								
Paraphrenia	♂	96						
	♀	170						
	both	266	511.5	7	1.4	678.0	8	1.2
Hebephrenia	♂	83						
	♀	103						
	both	186	353.5	8	2.3	357.5	8	2.2
Catatonia	♂	150						
	♀	82						
	both	232	437.0	12	2.7	367.5	13	3.5
429 Unsystematic Schizophrenias								
Affective Paraphrenia	♂	25						
	♀	95						
	both	120	231.0	5	2.2	270.5	36	13.3
Cataphasia	♂	120						
	♀	53						
	both	173	327.5	50	15.3	304.0	45	14.8
Periodic Catatonia	♂	67						
	♀	69						
	both	136	249.5	55	22.0	170.0	36	21.2
221 Cycloid Psychoses								
Anxiety-Happiness Psychosis	♂	38						
	♀	27						
	both	65	126.0	8	6.3	143.5	6	4.2
Confusion Psychosis	♂	41						
	♀	32						
	both	73	141.0	9	6.4	79.5	3	3.8
Motility Psychosis	♂	25						
	♀	58						
	both	83	156.0	8	5.1	170.0	7	4.1
131 Phasic Psychoses								
Manic-Depressive Illness	♂	20						
	♀	40						
	both	60	114.5	21	18.3	90.0	18	20.0
Monopolar Depressions and Euphorias	♂	26						
	♀	45						
	both	71	138.0	8	5.8	159.5	5	3.1
Total	♂	691						
	♀	774						
	both	1465	2785.5	191	6.9	2790.0	185	6.6

outer picture was modified here and there. While I was still in clinical practice in no case of systematic schizophrenia did treatment produce a fundamental improvement. Neither shock, convulsive, nor pharmacological treatment had any effect on the basic syndromes. Fundamental improvements with therapy occur only in unsystematic schizophrenias and cycloid psychoses. The condition of "latent" schizophrenia is not to be taken as contradicting the regular progressive course following the onset of systematic schizophrenia, for the pathological process never develops fully; in "latent" schizophrenia it is merely a "weakness of the system". Very rarely it does seem to happen that the disorder really begins and then stands still, without progressing towards a severe end-state. Here it is merely a question of the absence of further aggravation. In some cases of eccentric hebephrenia and manneristic catatonia we found mannerisms persisting as a form of compulsive acts. This may explain some of the "latent" schizophrenias.

In the literature the statement is often made that one-third of so-called schizophrenics recover, another third end with few symptoms and the last third lead to severe defect. The first are really cycloid psychoses, the second include unsystematic, and the third above all are systematic schizophrenias. Psychiatrists these days criticise KRAEPELIN for having given a too unfavorable prognosis in schizophrenia. What he actually did was to diagnose differently. As can be taken from his descriptions he included cycloid psychoses essentially under manic-depressive illness and assigned them a correct prognosis. The remaining cases had a largely unfavorable prognosis. He gave his dementia praecox cases a not too unfavorable prognosis, in fact better than in modern psychiatry which according to Eugen BLEULER and Kurt SCHNEIDER does not separate the prognostically favorable cycloid psychoses and the prognostically unfavorable schizophrenias. Of course the prognosis would be still better if cycloid psychoses were removed from the double grouping and considered as a separate entity.

The Question of Endogenous Mixed Psychosis

For many years the psychiatric literature has contained articles on **mixed psychoses.** It was assumed that schizophrenia as an entity was not infrequently combined with manic-depressive illness; the symptom picture and course of the disorder showed a mixture of both components. This was supposed to have been an explanation for the so-called **"atypical"** psychoses. Attempts were made to show that such patients had inherited one psychosis from one parent and the other psychosis from the other parent. But investigation did not show this to be true. The concept of mixed psychosis has gradually receded and in more recent times no longer plays an essential role. But the concept of **"schizoaffective"** psychoses has taken over and has led to attempts to distinguish these from schizophrenia and the phasic psychoses. In my own investigations the question of mixed psychoses played no important part as I was able to include most of the so-called atypical psychoses among the cycloid psychoses. For a period of time I assumed that, if different endogenous psychoses came together on a hereditary basis, one would exclude the other. But I had to correct this assumption. I can confirm that two psychoses in one person occur extremely rarely. The rarity depends, as I now assume, on the fact that different endogenous psychoses have different causes which generally are not found together in a single person at the same time. It has to be kept in mind that each separate form occurs very rarely, taking into account that the incidence of all schizophrenic disorders in the average population is less than 1%. Such rare disorders could hardly co-occur with another equally rare. I had to abandon the notion of mutual exclusion of different endogenous psychoses from the fact that such combinations may actually be observed. I did see a few in the course of many years.

Before presenting such cases I would like to mention one family described in the previous edition of this book as an example of my former assumption that two different forms of schizophrenia meeting together did not become mixed but one form excluded the other.

The mother had **affective paraphrenia,** her daughter (Hannelore) and her youngest son (Siegfried) had the same disorder. The father suffered from depression of old age and died during this depression after a prostate operation. He had worried greatly about his sick wife and children. One of the sons, our patient, had a form of systematic catatonia (proskinetic-parakinetic). A sister and a brother both had (systematic) simple parakinetic catatonia.

The presence of affective paraphrenia in Hannelore and Siegfried may be easily explained: their mother had the same disorder, their father had a depression of old age, both of them with genetic loading: in affective paraphrenia recessive heredity may be assumed. The presence of only one gene does not always stay hidden, it may be expressed as lability of affect; this happened in the father who with the stress from his sick wife became depressed in his old age. Of the seven children only two had affective paraphrenia, because the latter disorder becomes manifest much less often in men than in women – in this family there were two girls and five boys.

With the systematic catatonias of our patient and his two siblings the question arose whether or not they had been isolated in childhood. We will see (in the following chapter) the importance of lack of communication as a causative factor in the development of systematic catatonias. Because of his activities during the Nazi regime, after the end of the war, the father was taken into custody for two years and in the following years arrested several times. In the small locality where he lived he was treated, as he put it, as an "expellee". It is probably not wrong to assume that the isolation of the father during that time also affected the children. Perhaps the children had to face the contempt to which their father had been subjected. Two healthy brothers of the patient at the end of the war were 15 and 14 years old; their development had probably advanced to the point when isolation could no longer cause damage. The catatonic siblings at the end of the war were 13, 11 and 8 years old. Here the probability was greater that isolation still played a role. It remained unclear why the togetherness of the many siblings did not allow sufficient communication. In any case it may be presumed that the two disorders in the same family had different causes which were different preconditions for the different siblings.

Combinations of two psychoses both of which are polymorphic are difficult to differentiate, as there are transitions between the psychoses. A female patient in a excited phase presumably had a **confusion psychosis,** while in phases of the opposite pole she had **pure melancholia.** Here it was not only a question of a combination of the polymorphic confusional psychosis with the equally polymorphic manic-depressive illness, but rather with **pure melancholia.** Her father had a psychosis which after personal examination I found to be pure melancholia. Her sister according to her medical record had a "pure puerperal depression with lack of drive" which likewise might have been a case of pure melancholia. Her **mother** on the other hand during our visit was exceedingly talkative. To every question we asked her she replied with a long speech about her family and her opinion concerning the illnesses in the family. Very little factual material was gleaned from what she said, but again and again she talked of her own ideas. She had an "excited thought temperament", about which more is to be said later, and in this way might have transmitted her disposition to confusion psychosis to her daughter. The melancholia in the other phases might have been derived from the patient's father.

In one case there was a combination of **anxiety-happiness psychosis** with **systematic paraphrenia.** This female patient went through a phase in which she was extremely anxious and ecstatic, then she was well for 11 years, and then there was an insidious onset of confabulatory-phonemic paraphrenia. The fact that the first disorder was really a case of anxiety-happiness psychosis was confirmed in so far as this patient's sister had a severe anxiety psychosis later ending in suicide. We did not succeed in elucidating the basis of the patient's combined form of paraphrenia.

The combination of two systematic schizophrenias which have been described in detail might have been considered as mixed psychosis. In very rare cases a combination of **three forms** might be suspected. We will come across such cases in our description of childhood schizophrenia. A further consideration is that mixtures only occur within the separate categories, i. e. the paraphrenic, the hebephrenic, and the catatonic forms. Here too exceptions are very rare.

I saw a female patient with **eccentric hebephrenia** beginning at age 16. When I saw her again several years later the clinical picture was paranoid; the diagnosis was **confabulatory paraphrenia,** accompanying the eccentric hebephrenia which had continued uniformly throughout 10 years. Hebephrenic features were not suppressed: affect remained quite flat, monotonous speech was still noted. Her ideas were not expressed in the elevated cheerful mood otherwise characteristic of confabulatory paraphrenia.

Another female patient's psychosis began at the age of 20, diagnosable as **manneristic catatonia** from the medical record. Nineteen years later a paranoid psychosis developed which after personal examination I diagnosed as **hypochondriacal-phonemic paraphrenia.** In this case the patient's mother had had a gradually progressive psychosis leading to hospitalization lasting 34 years. The patient probably had a disposition towards developing paraphrenia from her mother. Because of her mother's illness she was brought up by her grandmother as a single child from the age of 2 years which perhaps played an important role in the development of her catatonia. In the last chapter of this book we will consider the danger of single children developing systematic catatonia.

These examples show that mixed psychoses do exist; individual forms do not exclude others. Their rare occurrence – even in their special forms – is explained by the fact that the different psychoses have quite different causes. Since each special form by itself is rare it can hardly be expected that the causes of two forms meet together in one person. In the last two cases both forms which were combined could be distinguished from one another quite readily as they originated at different ages of the patients' lives; each form started at an age characteristic for that form, i. e. paraphrenia later than hebephrenia and catatonia respectively.

Both these patients show at the same time that the claim that in the schizophrenias there is a constant change of symptoms, at any rate in the systematic forms, is not justified. Among several hundred patients I found only these two in whom a change of symptoms was clearly recognizable. Actually it was not a true change, rather there was one syndrome to which a second was added.

Etiology of Endogenous Psychoses

In spite of very intensive efforts made for decades with regard to elucidating the genesis of endogenous psychoses, very little is known. It is known that hereditary disposition plays a role; it is assumed that psychosocial factors are significant. But it is not known what the nature of the hereditary disposition is, nor which kinds of psychosocial factors are involved. This gross deficiency is, in my opinion, derived from the fact that since KRAEPELIN'S days etiological investigations have been carried out on only one schizophrenia and only one manic-depressive disorder (in the broad sense of the term). The separation at last of manic-depressive illness from monopolar depression has held good; whereas the **"schizoaffective psychoses"** now frequently mentioned have not led to any useful separation, as these include schizophrenic as well as phasic, especially cycloid psychoses. I have had the opportunity in recent years to spend much time conducting investigations and research, with the result that much more is now known than previously.

Above all there are **two circumstances** which have led to new insights. First there is a recognition that gross maternal deprivation, as observed by SPITZ in the case of children cared for in homes away from their mothers, can lead not only to severe mental retardation but also to early childhood schizophrenia. SPITZ describes catatonic symptoms, pointing out "certain unusual postures of hands and fingers", also "extremely bizarre hand and finger movements", and concludes that these children "present a picture of stuporous catatonia". Secondly there is the recognition that in the development from child to adult the influence of **siblings** plays a large role. The absence of siblings or an unfavorable constellation of siblings may give rise to defects in the development of affected children. The way I have classified endogenous psychoses has been an essential factor in arriving at these new insights. I wish to emphasize this very clearly, because if it is not given due consideration, there is a constant temptation to say, "it cannot be so", "other investigations have always had negative results". The findings are certain, they come not from investigating schizophrenia globally, but from special types which behave quite differently from, even in opposition to, one another.

According to my findings the **influence of siblings** on one another is so great that I must right away give a plausible explanation. The relevant events are of a **biological nature,** for they arise primarily in the case of the systematic schizophrenias where there is the least doubt of an organic basis, and which among all endogenous psychoses in their clinical picture and their course show the least relationship with neurotic phenomena. The question arises whether in the **animal kingdom** there is a parallel in the significance of siblings? HARLOW and HARLOW (1965) in their investigations of rhesus monkeys have ascertained that if the young animals were separated and later again brought together there was real damage in their development. But when one compares the duration of the togetherness of young animals and young humans one finds great differences. Young animals rapidly become independent and are looked after by their mothers until they are grown up. When the mother animal gives birth to a new animal the previous one is usually grown up and is often pushed out. Assuming that animals have a psyche young animals do not need any attention other than what their parents have given them. However when a second human child is born, the first is still immature. Considering that mothers used to give birth to second and subsequent children after only 1 -2 years the care of the former had to be shared with the latter, even more so as other siblings arrived. Maturity is reached as late as the age of 20 years, thus there is danger of inadequate development which does not occur from within but depends on the immediate environment of the child, above all on other people. Besides parents, the closest human environment of the child consists of siblings with whom the child is closely connected for years. It would be inconceivable if nature did not take advantage of this long togetherness in making use of alternative relationships through the participation of siblings. In these days of fewer children the situation might be different, but the biological basis continues to exist in spite of social changes. It should be mentioned here that these alternative relationships are not confined to siblings but more generally to other children. Friends are available to a large extent even in our day of small families.

Before I deal with the influence of siblings in greater depth I wish to deal with **the role of heredity** in the clarification of the diagnostic differentiation.

Significance of Hereditary Disposition

A hereditary disposition is likely if one finds the presence of psychoses in the patient's familiy. Table 33 shows the number of psychotic siblings of 1465 patients whom I have examined in the course of the last 15 years in various psychiatric hospitals. The **5 most important groups** to be differentiated are indicated. As can be seen in the table there is a great variance in the number of patients examined. This has nothing to do with the true incidence of the groups, but is derived from the fact that my examinations were primarily carried out on wards for chronic mentally ill patients. My intention was to observe chiefly cases of long duration. In the case of acute psychoses, the 5 groups appear with approximately equal frequency. This I could determine after I had examined 200 patients in the Berlin hospital during my time as Director. I found 21.5% of patients with systematic schizophrenia, 18.5% with unsystematic schizophrenia, 22.0% with cycloid psychoses, 18.7% with manic-depressive psychoses and 19.0% with monopolar affective psychoses. The different numbers in Table 33 therefore indirectly confirm the different prognoses of various types. The systematic schizophrenias have the least favorable course, therefore they were primarily found on wards for chronic patients (684 cases). Unsystematic schizophrenias with their principally remitting course have a much better prognosis, therefore there were fewer patients (429). Cycloid psychoses (221 cases) remit after every phase; one would perhaps expect still fewer cases, since manic-depressive and monopolar psychoses occur less frequently. The cycloid psychoses, as mentioned above, present more severe clinical pictures. Affective psychoses are more often treated in municipal and university hospitals than cycloid psychoses.

The number of psychotic parents and psychotic siblings vary to a large extent, as one can see from the Table. The unsystematic schizophrenias show a high hereditary disposition. It is especially remarkable that **periodic catatonia** has a very high incidence (22.0% parents, 21.2% siblings). Here I assumed a **dominant heredity.** It is true that the numbers are far from what would be expected theoretically, that is, 50%, but the gap is filled if a closer look is taken at the "healthy" parents and siblings. I saw many parents and siblings in their domestic milieu and found several abnormalities. Many of them are mentioned in the medical records of the patients. Thus I could to a large extent fill the gap between 20% and 50% (LEONHARD 1975). Frau VON TROSTORFF (1981) gave further evidence of dominant heredity. She found 17 cases in whom catatonia appeared in direct line in 3 generations, and 5 cases, in which it occurred in direct line even in 4 generations.

In the case of **affective paraphrenia** however there was evidence that this illness is inherited in a **recessive** manner. The evidence for this is that according to Table 33 many siblings had the illness (13.3%), whereas few parents were afflicted (2.2%). Furthermore many years ago (LEONHARD 1950) I described a family in which, of 18 affective paraphrenia cases, no less than 13 occurred in relatives. This frequency arose through inbreeding. Members of this family frequently intermarried in order not to imperil their reputation and their wealth.

In the case of recessive heredity the theoretically expected number of 25 psychotic siblings was far from having been reached, a plausible explanation being that this illness frequently does not manifest itself in men. As can be seen from Table 33, of our patients 95 were women and only 25 were men. Similarly in the close and also more distant relationship of our patients we found far more women than men. The tendency to develop this illness appears to be connected with the female constitution. Additional exogenous influences will be considered later.

Recessivity does not exclude the possibility that only a heterozygous gene has a particular significance. Not infrequently we noticed in healthy parents certain paranoid traits, as well as emotional lability with the tendency to reactive suicide. Both might indicate in diluted form the presence of affective paraphrenia. In two cases there was evidence of induced paranoia: the mothers had partially acquired the delusions of their daughters.

Cataphasia likewise has a hereditary basis as shown by the relatively large number of affected parents (15.3%) and siblings (14.8%). There is no question of recessive heredity as more parents than siblings were affected. Dominant heredity is likely since this condition was found nine times in the direct line of three generations; moreover an affected father with two different wives had two cataphasic children. Other causes, to be considered later, must play a rather important role in the case of dominant heredity in view of the numbers that were found.

Whether male or female constitution is involved at the onset of cataphasia we do not know. According to Table 33 we have far more men than women patients (120:53), but among the relatives, on the contrary, we found more affected women. Among the 50 parents listed in Table 33 there were 19 fathers and 31 mothers; among the 45 siblings there were 20 brothers and 25 sisters. We assume that these contradictory numbers reflect the fact that cataphasia in men is more severe and has a more chronic course, thus it is encountered more frequently in psychiatric hospitals.

In gross contrast to the unsystematic schizophrenias, the **systematic types** show a very scanty hereditary disposition. It was quite remarkable that in the case of systematic paraphrenia there were only 1.4% of affected parents, and 1.2% affected siblings. The percentage is hardly any higher than that found in the normal population (about 0.8%). I have been wondering all my life how the most severe forms of schizophrenia have arisen. This low hereditary disposition I ascertained during my first investigations of patients at the Gabersee Hospital.

It could be claimed that in the case of systematic schizophrenias hereditary did not play any part. But against such a claim was an important finding: affected relatives of patients with systematic schizophrenia, as far as we could judge, showed the same subtype, partly in pure form, partly in combination with other special forms. Furthermore in healthy relatives we found occasionally an at least latent syndrome of the same subtype. It might have been concluded that the subtypes could be inherited in their own form. But in view of the scanty hereditary disposition of systematic schizophrenia this conception could not be retained. I was therefore led to a different interpretation of the findings. There are other factors responsible for the origin of the illness, but the special configuration has an endogenous cause. The endogenous weakness of a system does not by itself produce the illness, but does underlie the precipitation of the illness by an exogenous trauma.

This hypothesis explains the fact that there are strikingly large numbers of combined schizophrenias. At the conclusion of a series of investigations we counted 220 simple and 89 combined cases among 309 cases of systematic schizophrenias. In view of the rarity of schizophrenia as a whole it is very unlikely that two different genes are found in one person and produce two different schizophrenias. On the other hand a weakness of the system which does not lead to illness can arise in many people and can occur in two systems at the same time. Thus I see the external cause as pathogenetic, and the endogenous disposition as pathoplastic.

Of course it could happen that the weakness of the system may at times reach beyond the limits of normality and contribute to the origin of the illness. In favor of this is the fact that combined schizophrenias show a higher hereditary disposition than simple ones. In previous editions of this book I have taken up the matter and given it a different interpretation. Among the combined forms there were 4.0% of affected parents and 3.5% of affected siblings, whereas in simple forms there were only 1.5% affected parents and 1.8% affected siblings. If in one family the weakness of the system occurs several times, the frequency of the illness rises visibly. This can only be explained on a pathogenetic basis.

Furthermore a particular form of schizophrenia, **parakinetic catatonia,** indicates that weakness of the system can by itself contribute to the genesis of the illness. My Frankfurt and my Berlin patients with this form of catatonia showed a higher hereditary disposition than the other forms. The corrected figures from my Berlin series contained five patients with parakinetic and parakinetic-combined catatonia from 102.5 ("corrected") parents, i.e. 4.9%; likewise there were 5 patients of 69.5 ("corrected") siblings, i.e. 7.2%. If parakinetic catatonia is excluded, the remaining catatonias contain 6 patients from 311.5% ("corrected") parents, i.e. 1.9%, and among 280.0 ("corrected") siblings there were 7 patients, i.e. 2.5%. Therefore these remaining patients with catatonia showed no essential difference from patients with hebephrenia. As far as heredity is concerned, the systematic schizophrenias apparently are quite similar in the three groups, as long as parakinetic catatonia is excluded. In the latter case a weakness of hereditary disposition apparently plays an essential role, judging from the higher percentages of affected patients. The reason why parakinetic catatonia is different from the others remains unclear. Perhaps it is questionable, but I find that parakinetic catatonia bears a certain similarity to Huntington's chorea, a disease which has a very clear heredity. I have already mentioned that the system attacked in parakinetic catatonia is only a step higher than the system affected by the chorea.

In the case of **cycloid psychoses** one finds only a scanty hereditary disposition, as shown in Table 33. There is no difference between the three different forms. This is of considerable significance in view of the findings from our research with twins. As we shall see later, there are great differences. In comparing anxiety-happiness psychosis with motility psychosis we see that the famous rule already enounced by GALTON as an important guideline in genetic research, has **no general validity,** i.e. the rule that in the case of monozygotic twins concordance points to heredity, whereas discordance implies exogenous origin. Anxiety-happiness psychosis and motility psychosis according to Table 33 are equally hereditary and non-hereditary, respectively. But the investigations of twins reveal differences. Among 6 monozygotic twins with anxiety-happiness psychosis only one partner was affected, whereas among 11 monozygotic twins with motility psychosis no less than 9 partners were affected. Eleven dizygotic twins with motility psychosis and 4 dizygotic twins with anxiety-happiness psychosis were all discordant. According to the rule of GALTON anxiety-happiness psychosis is supposed to show little heredity, whereas motility psychosis should show an unusually high degree of heredity, which however is not at all confirmed from the figures in Table 33. Affected parents and affected siblings with anxiety-happiness psychosis are found somewhat more frequently than in patients with motility psychosis. Even if the two psychoses were not separated but considered as a unitary cycloid psychosis, among 17 monozygotic twins we still found 10 concordant cases; thus a high hereditary disposition might have been assumed erroneously.

It is noteworthy that the findings of our research with twins in **monopolar phasic psychoses** contradict the rule of Galton. Although here the figures of hereditary disposition are less than those for the cycloid psychoses (according to Table 33) it was remarkable to observe that all 6 monozygotic twins affected with these conditions were without exception concordant. We shall look into this later. Social psychiatrists suggest sometimes that concordance in monozygotic and discordance in dizygotic twins could be explained by the fact that the former are treated by parents and educators in the same manner whereas the latter are treated differently. This is not an appropriate explanation, since in monozygotic twins more clearly than dizygotic twins there is often a "father's child" and "mother's child". Surely this shows the importance of different treatment.

Next to be considered are the different figures concerning the genetics of affective psychoses. There is much higher incidence of psychoses among the relatives of patients with **manic-depressive illness** than among those of patients with **monopolar types,** as can be seen from Table 33. The figures for the siblings of the former are 20.0%, the latter only 3.1%. Other authors found a difference as well, but not to the same extent; according to my opinion, these authors separated monopolar from bipolar forms only after the illness had run its course, whereas more accurately the pure forms of manic-depressive disease should be diagnosed from the clinical picture. Many patients with bipolar disease only show depressive phases. It had been pointed out before that in this connection there were regional differences between Frankfurt and Berlin. It can generally be ascertained from the clinical picture even during the first phase whether a depression is "pure", or whether it is part of manic-depressive illness, and thus the correct diagnosis can be made.

Finally there is an important finding if among the pure phasic psychoses the **euphoric** are separated from the **depressive** ones. Nowadays in general only the depressive forms are classified separately; manic forms are usually included in manic-depressive disease. But according to Table 34 euphoric forms have definitely fewer affected parents and siblings than depressive ones. Therefore, they ought to be distinguished even more definetely from manic-depressive illness with its high hereditary loading, and their inclusion with the depressive types is even less justified. I consider my conception of their independence confirmed in that I had the same findings in my earlier investigations.

Significance of Psychosocial Circumstances

Since from now on psychosocial causes will be considered it may appear that I was not going to maintain my view that the schizophrenias, especially the

Table 34 Psychotic Parents and Siblings in Patients with Monopolar Phasic Psychoses

	Number of Parents	"Corrected" Number of Parents	Affected Parents	%	"Corrected" Number of Siblings	Affected Siblings	%
Euphoric Forms	25	48.0	1	2.1	71.5	1	1.4
Depressive Forms	46	90.0	7	7.8	88.0	4	4.5
Total	71	138.0	8	5.8	159.5	5	3.1

systematic types, are diseases of the brain. But the psychosocial causes which I will point out are of a kind quite different from what has been assumed up to now; the development of brain mental processes is dependent on influences from the external environment. Systems that are not drawn into association with the environment do not mature adequately. This was already shown by the investigations by SPITZ (1945): children in a home suffering from lack of affection stood still in their development. Many actually died from this deficiency. Even **morphologically** deficient maturation from lack of stimulation from outside has been recognized, for instance by SEITELBERGER (1980):

"Not only the maturation of the sensory and motor systems but that of all parts of the brain which necessarily function as a unit is dependent on demands from outside. It has been proved that systems which have been deactivated through isolation do not mature morphologically and therefore remain dysfunctional. This failure cannot be remedied through later attempts to catch up once the critical period of maturation is past... The fine morphological differentiation of the nerve cells and their connections can be shown to diminish slowly up to the end of the second decade of life. They are accompanied by the processes of mental maturation of the individual."

I can rely on such observations, even though after the discovery of certain psychosocial causes I stay with my view that in the case of systematic schizophrenias certain brain systems fail to develop. Here there is no antithesis between psychosocial and organic factors. If the systems are not stimulated adequately, they suffer damage. This damage is so much the greater if there is already a certain constitutional weakness (this “weakness of the system” has been described in detail above). The damage is not immediately manifest because the schizophrenias begin infrequently in the early years of development. Probably the systems fail as they are exposed to the ongoing stresses of life.

In the social psychiatric literature much consideration is given to the presence of disharmony and conflict in the families of future patients. This may be significant for the development of neuroses, but I have no evidence that such circumstances play a role in the origin of psychoses. My findings – I shall demonstrate many – show that a deficiency in the relationship to the environment plays an important part, i.e. lack of communication. To start with here are the results of investigations with twins.

Lack of Communication in the Development of Systematic Schizophrenias

The text up to now has primarily dealt with systematic schizophrenias. What follows will be a consideration in greater detail of the question of communication.

Absence of Systematic Schizophrenias in Monozygotic Twins

Since in the case of the systematic schizophrenias hereditary factors could hardly be demonstrated nor could psychosocial causes of the kind assumed by social psychiatrists be relevant I eventually looked for a solution by **investigating twins.** As a matter of fact the result was a clarification, but of a kind completely different from what I had thought previously. I was looking for as many monozygotic twins with endogenous psychoses as possible in order to examine them personally with the aim of determining the exact type and to find out if the systematic schizophrenias were essentially concordant or discordant. With only slight hereditary disposition discordance might have been expected; but since apart from heredity no causes could be found, concordance had to be taken into consideration. The result was very surprising: I was unable to proceed with the comparison, as among monozygotic twins **not a single case of systematic schizophrenia could be found.** In the course of many years I have now personally examined 45 monozygotic twins with endogenous psychoses, as well as 24 affected twin partners. Neither among the 45 probands nor the 24 partners did I find a systematic schizophrenia. As this result became available I decided to look at dizygotic twins in order to examine them personally. I succeeded in examining 47 patients with endogenous psychoses. Among these there was no lack of systematic schizophrenias. In fact there were 12 systematic schizophrenias, 15 unsystematic schizophrenias, 16 cycloid psychoses, 2 monopolar and 2 bipolar phasic psychoses. Among monozygotic twins there were 16 unsystematic schizophrenias, 18 cycloid psychoses, 7 monopolar and 4 bipolar phasic psychoses.

Apart from the absence of systematic schizophrenias among monozygotic twins there was a striking frequency of cycloid psychoses among monozygotic

Table 35. Distribution of Diagnoses of Twin Pairs Examined by LEONHARD (LEONHARD did not analyze twins systematically, forgoing targeted calculations of concordance)

	Number of Monozygotic Twins	Number of Dizygotic Twins
Systematic Schizophrenias	0	12
Unsystematic Schizophrenias	16	15
Cycloid Psychoses	18	16
Monopolar Phasic Psychoses	7	2
Bipolar Phasic Psychoses	4	2
Total	45	47

as well as dizygotic twins. This finding is, as we shall see, derived from the fact that motility psychoses as well as anxiety-happiness psychoses occur relatively frequently in twins. Confusion psychoses, the third of the cycloid psychoses, occur in twins rather infrequently. Furthermore it was noteworthy that monopolar and bipolar psychoses were quite as rare among monozygotic as well as dizygotic twins. I assume that there must be other external factors at work. Because of the benign nature of these conditions family members do not volunteer much information to doctors and nurses, so that the presence of twins may have remained hidden or at least was not remembered during my questioning.

The observation that among monozygotic twins systematic schizophrenias could not be found was that much more certain because it was quite unexpectedly and puzzling. How could this observation be explained? Was there a biological significance? Was it possible that twins with a potential for developing systematic schizophrenia die before or after birth and therefore were not to be found? These possibilities were not to be taken seriously. Systematic schizophrenias in their protracted course hardly ever endanger their lives; and as long as they have merely a predisposition to the illness cannot exercise a negative influence on the life of a twin. One could go back to the origin of twins from one egg to try to find a biological explanation. I cannot think of such, and I became convinced of the importance of psychosocial influences, of a kind quite different from the opinions of social psychiatrists.

It is worth repeating that systematic schizophrenias are missing only among **monozygotic** twins. Among dizygotic twins they exist approximately in the frequency to be expected in theory. This demonstrates that the psychic influence on the origin of systematic schizophrenia occurs during a time when the resemblance between monozygotic twins is already quite significant. During infancy it is not yet evident how the partner will develop. Likewise in the early years of life the child is not yet differentiated to the extent that the presence of a partner of the same age is terribly significant. One has to think of an age in which mutual understanding plays a decisive role. This points to the later years of childhood which appear to be of great significance in the origin and the prevention of systematic schizophrenia. It may be mentioned here that the onset of the special form of catatonic schizophrenia may occur in the earlier years of childhood.

In any case a person appears not to become schizophrenic if he has another person of a completely similar existence beside him, that is, a person who understands him in his reactions to the highest degree, since he is, as it were, he himself. This close contact which arises out of necessity appears to offer a guarantee that the interaction between individual and environment proceeds adequately and the maturation of the psychic systems follows normally. Turning to people other than the twin does not seem to be so essential; communication may be sufficient if it is available at least with one person. In this case there is little risk of damage to the psychic systems.

In order to demonstrate convincingly the importance of contact between monozygotic twins there would have to be found a pair of whom one died in early childhood and the other later on developed systematic schizophrenia; or monozygotic twins would have to be found who were separated during early childhood and of whom one – or perhaps both – developed schizophrenia. Both observations would bring proof that monozygotic twinship was unimportant, that the only relevant factor would be that the twins grew up together. We cannot present any cases of the first or the second kind. Unfortunately we cannot check this in the literature, since no author has separated systematic from unsystematic schizophrenia; the former are of course missing among monozygotic twins. Without this separation our findings may perhaps be confirmed if further investigations are carried out with monozygotic schizophrenic twins in order to find out how frequently the course is protracted or how frequently the course occurs in phases or even longer periods. It is true that some unsystematic schizophrenias show a protracted course as has been described before, but statistically such an examination should show that among monozygotic schizophrenic twins the remitting and periodic courses predominate. Dizygotic twins could be used as controls. Among our series of monozygotic twins – 34 cases excluding phasic psychoses – a protracted course occurred in only two patients with paraphrenia and one with cataphasia. In all the other cases the course was clearly remitting or periodic. Among dizygotic twins – 43 cases excluding phasic psychoses – 11 of the 12 systematic schizophrenias showed a protracted course. Furthermore one case of affective paraphrenia and one of cataphasia showed a protracted course. In total, among dizygotic twins there were 13 protracted cases, among monozygotic twins only 4.

From this point of view I looked at the literature. Unfortunately twin researchers mentioned only one schizophrenia, not the more precise clinical picture and course; however case histories are presented permitting a certain conclusion. ESSEN-MÖLLER (1963) and TIENARI (1963) describe again and again remitting and periodic courses among monozygotic twins, hardly ever a chronic course suggesting systematic schizophrenia.

Further evidence that other authors did not include systematic schizophrenics among their twins is revealed through a **comparison of the sexes.** In most research series female twins preponderate. If I am correct the excess of females among monozygotic twins should be greater; or the unsystematic schizophrenias and cycloid psychoses occur more frequently in women than in men. Systematic schizophrenics, absent among monozygotic twins, do not show this excess. Here as well I re-examined the literature. LUXEMBURGER (1936) found no difference. SLATER (1953) stated that women exceeded men among dizygotic twins in the ratio of 100:53.3; among monozygotic twins the ratio was 100:41.3. KALLMANN (1946/47) stated that among the total number of twins the ratio of women to men was 100:83.3, among monozygotic twins 100:75.7. My expectation was therefore confirmed, for it is not important that LUXEMBURGER found no difference; he described only 17 monozygotic twins, whereas SLATER described 41 and KALLMANN 99.

The absence of systematic schizophrenias among monozygotic twins throws light on the fact that various twin researchers have found **very different figures of concordance and discordance.** If after a psychotic episode there were no or practically no residual symptoms some researchers would not take seriously peculiarities that are now present or had been present previously, that is they would perhaps call them "neurotic"; whereas other researchers preferred to assume that a genuine psychosis was or had been present. It is well known that researchers who have claimed little concordance still find some peculiarities among twin partners. As mentioned above, among twins cycloid psychoses appear quite frequently. These can be classified differently according to their benign course and the subjective attitude of the investigator. In Scandinavian countries in which several modern investigations of twins were carried out cycloid psychoses are largely considered to be "reactive psychoses", and are therefore distinguished diagnostically. If among monozygotic twins systematic schizophrenias were found in larger numbers such lack of clarity could not continue. For in these severe forms it can hardly be doubted whether the illness appears in the partner or not. We have seen that once systematic schizophrenias have become manifest the illness usually progresses to severe deterioration.

After all this I would like to see confirmed the absence of systematic schizophrenias among monozygotic twins. This is of the greatest significance, since it follows that, **to the contrary, schizophrenias of this type can be produced when lack of communication exists.** This conception of mine, already mentioned above, found its first concrete basis from research with twins. It had been suspected before that isolation of the person might be a precondition for schizophrenia. There was some evidence in favor of this notion, but definite proof was lacking. This was partly due to the fact that the emphasis had been on isolation in adults; partly also no clear result could be obtained as long as schizophrenia was considered as a whole instead of separating the different types. Our research with twins has produced such evidence in no other endogenous psychoses nor other forms of schizophrenia.

Lack of Communication in Systematic Schizophrenia of Childhood

Forms of systematic schizophrenia appearing in early childhood will have to be considered separately because it has to be shown by means of descriptions of the clinical picture that they actually exist, a fact which has been disputed. In the case of other schizophrenias of childhood it is evident that there is a significant **lack of siblings** resulting in a lack of communication. There was a strikingly high proportion of **single children.**

I shall now describe the clinical picture from personal investigations of 37 cases of systematic schizophrenia of childhood (25 male, 12 female), where the onset was between the ages of 3 and 15 years. There were 17 single children, i.e. 45.9% (half-brothers and half-sisters were not included). An earlier investigation produced fewer single children, because at that time children were included whose illness had begun before they were 3 years old, and in whose case the lack of siblings had no significance. Of 30 patients who in childhood developed unsystematic schizophrenia or a cycloid psychosis there were 7 single children, that is 23.3%.

TOMAN and PREISER (1973) found 21.8% of single children in the average population of BAVARIA, BECK and LEMPP (1965) in Baden-Württemberg 22%. The number of single children in the case of systematic schizophrenia of childhood is quite considerably higher, whereas in the case of unsystematic schizophrenias and cycloid psychoses it is similar to the average population. Lack of siblings means lack of communication, because single children are less together with other children than those who have siblings.

It is really quite remarkable that practically 50% of patients are single children. The number of those who grew up without siblings is even higher. Two patients, a boy and a girl, did have siblings, but they grew up as single children with their grandmothers. A girl whose illness began at the age of 6 years did have a sister who was 12 years younger; at the beginning of the illness and for a long time afterwards she was therefore a single child. A boy whose illness began at the age of 4 had a sister 4 years later. A boy whose illness began at the age of 6 had a brother 3 years later. All these cases could be counted as single children. A further observation is that when there is a great age difference between siblings the quality of communication between them is much diminished. This was the case where a boy had a sister who was 18 years older. Another patient had a brother who was 10 years older; another boy had a brother who was 16 years older; still another boy had a brother and a sister who were 16 and 13 years older, respectively. These observations confirmed **that a lack of siblings increases the danger of developing childhood schizophrenia.**

One of my patients did not fit in with these observations: he developed the illness at the age of 9 and had no less than 6 siblings. But this exception is clarified when one considers this child in greater detail. Whereas all the other children developed normally and did well in school the patient who was the youngest remained behind. He had rickets, did not start to walk until the age of 2 and talk at the age of 3. He did poorly in school and was often ridiculed by other children. One can certainly say that this both physically and mentally weak child could not develop a genuine relationship with his older stronger siblings, nor with his classmates; on the contrary he was a target for their mockery. I have often observed such "targets" in the case of neurotic and anxious children (1967) and find them not only in this case but in other cases of childhood schizophrenia. Isolation from other children is easily recognizable in the case of "targets".

I would have liked to have confirmed my findings in the literature. But on the one hand authors do not indicate whether schizophrenic children whom they have described had siblings; on the other hand they do not differentiate between systematic and unsystematic schizophrenia and cycloid psychoses. Children aged 10–14 years are often described in the literature; among these the systematic forms occur less often, because these are more numerous only in the younger age groups. Before I started my special investigations of schizophrenic children whose illness began in the first 3 years of life I had come across 13 such cases; 16 others developed the illness before the end of their 6th year. No psychoses other than systematic schizophrenia occurred in the first 6 years. Later on in childhood unsystematic schizophrenias and cycloid psychoses occurred much more often. Table 53 will later show the exact distribution of my cases of endogenous psychoses originating in the different age groups of childhood.

Siblings of Patients with Systematic Schizophrenias

The high proportion of single children with childhood schizophrenia induced me to look into the occurrence of single children among adult patients. Beyond this I became interested in the number of siblings that adult patients had. It turned out that not only the total number of siblings was decreased, but in part **certain groups** i.e. older, younger siblings, brothers, sisters were strikingly diminished in number.

Here I have to remark that it was primarily a question of a **relative** decrease in the number of siblings. When I speak of a **"lack"** this is meant in comparison with other groups of siblings. There is no absolute increase or decrease, for at different times and in different cultures the average number of children is totally different. For instance I cannot say whether paraphrenic patients have fewer siblings than healthy persons of the same age, of the same domicile and of the same social position, but it is quite clear that male paraphrenics have far more older than younger siblings. It appears to be a matter of disturbance of equilibrium. One can understand this perhaps if one considers that in the development of the child "older" and "younger" could be equivalent to "stronger" and "weaker", or "more or less experienced" or "more or less clever". For the development of the child it seems to be essential whether and to what extent the child feels superior or inferior towards his siblings.

This shows **that communication between siblings is quite different from that between monozygotic twins.** Between siblings it is not a matter of close understanding. I was able to make another observation: I looked into the question whether the lack of siblings is especially disadvantageous if one considers the age periods that are close to the patients. The opposite was shown: It was more a question of siblings who were quite a bit older or younger, whereas the mutual understanding between children who are close in age is surely much better. Later I shall consider in more detail the difference in communication between monozygotic twins and between siblings.

The first question was the number of single children with systematic schizophrenia beginning in adult age. Table 36 provides some information. The patients included in this table came from an older generation than patients with childhood schizophrenia. Since in former times there were more children it is not surprising that altogether there were less than 20% of single children. But there were some important differences.

The first observation was that **male patients** were far more often single children than female patients (25.8%:14.4%). Later we will note that male patients with systematic schizophrenia in general have fewer siblings than female patients. The possible explanation is that in earlier generations parents liked to have "a first born son and heir" after whose birth they decided to have no more children. If this is not true another explanation may be that because boys are more inclined to be introverted than girls they are at a higher risk if they are single children.

From Table 36 it can be seen that catatonics are single children more frequently than paraphrenics and hebephrenics (27.6% compared with 14.3% and 18.3% respectively). This corresponds to the greater frequency of single children with systematic childhood schizophrenia all of whom, as we will see later on, are catatonics.

It cannot be determined whether the proportion of single children with paraphrenia and hebephrenia is diminished. However Table 37 shows that the relative increase of single children with catatonia is due to the fact that the total number of siblings is smaller. Inspection of the right side of the table clearly shows that in contrast to the 302.3 siblings of paraphrenics and 249.5 siblings of hebephrenics there are only 207.3 with catatonia. The fact that hebephrenics have fewer siblings than paraphrenics indicates nothing definite, as our paraphrenics on the average were older than our hebephrenics and probably were born during the time when birth control was less common. The age of onset of paraphrenia, as we mentioned above, is about 10 years higher than that of hebephrenia. Moreover it is improbable that this great difference arises in only this manner, as the age difference is not great. Probably hebephrenics do have fewer siblings than paraphrenics, because with the former we find that in a certain group they have fewer male as well as female siblings, whereas with the latter only male patients have fewer. Certainly the number of children with catatonia is smaller, since there is very little difference between the age of onset of hebephrenia and that of catatonia

In analysing each group of siblings separately I would like to draw attention to the fact that the differences in so far as they are substantial are all significant.[21]

Table 36. Single Children with Systematic Schizophrenias

	Sex	Number of Patients		Single Children		Per 100 Patients	
Paraphrenia	♂	96	266	19	38	19.8	14.3
	♀	170		19		11.2	
Catatonia	♂	150	232	46	64	30.7	27.6
	♀	82		18		22.0	
Hebephrenia	♂	83	186	20	34	24.1	18.3
	♀	103		14		13.6	
All Males		329		85		25.8	
All Females		355		51		14.4	
Total		684		136		19.9	

[21] Cordial thanks are due to Mr. Littmann, psychologist in the Department of Psychiatry of the Charité, for his statistical evaluation of my data.

Table 37 shows that **male paraphrenics** have far fewer younger than older siblings (90.6:43.8 and 84.4:55.2). In the case of brothers the difference is exceptionally great ($p < 0.01$), but even in the case of sisters the difference is significant ($p < 0.05$). In the case of female patients the difference is evident only for their brothers (93.5:70.0).

Hebephrenics likewise have fewer siblings but the situation is different: **Male patients** are seen to have significantly fewer older brothers (38.6:65.1), and somewhat fewer sisters (47.0:63.9). **Female hebephrenics** show the opposite: they have fewer younger siblings, brothers (52.4:84.5) as well as sisters (52.4:88.3).

The differences between paraphrenics and hebephrenics have to be considered as highly significant. At one time I tried to explain them. The statistical significance of the findings confirms their objectivity; beyond this the consistency of the findings indicates that we are not dealing with a coincidence. I am not satisfied with my earlier explanations and believe now that the biological factor is more pronounced.

In **hebephrenia** – as this form of schizophrenia has been described above – there is a disturbance of affect. Normal maturation of affect occurs around the time of puberty. At this age there is an **internalization of affective life.** External expressions are no longer responded to with immediate or short-circuit-like affective reactions, but only after they have been processed internally. With this development man acquires a deeper attitude towards his emotional life. Emotional response becomes fixed for, as an adult, man continues to make value judgements in the same manner as learned in youth. Relearning later succeeds with difficulty. Nature apparently sees to it that a young person enters adulthood holding firmly established value judgements which he continues to maintain. At the same time he is looking for contact with people who are senior to him who may serve as an example. Therefore it may be a disadvantage if no older siblings are available. Examples are expected to be provided by parents and educators. I have already mentioned above the great significance of the relationship with other children. Moreover it is known that the adolescent often reacts with opposition to what grown-ups expect from him; he pays much more attention to what an older brother or an older sister considers good or bad. In this phase of development the influence of older siblings over younger brothers appears to me to be necessary. Nature has seen to it that the maturation of the corresponding brain systems depends on it.

However from Table 37 we see that in the case of **girls** it is the other way round, since there are fewer younger siblings. In order to make this observation comprehensible I have to express thoughts that perhaps in this age of emancipation of women may appear to be unusual, but which may nevertheless be valid as biological changes follow social changes rather slowly. What a woman considers good or bad, right or wrong, is for her much less solidly based than for a man; she is more adaptable. She often assumes the opinion of the man to whom she is closely related not because he is pushing her but because she herself has no inclination to oppose him with her own

Table 37. Number of Siblings of Schizophrenics (in Parentheses per 100 Patients)

	Number of Patients	Older Brothers	Younger Brothers	Older Sisters	Younger Sisters	Total
Paraphrenia						
Male	96	87 (90.6)	42 (43.8)	81 (84.4)	53 (55.2)	263 (274.0)
Female	170	159 (93.5)	119 (70.0)	138 (81.2)	125 (73.5)	541 (318.2)
Both	266	246 (92.5)	161 (60.5)	219 (82.4)	178 (66.9)	804 (302.3)
Hebephrenia						
Male	83	32 (38.6)	54 (65.1)	39 (47.0)	53 (63.9)	178 (214.5)
Female	103	87 (84.5)	54 (52.4)	91 (88.3)	54 (52.4)	286 (277.7)
Both	186	119 (64.0)	108 (58.1)	130 (69.9)	107 (58.6)	464 (249.5)
Catatonia						
Male	150	79 (52.7)	72 (48.0)	76 (50.7)	67 (44.7)	294 (196.0)
Female	82	50 (61.0)	42 (51.2)	47 (57.3)	48 (58.5)	187 (228.0)
Both	232	129 (55.6)	114 (49.1)	123 (53.0)	115 (49.6)	481 (207.3)

opinions. In the world of business she usually does not like to assume leadership roles; she prefers to leave this to others. In spite of all efforts women have seldom assumed positions of leadership such as should be expected from their numbers.

In contrast a woman is clearly committed and cannot be influenced by men in matters of **compassion and caring for weaker persons.** Consider how much deeper the emotional ties of a woman are for the care of children. Probably a girl in the development of her emotional life needs contact with weaker people, that is, with younger siblings. Well known is the devotion which girls often show towards younger siblings. Apparently nature sees to it that girls at the time of puberty and thereafter are looking for contact with younger siblings. The development of their emotional life and their cerebral systems underlying it may be damaged when younger siblings are missing. When turning my thinking to cerebral systems I refer to what I have said above.

In **paraphrenics** the central symptom is **disturbance of thinking,** exemplified by delusions and hallucinations. The delusions of affective paraphrenia which have an affective basis do not belong here but rather to an unsystematic form of schizophrenia. Maturation of thinking occurs in later childhood as the person does away with childish short-circuit reactions. During this time for males, as Table 37 shows, the lack of younger siblings is a disadvantage, especially brothers. Younger brothers are not present half as frequently as older brothers (43.8% compared with 90.6%). It might have been expected that as in hebephrenia the presence of older siblings might be more important, but the numbers show the opposite. In order to find an explanation I assume that a young man has already developed value judgements according to which he conducts his later life. It is a man's characteristic to want to try out his own opinions with others and act accordingly. The young male lover would like to pass on to his loved one the insights he has gained.

A husband rarely meets resistance in his family when he expresses his opinions, for the wife, as we have seen, does not find it important to express abstract opinions. Moreover, outside the family, a man often declares his opinions in a circle of friends, acquaintances and colleagues, trying to impress the others. In contrast, a woman may say "what this man or that one thinks, doesn't interest me at all". The man takes a position, tends to get involved in arguments, even if he personally is not directly concerned. All this indicates that in nature man has the inclination to give advice and direct others. There is a biological basis for this in that since early times a position of leadership has been assigned to him by means of which in the family and in groups outside of the family a sense of community was fostered. Changed social circumstances have apparently not yet led to a biological correction, as man even today generally still maintains a leadership role.

The development of thinking that goes in the direction of advising and guiding others does not come from within but results from interchange with the environment. Therefore people must be at hand to accept the opinions expressed by others. Among siblings it can only be the younger ones who, as we mentioned in discussing hebephrenics, turn to older siblings as role model. If there are no younger siblings psychological and biological maturation might be inhibited with a greater risk for systemic disorder to develop.

In thinking as in feeling it is a matter of the direction taken, as my comments should illustrate. Thinking by itself is certainly no less developed in women than in men, and emotions are no less developed in men than in women, but are of quite a different nature.

In **female paraphrenics** the situation is different from males, but not simply the opposite. Here too there are fewer younger brothers – I will deal with this a little later. There is no great difference in the number of sisters (81.2%:73.5%). From this I conclude that in female paraphrenics the number of siblings is without particular significance. I surmise that girls mature earlier than boys and that nature thus facilitates reproductive functions. For the changes necessary for their maturation nature has perhaps provided an association with a male partner. In order to elucidate this I must point again to circumstances in the lives of our ancestors. In those days a woman's tasks were quite different from today. At the time when there was no birth control she remained excluded from most occupations not only because women were not emancipated but also because after marriage a woman had to spend 2 to 3 decades bearing and bringing up children. To practice a profession as well would have been practically impossible. Hence it was necessary for a woman in her development to be prepared for family activities. From such considerations was formed my opinion that a girl was directed at an early stage towards a partner for her future family life, and that siblings were no longer needed in this phase of development.

Along this train of thought it should be added that in the case of sisters younger siblings are about equal in number, but in the case of brothers the number of younger siblings is clearly diminished (70.0%:93.5%). We shall see when we consider periodic catatonia that if the mother is absent the older sister is needed to take care of younger brothers, but not younger sisters. Perhaps girls should be biologically prepared in their thoughts for the special task of caring for younger brothers; therefore if there are none it would be a disadvantage for them.

If it is true that girls in their adolescence are looking for contact with male partners this would perhaps explain that paraphrenia occurs much more often in women than in men (see Tables 33 and 36). In times not so long ago girls generally were prevented from communicating with male partners. They were strictly supervised by parents and educators to prevent them from having immature sexual contact. I must add a remark here of general significance: if in my comments it has become apparent in different ways that biological factors are no longer in harmony with the demands of modern society, this does not indicate that a wrong turn has been taken, for ethical principles overrule biological challenges. Of course one cannot wish that women keep on having children year after year, as used to be the case; if they have none or only a few they should not be excluded from work in public life for which they now have time and energy. The possibility that having few or no children may favor the appearance of systematic schizophrenia requires that we must look for appropriate countermeasures. This will be considered later.

The supposition that for adolescent girls contact with siblings no longer plays a role appears to be contradicted by an observation which we made in our **twin observations.** Among monozygotic twins there were no paraphrenics, not even of the female sex. Close contact with a sister must be very effective here. Even dizygotic twins rarely had paraphrenia: there was only one, a woman who had a twin brother, compared with 6 catatonics and 5 hebephrenics. The contact between dizygotic twins seems to prevent the appearance of paraphrenia. The fact that the above mentioned number of 12 systematic schizophrenias among our dizygotic twins was somewhat less than theoretically expected was due to the rarity of paraphrenia. It therefore seems that communication with siblings in paraphrenia – in female patients as well – would be more significant than in hebephrenia and catatonia. I have already mentioned **that contact between twins is of a different kind from contact between siblings.** If siblings are not twins they are either older or younger, twins have no place here. Moreover the two forms of communication do not have the same background. In the first case the girl is looking for a male friend with potential for a close relationship such as with a monozygotic twin; in the second case she is looking for a male friend who responds to her biological demands appropriate to his age, whether greater or lesser. Probably nature assures the normal development of the girl by means of this double form of contact.

Now let us consider **catatonia.** Here there are no essential differences in the numbers of siblings of both male and female patients. But it may not be concluded from this that there exists no mutual influence, because the number of all siblings is greatly diminished.

Let me clarify to start with: catatonia is characterized by a **disturbance of the expression of the will.** The development of the will appears to be disturbed because there are altogether too few siblings. For its proper development it is significant that children compete among one another. For instance in serious or playful fights they want to prove to themselves or their comrades that they have won; or in front of adults they compete to show their prowess. By such endeavours the will is being developed. It is of no advantage if the child always has the last word or never wins. Therefore relatively younger, i.e. weaker, as well as relatively older, i.e. stronger siblings are a disadvantage. In both cases a disturbance of development of the will and deficient development of cerebral systems may result, which later may lead to catatonia. Since the equilibrium of the siblings may be shifted equally to one side or the other to their disadvantage, the statistics between older and younger siblings are balanced. The total number of siblings must be fewer, as we have actually found in catatonia. For both sexes the figures are the same. This shows that the development of the will starts in early childhood, boys and girls not being fundamentally different before puberty. This can be confirmed: small children fight for their toys; as early as the third year of life they go through a spiteful phase, thus demonstrating that a strong will is developing. If the development of the will precedes so much the other developments which have concerned us in hebephrenia and paraphrenia, it must be because man in the time in which he forms his value judgements and his thought processes must have learned to know the world in order to adapt to it with his new capacities. Children steadily learn to know the world, reacting to every novel event confronting them. Thus they collect experiences which they already have in good measure when maturation of their emotional life and thinking takes place.

Following all these findings I wanted to determine in more detail the influence of siblings on one another. I thought that the effect might be less important if the difference of age was several years, since siblings in this case spend less time together in everyday life. However when I checked this out, I found that the opposite pertained. The differences between older and younger brothers and sisters increased when I excluded siblings less than three years older or younger than the patient. This can be seen from Table 38. In order to make a direct comparison, in Table 39 are shown the figures from the previous two tables in such a way that the numbers of the older siblings are adjusted accordingly; they have a common numerator. The denominator shows the figures for the younger siblings. The difference between the younger and the older siblings can now be seen directly.

A slight decrease is the difference is found in the case of sisters of male paraphrenics; the younger ones are close to the older ones (55.2% compared with 56.7%). In the case of female paraphrenics the younger brothers do not differ much from older ones. But the increase in the difference mentioned above is very noticeable indeed in several places. Apparently what is decisive for the development in general is a greater age difference between siblings. In the case of male paraphrenics the 90.6% older brothers now contrast with still fewer younger ones (35.3% instead of 43.8%). In the case of male hebephrenics older siblings are now more frequent, especially sisters

Table 38. Older and Younger Siblings of Systematic Schizophrenics (counting those with and those without siblings less than 3 years older or younger, in parentheses per 100 patients)

		Older Brothers	Younger Brothers	Older Sisters	Younger Sisters
Paraphrenia	males with	87 (90.6)	42 (43.8)	81 (84.4)	53 (55.2)
	males without	70 (72.9)	25 (26.0)	55 (57.3)	37 (38.5)
	females with	159 (93.5)	119 (70.0)	138 (81.2)	125 (73.5)
	females without	107 (62.9)	91 (53.5)	111 (65.3)	90 (52.9)
Hebephrenia	males with	32 (38.6)	54 (65.1)	39 (47.0)	53 (63.9)
	males without	20 (24.1)	37 (44.6)	22 (26.5)	38 (45.8)
	females with	87 (84.5)	54 (52.4)	91 (88.3)	54 (52.4)
	females without	67 (65.0)	37 (35.9)	66 (64.1)	30 (29.1)
Catatonia	males with	79 (52.7)	72 (48.0)	76 (50.7)	67 (44.7)
	males without	58 (38.7)	46 (30.7)	63 (42.0)	39 (26.0)
	females with	50 (61.0)	42 (51.2)	47 (57.3)	48 (58.5)
	females without	38 (46.3)	26 (31.7)	36 (43.9)	33 (40.2)

Table 39. As Table 38, Numbers Expressed in Percentages of Equivalent Numerator

		Older Brothers	Younger Brothers	Older Sisters	Younger Sisters
Paraphrenia	males with	90.6	43.8	84.4	55.2
	males without	90.6	35.3	84.4	56.7
	females with	93.5	70.0	81.2	73.5
	females without	93.5	79.5	81.2	65.8
Hebephrenia	males with	38.6	65.1	47.0	63.9
	males without	38.6	71.4	47.0	81.2
	females with	84.5	52.4	88.3	52.4
	females without	84.5	46.7	88.3	40.1
Catatonia	males with	52.7	48.0	50.7	44.7
	males without	52.7	41.8	50.7	31.4
	females with	61.0	51.2	57.3	58.5
	females without	61.0	41.8	57.3	52.5

(now 81.2% instead of 63.9%), but even brothers as well (now 71.4% instead of 65.1%). Correspondingly the decrease of younger siblings in the case of female hebephrenics is more noticeable (now 46.7% instead of 52.4%, and 40.1% instead of 52.4%, respectively). **In this form the figures show even more clearly the great significance of the role of siblings.**

A correction of the explanation given at the beginning of this discussion appears to be necessary in the case of **catatonia,** as the figures no longer agree. Male catatonics have fewer younger sisters, female catatonics fewer younger brothers (31.4% compared with 50.7%, and 41.8% compared with

61.0%, respectively). I expressed the opinion that the disharmony caused by an absence of older siblings and the disharmony caused by an absence of younger siblings was evened out. Apparently this does not occur totally when the siblings are of the opposite sex. The deficiency of which I spoke is more noticeable if the older sibling belongs to the opposite sex. This can perhaps be understood if certain natural inhibitions between the two sexes are taken into account.

It is biologically difficult to understand how a certain constellation of siblings may be a disadvantage which may contribute to the development of a schizophrenic disorder, as has already been proposed by some authors. It cannot be correct that some children are by nature predisposed to this danger. Let us go back to ancient times during which biological systems were laid down. During early times man apparently lived together in herds, later in large families. In those times every child was together with other children from different parents. Under these circumstances it was unimportant if parents had many or few children. Nowadays the circumstances are completely different; it could be expected that systematic schizophrenia would occur much more frequently than is actually the case. But today there are sufficient possibilities for children to get together outside their own family. We also have to keep in mind that in the constellation of siblings we are dealing with only one causative factor; no matter how essential, it by itself cannot cause the disease.

I am now wondering what else may happen when the disorder actually appears. Heredity can hardly be considered seriously, as so few psychoses are found in the families of systematic schizophrenics. However I presume that certain characteristics which by themselves are normal may contribute to the illness under unfavorable conditions. By nature people vary in their capacity and readiness for contact. Some are more extroverted, others more introverted, as may be observed soon after birth. Weakness of contact and introversion – the expression autism combines both these characteristics – lead to a situation where some people do very little to seek contact with others (introversion) nor do they find it (weakness of contact). In this way the lack of communication from within is added to the outer lack from the unfavorable constellation of siblings. This observation appears to be confirmed, for it is a well known fact that schizophrenics long before the beginning of the illness are often quite autistic. This was shown clearly by Mrs. VON TROSTORFF (1970) in her study of systematic schizophrenia. I shall come back to this etiological connection between the characteristic features and the origin of systematic schizophrenia in the following chapter on early childhood catatonia.

From these considerations it follows that systematic schizophrenia has become more frequent in modern times. People in earlier times appeared to have been more extroverted, their lives were directed outwards and they were affected more directly by events in their environment. Moreover they had more children. These days it is often assumed that schizophrenia occurs equally frequently in less developed countries. However such estimates are quite uncertain. If these claims are true, this does not contradict an increase in systematic schizophrenia. It may be that other endogenous psychoses have become less frequent. Endogenous psychoses take a different course in different cultures. JILEK (1974) has made a thoroughgoing study of this problem and concluded as follows: "The frequency of transient psychotic episodes appears to have decreased in modern society together with the improvement of hygienic and nutritional conditions; however the chronicity of symptom formation appears to be more accentuated in the Western world than in Africa". My conception of this is that the systematic schizophrenias with their insidious progression have increased in our part of the world, whereas the cycloid psychoses have become less frequent. I shall return to this fact later on.

There is another question: if the supposed developmental disturbances of childhood and adolescence could lead to a schizophrenic disorder, the question may be asked why the illness manifests itself so much later. Only in the case of catatonia of childhood does the illness follow soon after the lack of communication. In other cases it may be assumed that the developmental disturbances at first produce only latent damage in the system which is involved. The actual breakdown occurs in the wake of the greater and more persistent challenges of adult life. Moreover, it may be stated that in early adulthood people are expected to show more activity and affect, whereas older people are supposed to be more thoughtful. Thus the sequence in which the different forms of systematic schizophrenia manifest themselves are connected not only with the way in which the early developmental disturbances occur but also with the challenges of later life.

Furthermore our findings clarify a problem concerning my **classification.** Systematic schizophrenias may be associated as combined systemic illnesses. But the only combinations are within catatonic forms, within hebephrenic forms, and within paraphrenic forms. Combinations from different groups are extremely rare. Here I must add that it is not a matter of the question of the mixture of symptoms, but of two circumscribed subtypes of systematic schizophrenia. By means of limitation of combinations of associations within the same subtype diagnosis is greatly facilitated; theoretically at first I could not understand this limitation. Now however it has become clear: every constellation of siblings is disadvantageous at a certain period of time only in the sense of one form of schizophrenia, because the maturation of the different psychic functions follows at different phases of life. It should happen very rarely that in several phases of development there was a lack of siblings and also of friends necessary for good communication.

Prophylaxis of Systematic Schizophrenias

As long as no cause of schizophrenia was known nothing could be done to prevent it. From the point of view of social psychiatry it was thought that the appearance of the disorder could be limited if the child could have a harmonious life and could be spared deprivations and conflicts. However concrete proof that this might be successful was not found. Nor did my findings show anything definite. However the lack of meaningful contact with siblings and with other children was shown to be highly significant. Therefore prophylaxis is possible. Since we do not wish to return to families with many children our efforts have to be directed to provide rich contact with other children. This already happens in kindergartens. A disadvantage of the latter consists in that here children of the same age are grouped together, whereas the child should have contact with older and younger children. In some places this has been achieved, which is surely an advantage. In school too children of the same age come together even if here there is a greater possibility of finding comrades other than classmates. Since it is difficult to make changes in kindergartens and schools special attention should be given to children's clubs and to youth clubs bringing together older and younger children.

In so far as the child is not part of a larger group but can look for a comrade by himself one can rely on his instinctive choice. I am convinced that a child looking for a comrade of a certain age according to his own age and his stage of development is driven by an inner impulse in the right direction. A boy in puberty who, as indicated above, in conformity with his biological development needs a comrade as a model will associate himself with an older comrade; whereas a girl of the same age, because of her different conformance to biological development will want to take care of a younger child. These children must be given the opportunity of seeking and finding the comrade whom they need biologically; they should be enabled to associate with many children of different ages so that they have a choice. Their comrades should not be chosen for them. Many parents and educators in this respect do not do what is beneficial for the child; they believe that they know best which comrade is suitable for their child, which one would have the best influence. They are of course right when the child has to be prevented from comrades who set a bad example, but in that respect they only prevent a bad psychological development. They may at the same time inhibit the biological development, in so far as no other friend is available. Some children stubbornly hold on to a friend of whom their parents do not approve; they appear to be dominated by an unconscious striving which biologically is more correct than the demands of the parents.

Especially with single children who can only find their comrades outside of the family it is important that they are given the opportunity to make their own choices. It is the mothers of single children who often do not do the right thing. They want to protect the child from alien, that is as they think, harmful influences, and as a consequence tie the child to themselves. For the single child this would present a serious danger if it did not have the freedom to choose a friend it would need in the various phases of development. A single child with sensible parents under certain circumstances may actually be in a relatively favorable situation. If siblings are available these, through their constant presence, act as comrades even if at their age they may not satisfy its biological needs. The child may thereby be prevented from looking for a true friend outside of the family. Probably single children take advantage of such a situation, otherwise far more of them would become affected than actually happens.

Furthermore even if the second form of contact, which plays such a big role in the case of monozygotic twins, is taken into account the child must be given the opportunity to choose a friend by itself. Here it is not relevant to speak of an instinctive drive. Truly psychological factors allow good contact between children who understand each other, whose thoughts and feelings are similar. In their close spiritual rapport they protect each other from the danger of losing the necessary touch with the environment. As with twins it is sufficient if at least one person is available for unrestricted contact. Thus the necessary interchange with the environment is guaranteed.

On the whole it may be stated that the task of parents is not to tie the child too much to them, but to give enough freedom so that it can make contact with other children. I am not saying anything new, for it has been known for a long time that an overprotective mother damages the child if she wants to take care of it all the time. However it was not yet known whether the behaviour of such a mother might contribute to the manifestation of severe systematic schizophrenia or not. Here I want to avoid a misunderstanding: I want to underline again that I am keeping biological development in mind; for psychological development close contacts with parents for the sake of setting good examples and education in general are necessary.

Exogenous and Constitutional Causes of the Unsystematic Schizophrenias

The unsystematic schizophrenias have, as we have already seen, a genetic foundation. Nevertheless external causes have to be present at the time when the illness manifests itself. Here too the influence that children exert on one another turns out to be significant.

Periodic Catatonia

First of all, Table 40 shows that patients with periodic catatonia in comparison to those with the other unsystematic schizophrenias have strikingly **fewer siblings.** Whereas patients with affective paraphrenia have 284.2 and patients with cataphasia 234.3 siblings per 100 patients, those with periodic catatonia have only 179.4. We remember that likewise among the systematic schizophrenias the catatonics have the lowest number of siblings. The reasons for these findings are however different. One possible explanation might be that periodic catatonics have fewer children because they become ill earlier and therefore have less time for reproduction. To a small extent this is confirmed, as the probands with healthy parents according to Table 41 have more siblings than those with affected parents (188.9 compared with 165.5). But this is only the case with men, not with women as shown by the table. This shows that there must be another reason, perhaps only one other reason.

Table 40. Number of Siblings of Unsystematic Schizophrenics (in Parentheses per 100 Patients)

	Number of Patients	Older Brothers	Younger Brothers	Older Sisters	Younger Sisters	Total
Affective Paraphrenia						
Male	25	21 (84.0)	8 (32.0)	21 (84.0)	15 (60.0)	65 (260.0)
Female	95	76 (80.0)	55 (57.9)	88 (92.6)	57 (60.0)	276 (290.5)
Both	120	97 (80.8)	63 (52.5)	109 (90.8)	72 (60.0)	341 (284.2)
Cataphasia						
Male	120	81 (67.5)	62 (51.7)	76 (63.3)	57 (47.5)	276 (230.0)
Female	52	31 (59.6)	29 (55.8)	39 (75.0)	28 (53.8)	127 (244.2)
Both	172	112 (65.1)	91 (52.9)	115 (66.9)	85 (49.4)	403 (234.3)
Periodic Catatonia						
Male	67	33 (49.3)	25 (37.3)	18 (26.9)	27 (40.3)	103 (153.7)
Female	69	34 (49.3)	38 (55.1)	29 (42.0)	40 (58.0)	141 (204.3)
Both	136	67 (49.3)	63 (46.3)	47 (34.6)	67 (49.3)	244 (179.4)

What matters is whether the father or the mother was affected; only in the latter case the number of children is diminished. If the father is affected the number of children (181.3) is almost the same as in the case of healthy parents (188.9). Finally one may observe that the influence of the affected mother affects only male patients (167.6 compared with 126.1), and not female patients (206.8 compared with 206.3). This latter finding arises from the fact, as we shall see later, that male patients are much more affected by the illness of their mother than female patients. This can be seen from Table 41 taking the following into consideration: the total number of our male and female probands is approximately the same (male 67, female 69), but according to the table female probands with healthy parents exceed males by 44:37, whereas male probands with affected parents exceed the females by 30:25.

Thus it can be seen that the smaller number of children in the case of catatonics is not to be explained by the earlier onset of the illness. There is a different reason. The question is whether smaller numbers of siblings favor the development of the illness. If this is the case, there should be relatively fewer patients when there are more siblings than when there are fewer. The average number of siblings in the case of periodic catatonia is 1.79 as Table 40 shows. In order to find an answer to the question I separated the families with siblings above the average from the families with siblings below the average. In practical terms this means that 1.5 siblings belong to the first group, whereas 2.0 siblings belong to the second. The number of siblings had to be "corrected", as a comparison had to be made between the number of healthy and the number of affected siblings. So what was decisive was whether the siblings had already passed beyond the age of onset of the illness or they might still become affected. Table 42 shows the result. In the case of larger numbers of siblings the incidence of illness is 17.6, whereas in the case of the smaller ones 29.4. Therefore the following is confirmed: **the probability of the illness becoming manifest increases as the number of siblings decreases.**

We were able to affirm the significance of lack of siblings when we paid attention to siblings of our **patient's parents.** We compared the number of siblings of affected parents with the number of siblings of healthy carriers of the hereditary disposition in order to find out whether the latter had more siblings than the former. In so far as possible, Mrs. VON TROSTORFF counted the number of siblings of the patients' fathers and mothers. She succeeded in obtaining reliable data in 115 cases (59 male and 56 female probands). We had to know from which parent the disorder came; in most cases the information was adequate. Even if the parents were both healthy, one of them was often quite abnormal; in other cases among siblings or other relatives of the carrier of the hereditary disposition psychoses or severe abnormalities were found. In the other line however nothing significant was found and it was unnecessary to consider it further. In the case of dominant heredity a unilateral genetic disposition would have been expected. But there were some cases with abnormalities on both sides (4 male and 2 female probands). These had to be excluded from our investigation. Furthermore there were some cases where neither in the paternal nor in the maternal line any definite abnormalities were discovered (6 male and 5 female probands). Among the 98 remaining probands we could check the number of siblings of affected parents and the number of healthy parents with hereditary disposition. Table 43 shows that the affected parents had significantly fewer siblings than the healthy carriers of the hereditary disposition (306 compared with 375). Thus the expectation was confirmed. Once again the statistics **show that the number of siblings does have an influence on the incidence of periodic catatonia.**

My next question was to determine the exact circumstances producing this effect. First of all I believe that the reason is not lack of communication as in the case of the systematic schizophrenias; investigations of schizophrenic twins show no evidence of this. Going back to Table 40 we observe again the striking constellation of siblings. There is a lack of **older sisters.** It is not a big lack, but it is bigger if we include only probands with affected parents, as is seen in Table 41. In the case of affected parents the number of older sisters of male probands is only 13.3 compared with 36.7 younger sisters. The absolute numbers are small, but the result in consideration of the large difference is significant ($p < 0.05$). In the case of female patients, there is no similar difference (44.0 compared with 48.0). Therefore we see that the lack of older sisters becomes important **only when one parent is affected, in this case also only with male probands.** Nothing new is found if the mother's illness is taken into account. Concerning the father's illness the figures are so small that they cannot be evaluated. But it is striking that according to Table 41 older sisters of male patients with affected fathers are missing altogether.

Table 41. Number of Siblings of Periodic Catatonics with Healthy and Affected Parents (in Parentheses per 100 Patients)

	Number of Patients	Older Brothers	Younger Brothers	Older Sisters	Younger Sisters	Total
Both Parents Healthy						
Male	37	18 (48.6)	14 (37.8)	14 (37.8)	16 (43.2)	62 (167.6)
Female	44	19 (43.2)	26 (59.1)	20 (45.5)	26 (59.1)	91 (206.8)
Both	81	37 (45.7)	40 (49.4)	34 (42.0)	42 (51.9)	153 (188.9)
One Parent Affected						
Male	30	12 (40.4)	13 (43.4)	4 (13.3)	11 (36.7)	40 (133.4)
Female	25	14 (56.0)	14 (56.0)	11 (44.0)	12 (48.0)	51 (204.0)
Both	55	26 (47.3)	27 (49.1)	15 (27.3)	23 (41.8)	91 (165.5)
Mother Affected						
Male	23	8 (34.8)	8 (34.8)	4 (17.4)	9 (39.1)	29 (126.1)
Female	16	10 (62.5)	8 (50.0)	7 (43.8)	8 (50.0)	33 (206.3)
Both	39	18 (46.2)	16 (41.0)	11 (28.2)	17 (43.6)	62 (159.0)
Father Affected						
Male	7	4 (57.1)	5 (71.4)	0 (0.0)	2 (28.6)	11 (157.1)
Female	9	4 (44.4)	6 (66.7)	4 (44.4)	4 (44.4)	18 (200.0)
Both	16	8 (50.0)	11 (68.8)	4 (25.0)	6 (37.5)	29 (181.3)

Table 42. Affected Siblings of Periodic Catatonics from Larger and from Smaller Sibships

	Number of Patients	"Corrected" Number of Siblings	Number of Affected Siblings	%
Patients from Larger Sibships	42	119.0	21	17.6
Patients from Smaller Sibships	94	51.0	15	29.4
Total	136	170.0	36	21.2

Table 43. Siblings of Healthy and of Affected Parents of Periodic Catatonics (in Parentheses per 100 Parents)

		Number of Patients	Number of Siblings
Healthy Parents	Parents of Male Patients	22	77 (350.0)
	Parents of Female Patients	26	103 (396.2)
	Both	48	180 (375.0)
Affected Parents	Parents of Male Patients	27	84 (311.1)
	Parents of Female Patients	23	69 (300.0)
	Both	50	153 (306.0)

It could be that not only the illness of the mother but also the illness of the father may be significant when at the same time there is no older sister. In any case the conclusion is that the mother's illness increases the danger of the manifestation of a psychosis if at the same time no older sister is available.

It should not come as a surprise that older siblings, especially older sisters, are expected to help in caring for younger siblings. Boys are more likely to go their own way. Girls are not necessarily expected to play this role by their parents, but they have a better attitude and also greater inner readiness to look after younger siblings than boys have. In my opinion girls are biologically chosen for this role. The double lack of mother and older sister has an effect on only male children because of the greater capacity for resistance shown by girls. Another possibility is that younger sisters of patients in the absence of a mother and of an older sister may at times assume the role of a mother and are in that way less likely to be affected.

I had to leave the question unanswered whether **affected fathers,** apart from heredity, might be partly responsible for appearance of periodic catatonia. Against this possibility is the observation often made in social psychiatric circles that schizophrenics have affected mothers much more frequently than affected fathers. This applies especially to periodic catatonia, as indicated by many statements in the literature concerning the parentage of the majority of schizophrenics. ESSEN-MÖLLER (1941) gave a different explanation. Women on the average marry earlier and on the average develop a schizophreni disorder later than men so that there is a greater likelihood for them to have affected children. But the great difference which we found in our cases cannot be explained in this manner. Among the 55 affected parents listed in Tables 33 and 44 there were no less than 39 mothers but only 16 fathers. In the case of cataphasia which shows an equally strong hereditary tendency we find a much smaller difference between affected mothers and affected fathers. Probably the figures correspond to those of ESSEN-MÖLLER. HESTON (1966) however contradicted the notion that the pathological behaviour of the mother could lead to the illness, for the children with schizophrenic mothers who grew up in a children's home developed the illness as often as children who stayed with their schizophrenic mothers. But if the social psychiatric thesis is modified in a certain way we see that HESTON'S findings no longer contradict the facts. Presumable it is not a matter of the pathological behaviour of the mother, but her total absence, be it because she is hospitalized or because of her illness she was unable to be a proper mother. Children who grew up in a home likewise had to do without genuine maternal care. Later when we are considering early childhood schizophrenia we will see that staying in a home cannot replace the mother. Therefore the important thing is the absence of the mother, rather than her schizophrenic behaviour. This follows from the observation that older sisters may compensate for this lack; they of course cannot prevent the pathological behaviour of the mother, but for younger siblings they can replace the missing mother.

Another observation that we made confirmed that it was not the mother's pathological behaviour that mattered. Affected mothers do not have more affected children than healthy mothers with hereditary disposition; on the contrary, they have fewer. To our surprise we noticed this from the figures in Table 44. The general incidence of periodic catatonia among siblings according to Table 33 is 21.2%. According to Table 44 the incidence rises in the case of patients with healthy parents to 23.7% and falls in the case of affected parents to 17.1%. There must be something specific here for periodic catatonia, as when we come to consider cataphasia and the cycloid psychoses we observe that here patients with affected parents do have more affected siblings. This odd finding requires explanation.

By now we have found the following: the number of siblings of patients with periodic catatonia is small. There is a lack of older sisters, but this applies only to male patients; it also applies if at the same time the mother is affected; affected parents in comparison with healthy carriers of the hereditary disposition do not have more, but have fewer affected children. The large contradiction derived from these findings can be explained if it is borne in mind **that an unfavorable influence on the proband need not be at the same time an unfavorable influence on his siblings.** Where there is lack of a parent among the daughters only the oldest is at risk, as all others girl have an older sister; and among the brothers only those are at risk who are older than their oldest sister. The patients with affected parents who had no older sister were therefore more at risk than their siblings. If however in the case of healthy parents there are other reasons leading to manifestation of the disorder siblings would probably be equally affected. The chances of them becoming affected is just the same as with patients. Here is an example to clarify this: if among 10 patients 5 do not have an older sister and if they only became affected under these conditions then their siblings can expect to remain healthy. This is the case if one parent is affected. The other 5 patients whose parents are healthy are all equally at risk.

This interpretation may be confirmed if male patients who, because of lack of mother and older sister are vulnerable, have fewer affected siblings than female patients. According to Table 45 this is true; male patients have 18.9%, female patients have 22.7% affected siblings.

Furthermore Table 45 shows strikingly that male patients have so very many fewer siblings than female patients. This probably depends on the fact that female patients in all cases of endogenous psychoses have more siblings. But the difference is not particularly great. It is greater in the case of periodic catatonia because the male but not the female patients have relatively fewer older sisters.

From the fact that only boys lacking a parent are vulnerable it may not be concluded that periodic catatonia should occur more frequently in men than in women. In the case of the latter there may be other factors at work. In our investigations we found approximately the same number of female as of male patients with periodic catatonia. The question may be asked which circumstances could be more damaging for females. Perhaps it has to do

Table 44. Affected Siblings of Periodic Catatonics with Healthy and Affected Parents

	Number of Patients	"Corrected" Number of Siblings	Number of Affected Siblings	%
Both Parents Healthy	81	105.5	25	23.7
Father or Mother Affected	55	64.5	11	17.1
Mother Affected	39	41.5	7	16.9
Father Affected	16	23.0	4	17.4

Table 45. Number of Affected Siblings of Male and of Female Periodic Catatonics

	Number of Patients	"Corrected" Number of Siblings	Number of Affected Siblings	%
Male	67	68.5	13	18.9
Female	69	101.5	23	22.7
Both	136	170.0	36	21.2

with female constitution where a hereditary disposition by itself may give rise to the psychosis. It could also be that girls are generally more sensitive than boys. This fits in with the fact that M. BLEULER (1972) found that female schizophrenics had suffered more shocks before the onset of their illness.

Looking at our findings gives rise to the more specific question what effect lack of siblings has on male and female patients, and in the case of male patients how they are affected by lack of mother and older sister. In my opinion the presence of siblings provides a certain protection for the child against shocks from outside. Even the father would play a certain role here although we could not ascertain this from our findings. A **strong protection** of the child within the family, keeping anxiety producing influences at a distance help to prevent the manifestation of inherited periodic catatonia. The larger the number of siblings with whom the child is in contact, the more likely the child will remain healthy.

The general statement could be made that insufficient care of the child may be harmful even if parents and siblings are available but if they do not fulfil their task to protect the child. In the case of one family in which 4 of 10 siblings became ill and where the probability of periodic catatonia developing should have been diminished though the presence of so many siblings, we found the following circumstances: the father was an alcoholic who spent most of his time in bars. He often beat the children; if a smaller child was crying he hit it or he hit an older child which he held responsible for the younger child's noisiness. When the father came home the whole family became quiet, and the children hid in a corner. The mother tried to intervene but was pushed away and often beaten. One can understand that in such a family even though both parents and many siblings were available there was no sense of security.

In **investigations of twins** we could not at first confirm our interpretation. The observation is often made in the literature that the second born of monozygotic twins is more likely to become ill, especially if at the same time his body is weaker and if he is dominated by his partner. If this was true it could be argued that the weaker twin had a greater need of protection than the stronger. But among our 6 monozygotic twins with periodic catatonia the first born 5 female twins were either the only ones affected or were more severely affected than the second born. The affected male twin was the second born, but he dominated the partner who submitted but remained healthy. I believe that the explanation for this contradiction is that it is not a matter of the need of the child, but on the degree of protection provided from outside. All children require protection, the stronger twin not less than the weaker one. But the latter receives constantly additional care from his stronger partner. He keeps on feeling well protected whereas his partner is more exposed to the environment and has to rely on himself. Thus it appears that the stronger twin grows up with less protection than the weaker one.

We have to make the following assumptions: since the factors influencing the development of periodic catatonia are at work in childhood, there is a considerable time interval before the actual manifestation of the illness, therefore there must have been at first a sensitization of the nervous system in addition to the hereditary disposition for the symptoms to appear. It is not clear if stressors and anxieties in **adulthood** may be contributing factors. Our investigation showed no evidence for this.

For the sake of **prophylaxis** we have the following clues: from the point of view of heredity it remains to be seen what can be done. But to protect the child from excessive anxiety is a clear-cut task. In systematic catatonia the lack of children in our times may be compensated for through contact with other children; in periodic catatonia it remains a definite disadvantage. The presence of siblings helped to provide security where the family holds together and where other children cannot enter the picture. Only a community of children may help if the goal is to protect every single one. Unfortunately children are not always so considerate, on the contrary the stronger ones are liable to make the weaker ones "targets" of their aggressive attitudes. Here it is especially important that parents and educators be vigilant in order to prevent a vulnerable child from suffering further damage.

Affective Paraphrenia

In the case of affective paraphrenia I assume, as already indicated, that we are dealing with recessive heredity. There is a constitutional basis which increases the probability of the illness becoming manifest; this is acknowledged by the fact **that women are affected very much more frequently than men.** Table 33 showed 95 women and only 25 men. Among the relatives as well more women than men were affected. Table 33 showed 23 women and only 13 men among the affected siblings. The incidence of 13.3% among siblings would be considerably higher if men were affected as frequently as women. Looking for an answer as to why this should be so the **greater affective lability** of women will have to be taken into account. Perhaps it follows from this that not only may women's lability be of significance, but a lability of an individual sort as well.

I see a confirmation of this opinion from an observation of a **pair of monozygotic twins.** The twins were the sisters Felicitas and Lea. The illness of Felicitas began 10 years earlier and its course was much more severe. It might be thought that the illness of Lea was induced by Felicitas. The latter was always more serious than her sister, often clearly depressed. With every misfortune she lost her enjoyment of and interest in life. Her sister Lea on the contrary was usually vivacious and cheerful; she was pleased to be alive. One may assume that the affective lability of Felicitas was one of the reasons why her affective illness began earlier and had a more severe course.

There may be a relationship between this explanation and the findings of Table 40. Here it is noted that there are **far fewer younger siblings than older ones.**

In adding the number of older siblings as well as the younger ones according to this table, the overall ratio comes to 206:135 (male 42:23, female 164:112). Affective paraphrenics are often the youngest of the siblings. Among the 120 patients 36 were the youngest, 17 the oldest. Fewer younger siblings are to be expected in these days of smaller families, but this in no way explains such great differences. There must be internal causes especially involving affective paraphrenia. The problem may be solved as follows.

A child in its emotional life is much more labile than a grown-up. Normal maturity of younger children is delayed when they are dominated by older children. The development of the youngest child is often inhibited by parents through the way in which it is treated, thus slowing maturation. It is well known that the youngest children often stay childish much longer and in later life remain strongly dependent on their parents. Some may even be described psychiatrically as infantile. Thus it may happen that younger and youngest children in their emotional life remain more labile than older ones and in this way show a greater liability to develop affective paraphrenia. The way these children are treated results in repression of their own life; direct repression may have similar consequences.

Here is an example of a mother who through strict and repressive control of her **monozygotic twin daughters** inhibited their development. Both sisters became affective paraphrenics. They could not live their own lives as their mother supervised every move they made. When the girls were grown up it never occurred to them to undertake anything without their mother's permission. During adolescence they had no opportunity for contact with the opposite sex. Later in life one of the daughters, now more than 30 years old, and divorced, wanted to marry another man. Her mother did not want him and beat her daughter. The other twin sister was even more submissive. She became ill 22 years before her sister with affective paraphrenia. The twins having been treated in this way could never develop independent personalities. They remained childishly dependent and suggestible and emotionally quite labile.

In the literature it is often mentioned that a child who later became schizophrenic did not progress into the role expected of the child from his age and his sex. If the concept of "role" is more clearly defined, keeping in mind the insufficient maturation of affectivity, I would have to agree with this opinion but only in the case of affective paraphrenia.

The **prevention** of the development of affective paraphrenia is facilitated if in spite of the fundamental hereditary disposition care is taken that the bonds between the child and the parents and older siblings are loosened and the child is allowed to develop its own independence. This fits in with the efforts of psychotherapists when they try to separate infantile personalities from parental domination. However this should have been started earlier before the harmful development occurs.

Cataphasia

Apart from heredity there are other causative factors at work in cataphasia, the third form of unsystematic schizophrenia.

Table 40 shows that like affective paraphrenia, though not to the same degree, older siblings are more numerous than younger ones (227:176). The difference for female patients is small (70:57), for male patients greater (157:119). The question therefore arises as to why male patients with cataphasia have a relatively frequent absence of younger siblings. The following is a possible answer: in cataphasia there is a special form of thinking disorder, above all that of abstract thinking. Patients often speak about objective circumstances in an orderly manner, but quickly become perplexed when the conversation turns to theoretical questions. Perhaps younger siblings with their extrovert attitude and their concrete thinking exert a favorable influence over the older ones. If this is true a lack of younger siblings would affect particularly the male sex, since men more than women tend to become concerned with theoretical questions.

Our interpretation was supported when we checked the origin of our patients from urban and rural regions. In the past investigations along these lines showed nothing certain, because schizophrenia had always been taken as a whole. Table 46 shows the results of our investigations. The total numbers are a bit smaller when compared with Table 33 as we could not determine the place of birth of all patients. As can be seen, patients with cataphasia come relatively rarely from the country (32.5% compared with 49.5% of patients with affective paraphrenia and 43.7% of patients with periodic catatonia). However they come relatively frequently from a metropolis (37.9% compared with 25.2% and 27.4% of the two other forms, respectively). The findings are significant ($p < 0.05$). In connection with my train of thought I wish to point out that in the country concrete thinking is more prevalent than in the big city; in the latter, abstract thinking is more in demand not only at work but also in activities of everyday life. Thus I have reached the opinion that the probability of cataphasia becoming manifest is greater if there is a higher demand for abstract thinking in childhood and youth,

Table 46. Place of Birth of Patients with Unsystematic Schizophrenias (in Parentheses Percentage)

Place of Birth	Sex	Affective Paraphrenia	Cataphasia	Periodic Catatonia	Total
Under 2,000 Inhabitants	♂	11 (47.8)	39 (32.8)	26 (40.0)	76 (36.7)
	♀	44 (50.0)	16 (32.0)	33 (47.2)	93 (44.7)
	Both	55 (49.6)	55 (32.5)	59 (43.7)	169 (40.7)
2,000–20,000 Inhabitants	♂	4 (17.4)	25 (21.0)	16 (24.6)	45 (21.7)
	♀	15 (17.0)	13 (26.0)	8 (11.4)	36 (17.3)
	Both	19 (17.1)	38 (22.5)	24 (17.8)	81 (19.5)
20,000–100,000 Inhabitants	♂	1 (4.4)	10 (8.4)	7 (10.8)	18 (8.7)
	♀	8 (9.1)	2 (4.0)	8 (11.4)	18 (8.7)
	Both	9 (8.1)	12 (7.1)	15 (11.1)	36 (8.7)
Over 100,000 Inhabitants	♂	7 (30.4)	45 (37.8)	16 (24.6)	68 (32.9)
	♀	21 (23.9)	19 (38.0)	21 (30.0)	61 (29.3)
	Both	28 (25.2)	64 (37.9)	37 (27.4)	129 (31.1)
Total	♂	23	119	65	207
	♀	88	50	70	208
	Both	111	169	135	415

whereas this disorder, in spite of its hereditary disposition, does not usually become apparent if in childhood the thinking is predominantly directed into concrete channels.

If this interpretation is correct the question may be asked whether or not such an influence is possible in adulthood as well. In looking for a solution we tried to determine the **place of residence** of patients at the beginning of their illness apart from their place of birth. Taking into account the movement of the population from the country into the city we found that more people lived in large cities than a few decades beforehand. Accordingly among our unsystematic schizophrenics 132 resided in large cities, but only 119 had been born there. In affective paraphrenia the number of patients rose from 28 to 38, in periodic catatonia from 37 to 44. But in cataphasia there was no increase. Sixtyfour patients were born in large cities, and almost the same number, that is 65, lived in large cities at the beginning of the illness. The increase of the population due to moving from the country into the city showed no greater number of persons who developed cataphasia. It appears that people who do not grow up in the big city but only move there as adults are no longer subject to the harmful influence of the big city in the origin of cataphasia.

If in large cities excessive demands on thinking are a factor in the origin of cataphasia the incidence of this illness must have **increased in modern times;** for the rapid development of technology makes much higher demands on man's thinking than in earlier times. Unfortunately I cannot compare the number of my Frankfurt cases of cataphasia with that of my

Berlin cases, as in Frankfurt I had not yet come to know the inhibited form of cataphasia. However it is significant that in Frankfurt, that is about 3 decades earlier, I found only 17 patients with cataphasia besides 26 with affective paraphrenia and 23 with periodic catatonia, the inhibited form of cataphasia being much less common than the excited form. In the Berlin series there were 173 patients with cataphasia compared to 120 patients with affective paraphrenia and 136 with periodic catatonia. Therefore here in Berlin cataphasia occurs far more frequently.

Another finding gives a better result. When we first evaluated statistically the Berlin investigations there were 126 patients with cataphasia compared with 94 with affective paraphrenia and 105 with periodic catatonia. Additional investigations showed greater numbers and attained the values previously given. The result was that the increase from the new cases was 37.3% for cataphasia, 27.7% for affective paraphrenia, and 29.5% for periodic catatonia. Thus cataphasia occurred relatively more often in patients born later than those born earlier. The average year of birth for the earlier ones was 1923, for the later ones 1934. Hence this result shows that cataphasia has become more frequent in recent times.

From the fact that thinking processes in cataphasia may be damaged from excessive demands in the big city it may be assumed that **parents with cataphasia** have a harmful influence on their children not only because of their hereditary disposition but also through their own thinking disturbance. Therefore one would expect that affected parents would have more affected children than healthy parents with hereditary disposition. In periodic catatonia this was not so, but in cataphasia it was. Table 33 showed 173 patients with 50 affected parents, 19 fathers and 31 mothers. Since in the case of 5 patients both parents were affected, these 50 cases applied to 45 patients. As Table 47 shows, among siblings of patients with affected parents there were 26.8% affected siblings, whereas among the siblings of patients with healthy parents there were only 11.3%.

The difference between the number of affected siblings of healthy and that of affected parents cannot be explained on a hereditary basis. In the case of **dominant heredity** one of the parents is a full carrier, therefore should have the same number of affected children as an affected parent. In the case of recessive heredity healthy parents would have only half as many

Table 47. Affected Siblings of Cataphasic Patients with Healthy and Affected Parents

	Number	"Corrected" Number of Siblings	Affected Siblings	%
Patients with Healthy Parents	123	231.0	26	11.3
Patients with Affected Parents	45	74.5	20	26.8
Total	168	305.5	46	15.5

affected children (25%) as a healthy parent with an affected one (50%), but in these cases affected siblings should occur more often than affected parents, which is not the case according to Table 33. It might be assumed that constitution played a certain role. Hence the situation would be compatible with recessive heredity; there should be more affected siblings than affected parents. In fact in cataphasia, according to Table 33 there are a few more affected parents than affected siblings (15.3% compared with 14.8%). As far as cataphasia is concerned I concur with the opinion of social psychiatrists that affected parents have an unfavorable influence on their children in their development.

The disadvantage coming from parents with cataphasia becomes considerably greater when both parents are affected, which was the case with 5 patients. A special form of heredity would not explain this. Among our patients with periodic catatonia there was only one where both parents were affected; among our affective paraphrenics there were none.

Another question is whether the influence of the father is the same as that of the mother. Among affected parents we found 18 fathers and 31 mothers. The difference may be explained according to the calculations of ESSEN-MÖLLER mentioned above. Accordingly the influence of the father and the mother is similar. This is understandable as thought processes develop during a period of time when the care of children is no longer provided primarily by the mother.

Furthermore Table 47 shows that affected parents have fewer children than healthy parents (1,6 compared with 1,9). This was to be expected, and corresponds approximately to the situation for periodic catatonia.

Exogenous and Constitutional Causes of Cycloid Psychoses

In the so-called schizophrenias, that is using the term "schizophrenia" in the wide concept, as it is quite customary these days, about one third of all cases belong to the cycloid psychoses.

From Table 48 it can be seen that healthy parents have more children than affected parents (1.8 compared with 1.4). Furthermore patients with healthy parents have far fewer affected children than patients with affected parents (3.1% compared with 15.4%). It is not a question of recessive heredity – more parents than siblings are affected – therefore it seems from the figures that in cycloid psychoses heredity plays no important role.

It is clear that the illness of parents leads directly to that of their children. The difference in spite of the small number of affected parents is significant ($p < 0.05$). Therefore we have to look so much the more for external causes. Let us examine the **sibships.**

Table 49 shows large differences in the number of siblings of patients with the three cycloid psychoses. The numbers are not "corrected" since it is now not a question of heredity but one of mutual influence. Patients with anxiety-happiness psychosis and motility psychosis have many siblings (286.2 and 300,0), whereas patients with confusion psychosis have relatively few (186.3).

Female patients with confusion psychosis have especially few siblings (175.0). In reverse, female patients with anxiety-happiness psychosis and motility psychosis have very many siblings (311.1 and 329.3). Patients with confusion psychosis have unusually few older sisters, whereas patients with anxiety-happiness psychosis have many more younger sisters, and patients with motility psychosis have many more older.

A comparison between confusion psychosis and motility psychosis may be of interest. The question arises as to why the presence of many siblings contributes to the development of **motility psychosis,** whereas lack of siblings increases the likelihood of **developing confusion psychosis.** In the investigations of twins (LEONHARD 1976a) I pointed to an interesting parallel: among both monozygotic and dizygotic twins there were relatively many cases of motility psychosis, whereas only one single case of confusion psychosis was found (in a pair of dizygotic twins). Being twins as well as having many siblings increases susceptibility for motility psychosis to develop and decreases susceptibility for confusion psychosis to develop. **Close contact between twins appears to correspond to the increased contact when many siblings are available.**

In this connection – unlike systematic schizophrenias – monozygotic and dizygotic twins are similar; it follows that the influence begins quite early in childhood at a time when the psychic similarity or difference between twins does not yet play an essential role. Should the influence begin in early infancy it could – apart from twins – come only from older siblings, as a younger sibling has not yet arrived. It does not make much difference whether the influence is limited to the period of early infancy or to the first years of life; for if during this time a younger sibling is already present, because of its helplessness it cannot have an influence on another child. In fact further observations confirm that the influence occurs in early childhood, for the difference in the number of siblings in confusion psychosis and motility psychosis is produced only through **older siblings.** According to Table 49 the ratios in these two psychoses of older siblings is 172.3:78.1, and that of the younger siblings 127.7:108.2. This seems confirm that the first years of life are decisive. Anxiety-happiness psychosis which in the number of siblings corresponds more or less to motility psychosis does not show the rel-

Table 48. Affected Siblings of Patients with Cycloid Psychoses with Healthy and Affected Parents

	Number	"Corrected" Number of Siblings	Number of Affected Siblings	%
Patients with Healthy Parents	197	360.5	11	3.1
Patients with Affected Parents	24	32.5	5	15.4
Total	221	393.0	16	4.1

Table 49. Number of Siblings of Patients with Cycloid Psychoses (in Parentheses per 100 Patients)

	Number of Patients	Older Brothers	Younger Brothers	Older Sisters	Younger Sisters	Older Siblings	Younger Siblings	Total
Anxiety-Happiness Psychosis								
Male	38	22 (57.9)	28 (73.5)	34 (89.5)	18 (47.4)	56 (147.4)	46 (121.1)	102 (268.4)
Female	27	15 (55.6)	19 (70.4)	18 (66.7)	32 (118.5)	33 (122.2)	51 (188.9)	84 (311.1)
Both	65					89 (136.9)	97 (149.2)	186 (286.2)
Confusion Psychosis								
Male	41	13 (31.7)	22 (53.7)	22 (53.7)	23 (56.1)	35 (85.4)	45 (109.8)	80 (195.1)
Female	32	15 (46.9)	16 (50.0)	7 (21.9)	18 (56.3)	22 (68.8)	34 (106.3)	56 (175.0)
Both	73					57 (78.1)	79 (108.8)	136 (186.3)
Motility Psychosis								
Male	25	19 (76.0)	17 (68.0)	9 (36.0)	13 (52.0)	28 (112.0)	30 (120.0)	58 (232.0)
Female	58	65 (112.1)	38 (65.3)	50 (86.2)	38 (65.5)	115 (198.3)	76 (131.0)	191 (329.3)
Both	83					143 (172.3)	106 (127.7)	249 (300.0)
Total								
Male	104							
Female	117							
Both	221	149 (67.4)	140 (63.3)	140 (63.3)	142 (64.3)	289 (130.8)	282 (127.6)	571 (258.4)

ative increase of older siblings. The latter are even somewhat less in number than younger siblings. Apart from confusion psychosis and motility psychosis only the pure phasic psychoses indicate that the first years of life are critical.

To continue the comparison of numbers of siblings here are the average numbers for the cycloid psychoses added together: 221 patients have 571 siblings. If there was an equal distribution in the four age groups there would be 142.8, that is 64.6 siblings per 100 probands. Using this formula enabled me to recognize deviations upwards or downwards. Having described my findings I have tried to explain them. Since my interpretations cannot be truly proved to be correct they are not binding and may be replaced by better, less complicated, explanations. However the findings remain. The differences shown in spite of small numbers are significant in so far as they are brought to light ($p < 0.05$).

Starting with anxiety-happiness psychosis, we are dealing with a pronounced affective disorder, therefore disturbances of the emotional development should be investigated. Ordinarily the emotions undergo a deepening during childhood. A small child cries easily and readily, but the emotions during this time are not deep. With the passage of time this changes; older children are capable of having deep emotions, they can become very enthusiastic, or they can despair as far as suicide. During this period of deepening emotions disturbances should be looked for which are of significance in the development of anxiety-happiness psychosis.

It is striking that **female patients** with anxiety-happiness psychosis have nearly twice as many younger sisters (118.5) as the average (64). I presume that **tensions in the emotional life** caused by younger sisters have a role to play with the older ones. An increase of younger sisters raises the tensions. Parents expect their daughters to do housework; that was certainly the case in former times when there were more children in the family. Older girls were obliged to help with the care of younger siblings, a disadvantage to them when it seemed that their mother preferred the younger ones. The tensions produced in the older girls had an unfavorable influence in the development of their emotional life and thus contribute to the genesis of anxiety-happiness psychosis. The fact that brothers were not expected to do so much work did not play an essential role in the minds of the sisters as they had a different attitude towards life. If they were preferred by their mother this worried the girls very little. At any rate this was the case at the time when the emancipation of women had not yet progressed. Accordingly the average of 70.4% with younger brothers was hardly exceeded.

In the case of men such situations of jealousy do not play an essential role. The figures concerning **male patients** show very little difference in Table 49. This does not of course exclude the possibility that tensions of a different kind affecting boys may contribute to the genesis of anxiety-happiness psychosis. That this is so may be gleaned from the fact that male patients also have relatively many siblings, for in large families there has to be more sharing and individuals receive less attention. That way tensions may arise more easily. I would like to draw attention to something else shown in Table 49 concerning male patients, but first I wish to point out that the significance of tensions in the origin of anxiety-happiness psychosis were revealed by our **investigations of twins.**

Of the discordant monozygotic twins it was always the mentally weaker one who through his submissiveness tended to become jealous. We have noticed that this did not apply in periodic catatonia. Concrete indications of the presence of jealousy were found among our twins with anxiety-happiness psychosis. In one case the patient was not only submissive to his twin brother but also was valued less by his parents. Similarly a girl patient whose healthy twin sister was accepted in an advanced school suffered greatly when she was not accepted. During childhood another female patient was often physically ill which, though not severe, had a negative influence on her performance. Her sister was always healthy and dominated her twin. Still another patient at the age of 12 years suffered a fractured spine and for many years had to sleep in a plaster cast. Many tensions arose in this disabled twin; there was much fighting between the girls which is generally quite rare with monozygotic twins. Clearly tensions arose only in the affected twin. Therefore the discordance is understandable.

From the figures for **male patients** with anxiety-happiness-psychosis it can be seen that there are more older sisters (89.5) and fewer younger sisters (47.4). Here I would like to add some thoughts concerning the age differences between husband and wife. Usually a man chooses a younger woman, and a woman chooses an older man. The fact that the capacity for reproduction disappears earlier in women than in men is not decisive in the choice of partner; lovers do not think what will happen 20 or 30 years later. But the capacity for reproduction is biologically determined; this may be the reason why the choice of partner has a biological basis with regard to different ages between men and women. These thoughts led me to surmise that in nature an age difference is foreseen in the relationship between the sexes starting already in childhood at a time when sexual feelings are not yet evident; disharmony may arise if the son has many older and few younger sisters. Female patients however have fewer older brothers than younger ones (55.6 compared with 70.4); girls have been shown repeatedly to cope better with disharmony than boys.

In looking for explanations of the findings with **motility psychosis and confusion psychosis** I will start with the statement that in the former there is a disturbance of psychomotor functioning, in the latter a disturbance of thinking. It seems that patients with motility psychosis who have many older siblings have an impairment in development in the one psychological area, whereas with lack of siblings in the case of patients with confusion psychosis there is impairment in the other. In dealing with systematic schizophrenia we have already discussed the disturbance in psychomotor functioning and in thinking, but there we considered the importance of higher mental processes by means of which the child matures into adulthood. The developments which we are considering here take place on a much lower psychological level. These developments, I assume, lead to a condition where external impressions are no longer accepted just passively, but are reinforced through thinking. Thus they are available for conditioned reflex processes appearing simultaneously, in that in parallel to the first thought processes there is the development of psychomotor functioning.

For proper thinking many stimuli from outside must be advantageous as it develops in a changing environment. Should stimuli be reduced because few siblings are available the danger arises of a disturbance developing leading to confusion psychosis. On the other hand if there are many stimuli because of the presence of many siblings, although for proper thinking this is not a disadvantage it may lead to an excessive demand on psychomotor functioning, with the risk of a reflex and automatic disturbance in line with motility psychosis. This is my explanation for the fact that there is a great difference between the number of siblings, more importantly of older siblings: only 78.1 older siblings of patients with confusion psychosis, compared with 172.3 older siblings of patients with motility psychosis. Girls being more labile than the more robust boys makes them more susceptible to excessive demands or excessive stimulation. Thus we find that motility psychosis, as Table 33 showed, is much more frequent in women than in men.

May we point to a parallel situation existing between the presence of many siblings and that of twins: excessive demands in both cases play a role as we see in the case of the two monozygotic discordant twins mentioned above. In both cases the affected twin on whom excessive demands had been made was the second born and physically and mentally weaker than his partner. Furthermore in one case the affected twin suffered from asphyxia at birth and perhaps suffered brain damage. The other monozygotic twins did not show such differences; probably each one had the same influence on the other with result that both became ill. Thus, as shown above, **Galton's law** is not valid in the case of twins with motility psychosis. The concordance here does not arise through heredity, but from the fact that each twin produces the illness in the other. Dizygotic twins are different, here an excessive influence of the one over the other makes the discordance understandable.

Important peculiarities emerge if **male and female patients** are considered separately. Female patients have an unusually high number of older brothers (112.1), the reverse is true for older sisters. It is to be borne in mind that boys who are more robust may cause overstimulation more readily than girls who are more gentle. On the other hand a smaller number of older sisters has quite a special effect on female patients with confusion psychosis (21.9). Stimulation of the thought process is evidently more appropriate and consequently more effective from girl to girl. Hence there is a smaller number of brothers in the case of male patients (31.7).

Contrary to expectations in **male patients** with motility psychosis the number of older sisters is not higher, but considerably lower (36.0). This can be understood from the fact that weaker girls cannot produce overstimulation in their brothers. But why is there a smaller number of sisters? When we were considering periodic catatonia we saw that the oldest girl in a series of siblings played a special role. Evidently there is a biological provision for them to exercise a motherly role. We saw that if the real mother was missing the oldest girl may replace her. The present finding shows that the oldest girl of the siblings is expected to act as a mother substitute. The baby's reaction to its mother's loving care is at first a motor response, with an increasingly affective reaction in the background. For the normal development of motor processes the interchange between mother and the small child is surely necessary. An older sister appears to be biologically available to support the mother in her task. Thus I see the fact that lack of an older sister impairs motor development and predisposes to the appearance of motility psychosis. In cases where this lack affects only boys a parallel is found in periodic catatonia, for there the lack of an older sister likewise is significant only for boys. In the case of girls it appears that their development proceeds generally in a more autonomous manner, that is with less dependence on connection with the environment.

In addition to all the above the constellation of siblings turned out to be highly significant in the case of the cycloid psychoses. It may be asked what circumstances contribute to their emergence as hereditary disposition appears to be relatively unimportant.

This question leads to a consideration of the **temperament** of patients and their parents. As Frau von Trostorff (1966) showed in greater detail in cycloid psychoses, we often found peculiarities of temperament, somewhat resembling the symptoms of the illness. Patients with motility psychosis when in remission often showed a **"motion temperament"**; they were described as having vivid expressive movements and vivaciously addressing other people; or on the contrary, because of the bipolarity of the illness their expressivity was clearly quite poor. Patients with confusion psychosis when in remission often showed an excited or inhibited **"thought temperament"**, either lacking concentration or slowing down in their conversations with others. Patients with anxiety-happiness psychosis when in remission often showed an **"anxiety-happiness temperament"**, that is, there were exceedingly strong and sudden fluctuations in their emotions. So too among many relatives of these patients we found such temperaments. By themselves they do not show unfavorable fluctuations in temperament; the opposite may be observed. The larger number of expressive movements found in patients with motility psychosis when in remission often increases their capacity for contact in a delightful manner; slower thinking found in healthy periods of patients with confusion psychosis when in remission is usually more thorough; a certain exuberance increases the depth of human emotion. However temperaments of this sort must have more serious consequences when the constellation of siblings is unfavorable. To be favorable a constellation has to be in harmony with constitutional make-up. There is certainly no danger when, for example, the constellation of siblings tends towards anxiety-happiness psychosis if the constitutional individuality is characterized by emotion temperament or a thought temperament.

Such connections are generally of great interest in the question of the origin of mental disorders. Peculiarities of temperament are normal, constellations of siblings are commonplace, but with unfavorable combinations where two tendencies point in the same direction a psychosis may occur. Health may be turned into sickness through an unfavorable combination of circumstances.

Since the two components mentioned above which may contribute to the development of a cycloid psychosis are very common one can understand that these psychoses occur quite frequently. In **simpler more primitive cultures** they occur probably even more frequently. It was shown that with anxiety-happiness psychosis and motility psychosis many siblings were available, whereas with confusion psychosis there were few. It can therefore be assumed that in cultures where many children per family are common the first two psychoses occur relatively often, whereas the third is rare. Presumably cycloid psychoses occur much more frequently in people with many children than in our own culture with fewer children. Previously it has been mentioned that systematic schizophrenia occured more frequently per family in our culture. It follows that in the wide concept of schizophrenia in which the cycloid psychoses are included the total incidence is similar in the various cultures.

Taking into account different causative factors in the genesis of cycloid psychoses points to different kinds of **prophylaxis.** In order to prevent anxiety-happiness psychosis, children should not be exposed to undue tensions. Parents and educators bear a particular responsibility here. Frequently the behaviour of parents produce rather than prevent these tensions. Children are not infrequently exposed to conflicts between their parents. Moreover the child's jealousy may be nourished by the fact that parents do not always treat all their children in the same way; often one child is preferred, another is placed at a disadvantage. The parents may dispute this, but objectively it is often observed. To protect a child from being disadvantaged may not be enough. Children vary greatly in their reactions, this should be taken into careful consideration. This is described in more detail in my book: **"Kinderneurosen und Kinderpersönlichkeiten"** (LEONHARD 1967). It must be kept in mind that a persistently sensitive child tends to become jealous. Certain other **"accentuated" characteristics** (LEONHARD 1970b)lead to the kind of tension which increases the danger of anxiety-happiness psychosis developing later. Anxious and emotional children are more easily affected and their anxiety is increased, especially when they witness parents quarreling. They react with greater sensitivity than others even to everyday events; educators should be aware of this when dealing with an anxious or emotional child. Educators who have empathy for the child's mind do not have to be told this, but for others it should be stressed. Another situation applies to children who are excessively precise. They are very conscientious and suffer greatly when they have made a mistake. The duties in school are often a heavy burden for them especially if they find it difficult to carry out their tasks. Such children when they make mistakes should be consoled rather than scolded. I cannot verify that pedagogues take into sufficient consideration the various ways that children react. They do try to avoid the mistake of taking sides, a mistake which many parents make, but in a deeper sense their treatment is not thereby justified. A small penalty which hardly touches a robust boy may deeply hurt a sensitive child. This does not mean that I am in favor of different punishment for similar mistakes: outwardly the punishment may be the same, but it has to be meted out to one child with quite different words and a quite different emotional tone of voice from that meted out to another child. In this way not only the development of neuroses may be prevented, but it helps to prevent cycloid psychoses as well. Patients with anxiety-happiness psychosis have often been made very anxious as children. Neuroses and psychoses here have exceptionally a similar basis.

For the prophylaxis of **motility psychosis** and **confusion psychosis** the difference in number of available siblings creates a difficult task. Should there be many siblings careful attention should be given to prevent motility psychosis; should there be few siblings care should be taken to prevent confusion psychoses. Surely some good results may be obtained if the people concerned were able to make carefully considered adjustments in their relations with the sensitive child. What matters above all is the presence or absence of older siblings. If they are present, they have to be kept at a distance to prevent excessive stimulation of the young sibling. Should older siblings be missing, other older children from outside the family may be of help. In both psychoses the presence of an older sister with the mother is beneficial. If an older sister is not available it becomes that much more important for the mother to provide sufficient care for her baby and small child. It is not simply a matter of physical care, but also a rich mental involvement with the child. It was an advantage in the former German Democratic Republic that there were good possibilities for a woman who had an outside position to devote enough time to her baby. She received sufficient financial support enabling her to interrupt her work for a whole year without losing her job.

Exogenous and Constitutional Causes of Manic-Depressive Illness

In recent years we did not have many cases of manic-depressive illness for our investigations. The reason for this was as already mentioned that our activities were primarily on units with chronic patients. Nevertheless the findings complement our earlier results.

In looking for reasons for the fluctuations of the symptoms of manic-depressive illness the first fact to be noted is that among the patients **affected woman** are much more common than **affected men.** Among our present patients there were 40 women and 20 men. Among the 18 affected siblings listed in Table 33 14 are female and only 4 are male. Thus it seems that the female constitution contributes to the appearance of manic-depressive illness. Women's more marked affective excitability presumably produces a greater propensity towards affective illness. We have already described a similar situation in affective paraphrenia.

There may be another cause for fluctuations in symptoms in that patients with affected parents have more siblings than patients with healthy parents.

According to Table 50 there are 15.5% of affected siblings whose parents are healthy compared with 28.1% whose parents were affected. I presume that the illness of the parents produces a restlessness in their children resulting in an affective lability which may later contribute to the manifestation of the illness. But concrete proofs for this statement I do not have.

In examining the role of siblings the small numbers gave rise to accidental fluctuations with no exact criteria to show whether or not there were too many or too few siblings. However the **total number of siblings** of patients with manic-depressive illness was strikingly small. If no "correction" is made, there were 193.4 per 100 patients, whereas according to Table 49 in the cycloid psychoses there were 258.4 siblings per 100 patients, in spite of the fact that confusion psychosis with its number of siblings was included. From these findings I do not wish to come to any final conclusions regarding the genesis of the illness, but I assume that patients with manic-depressive illness on the one hand have a limited number of children because of the many phases of the illness, and on the other hand being aware of the risks of heredity decide to have few children.

For symptoms of manic-depressive illness to develop there is a constitutional peculiarity related to temperament. Patients with this disease exhibit various transitions to hypomanic, dysthymic and cyclothymic temperaments showing no exact borderlines with cheerful, serious, and labile temperaments. Of course such trends – which may be grouped together as **affective temperaments** – in general have no connection with manic-depressive disease. But there may be exaggerations when people become distinctly hypomanic, dysthymic or cyclothymic. In an investigation I undertook with KORFF and SCHULZ (LEONHARD et al. 1962) we found that 17.7% of parents of patients showed affective temperaments to a degree which could be considered abnormal, and the same in 11.7% of siblings. In addition there were many affective temperaments within normal limits, in fact no less than 31.3% in parents and 25.6% in siblings. There is a temptation to consider all these characteristics as mild forms of the disease and call them latent manic-depressive psychosis. Adding the percentages to that of the manifestly ill persons 50% of siblings would theoretically be expected in dominant heredity. But in the case of parents there would be 67.3%. Therefore it is quite

Table 50. Affected Siblings of Manic-Depressive Patients with Healthy and Affected Parents

	Number	"Corrected" Number of Siblings	Affected Siblings	%
Patients with Healthy Parents	35	58.0	9	15.5
Patients with Affected Parents	21	32.0	9	28.1
Total	56	90.0	18	20.0

improbable that these temperaments are all expressions of manic-depressive illness. This would only apply in those temperaments where the degree reaches abnormal proportions. Our results led to an explanation that I still accept today: the probability of manic-depressive illness becoming manifest is increased if in addition to hereditary disposition there is an independent affective temperament. In this way may be understood the juxtaposition of the illness and the normal temperaments in the families of manic-depressives. Even theoretically this explanation should apply, as illness and temperament, in spite of their fundamental differences, do have a similar psychic basis, reminiscent of the point of view which I mentioned in discussing the genesis of cycloid psychoses; however the characteristics of temperament differ here.

Investigation of **twins** gave us another opportunity to observe the temperaments of manic-depressives. It is known that monozygotic twins in spite of their great similarities are often different in their temperament. Usually the differences are insignificant so that they might be ascribed to influences from outside. If the physically stronger twin becomes the leader he may also appear as more vivid. But sometimes there are opposite temperaments. I have already mentioned a pair of twins with affective paraphrenia where one partner tended to be cheerful whereas the other was more serious and slightly depressed. In manic-depressive illness one finds this more frequently as there is a much greater tendency towards a bipolar course than in affective paraphrenia. I assume that major differences in temperament shown by monozygotic twins can no longer be understood as normal variants but indicate the presence of bipolar disease.

The monozygotic twins Leni and Martha G. were quite different in temperament. Leni was always quiet and a bit lame, whereas Martha was usually cheerful and even jolly. Leni hat her first depression at the age of 38 years. Twenty years later she was depressed again. After that she had both manic and depressed phases. Martha had her first depression at the age of 64, and her second at the age of 66 years when she committed suicide. I suppose that here the opposite temperaments were already an expression of bipolar disease, that is they should be thought of as latent manifestations of the illness. Corresponding to the fact that Leni's depressive character appeared to be more pathological than the cheerful character of her sister, her illness began many years earlier. It is not necessary that temperament and illness should be associated with the same pole. The cheerful Martha only had depressions during her life, the more serious Leni also had manic phases.

If temperament, as is probable in this particular case, is an expression of the illness itself, it cannot be counted as an etiological factor. But otherwise I suppose it contributes to the manifestation of the illness in that it supervenes as independent characteristic to the hereditary disposition.

The significance of exogenous factors thereby remains untouched. If anxieties are produced in the child through illness of a parent the question arises whether irritating influences may also contribute to the appearance of a phase of the illness in adulthood. For some this seems self-evident, for instance in English psychiatry the opinion is sometimes expressed that endogenous and reactive depressions cannot be distinguished. German psychiatry spoke in the past about "provoked" melancholias and manias; today little is said about this, as it is felt that the reactive factor rarely plays a role in the endogenous illness. From my findings I cannot make any definite statement about this.

Prophylaxis of manic-depressive illness may be facilitated by means of understanding that during childhood external influences play an essential role, as affected parents have more affected children than healthy people with hereditary disposition. Everything should be done to protect children in families with manic-depressive illness from strong anxieties and by exposing them to a stress-free environment. The necessity of avoiding family conflicts in the presence of children has already been dealt with in the discussion of anxiety-happiness psychosis, which of course is somewhat related to manic-depressive illness. This recommendation must be repeated here. If the mother herself is affected her absence brings for the child other anxieties making it difficult to do the right thing. What was stated for the prophylaxis of anxiety-happiness psychosis should be stressed here as well, keeping in mind that manic-depressive illness has a hereditary nature, which is essentially not the case in cycloid psychosis.

Exogenous and Constitutional Causes of Pure Phasic Psychoses

Regarding the pure phasic psychoses, that is, **pure mania** and **pure euphorias,** as well as **pure melancholia** and **pure depressions,** for similar reasons as in the case of manic-depressive illness, not many new cases were available for my investigations. Some of the results therefore are not conclusive, but together with several other results pointing in the same direction help to show them to be valid. Thus in Table 33 it can be seen that patients with pure phasic psychoses have far more affected parents than affected siblings (5.8% compared with 3.1%). This may not be particularly significant as there were only 71 patients. But in Table 34 we saw that this ratio was found to be the same in both the depressive and euphoric forms of the illness; in both cases parents are affected far more often than siblings (2.1% compared with 1.4% and 7.8% compared with 4.5%, respectively). In Table 51 pure melancholia and pure depression have been separated. In both cases again affected parents are more frequent than affected siblings (12.5% compared with 7.5% and 5.2% compared with 3.3%, respectively). This is certainly a reliable finding. Moreover similar results were obtained in the earlier investigations.

The psychoses of relatives according to expectations showed the **same polarity** as the patients. Depressed patients had only depressed relatives. In the case of the two relatives of euphoric patients shown in Table 34 the mother of a confabulatory-euphoric patient was also confabulatory-euphoric and likewise a manic patient had a manic sister. Monopolarity is confirmed in that the relatives have the same condition as the patients.

With regard to the pure forms, in view of the low hereditary disposition external or constitutional causes must be particularly decisive. In our experience with the other endogenous psychoses we were always inquiring into the **number of siblings.** It was shown that euphoric forms had considerably greater numbers of siblings than depressive forms. According to Table 52 there are 344 siblings per 100 euphoric patients, but only 258.7 per 100 depressive patients. It should be kept in mind that patients with euphoric psychoses often have parents with a hypomanic temperament which may increase their chances of having children. But we do not have to pursue this idea, as the following observation shows that other causes must be at work. If older and younger siblings are separated, Table 52 shows essentially that only the older siblings are a factor to explain the difference. This is quite large (192.0 compared with 123.9). In the case of younger siblings the difference is not significant (152.0 compared with 134.8). If it had been a question of the fertility of the parents the different frequency of older and younger siblings could not be understood.

From this numerical relationship I take it that the probability of euphoric forms appearing rises if there is an **increase in older siblings,** whereas in the case of depressive forms it rises through a **decrease of older siblings.**

If this is so, we should have a look at a **transcultural observation.** Among more primitive people with their larger number of children per family, mania occurs much more frequently, as Kraepelin found out in Java. This has been often confirmed. In our area the euphoric forms of pure phasic psychoses are even more rare, as shown in Tables 34 and 52. As they tend to become chronic we were able to observe them relatively often. If it is believed that pure mania and pure euphorias are rare here, it should be determined whether mania in more primitive people belongs to the pure

Table 51. Affected Parents and Affected Siblings in Pure Melancholia and Pure Depressive Psychoses

	Number of Patients	"Corrected" Number of Parents	Affected Parents	%	"Corrected" Number of Siblings	Affected Siblings	%
Pure Melancholia	16	32	4	12.5	26.5	2	7.5
Pure Depressive Psychoses	30	58	3	5.2	61.5	2	3.3

Table 52. Number of Siblings of Patients with Monopolar Psychoses (in Parentheses per 100 Patients)

	Number of Patients	Number of Older Siblings	Number of Younger Siblings	Total
Euphoric Forms	25	48 (192.0)	38 (152.0)	86 (344.0)
Depressive Forms	46	57 (123.9)	62 (134.8)	119 (258.7)

form or whether manic-depressive disease with manic phases occurs more often than here. With regard to the latter there are great differences independent of cultural circumstances. As pointed out before, the Frankfurt patients showed manic phases much more often than the Berlin patients. If, in general, only **older siblings** make a difference, my explanation for this is similar to what is seen in confusion psychosis and in motility psychosis, that is that the influence is present in the first years of life at a time when younger siblings are either not yet available or if available in their helplessness are unable to have any influence on their older siblings.

It is surely the development of the **emotional life** to which the influence applies, for the disturbance in pure euphorias and pure depressions affects emotional side of the psyche, in pure mania and pure melancholia the, apart from the impaired thought process and psychomotor function, the disturbance especially affects the emotions. This had already been mentioned in the discussion of the development of this psychic sphere; now we are dealing with the first appearance of emotions that can already be observed in the newly born. Some of these emotional characteristics must have started to develop even before birth, as they are present immediately after birth. The most important impulses (demand for nutrition, for fluids, for oxygen, for bodily warmth) are visible quite soon after birth.

In order to explain the relatively **large number of children** in the case of euphoric forms and the relatively **small number of children** in the depressive forms I suggest – tentatively, as a better explanation may be found – that the stimulations from the relatively many siblings may lead to an excess which may contribute to the origin of mania; whereas with relatively small numbers of children a lack of stimulation may lead to a depressive psychosis. The vivacity and overactivity seen in mania would point to "too much", whereas the inhibition of depressions would point to "too little". I have given a similar explanation to the fact that patients with motility psychosis have more siblings, whereas patients with confusion psychosis have fewer. In those cases events in early childhood were also significant – I did not find this with any other endogenous psychosis – therefore a similar interpretation may be justified.

An inner tendency towards developing the illness is revealed as in the case of manic-depressive illness through the frequent presence of an affective temperament in the family. In the work of KORFF and SCHULZ mentioned above among 263 parents and siblings of patients with pure depression 34.2% showed an affective temperament. With respect to the low hereditary disposition of the illness the presence of an affective temperament is probably here even more essential than in manic-depressive illness. In connection with the more vivacious and labile temperament of women the pure forms are represented much more frequently in women than in men. Table 33 showed a ratio of 45:26.

While a manic-depressive with only depressive phases may have a hypomanic temperament, in the pure forms there is a **parallelism between illness and temperament.** Should that not be so in a certain case it is probable that a manic-depressive illness was wrongly diagnosed as a pure form. I do believe that bipolar affective disorder can be diagnosed from the clinical picture even if there have been only depressive phases, but in rare cases it may look so much like a pure form that there may be a mix-up. This had to be kept in mind when we found in cooperation with KORFF and SCHULZ (see above) that there were cheerful temperaments among the parents and siblings of patients with depression. Not many, that is only 2.7% of those counted, had a clearcut diagnosis of hypomania. In a later investigation (LEONHARD 1963) among the parents and siblings of pure euphoric patients no one with a definite dysthymic personality was found although some were morose. A slight tendency of the temperament towards the opposite pole does not mean that the diagnosis is wrong. Parallelism can be expected only in the very special situation in which the pure form manifests itself. A person with self-tortured depression as expected has a self-tortured temperament, but in all other areas of the emotional sphere there needs to be no deviation from the norm; there may even be a tendency towards cheerfulness in the form of mixed emotions.

The temperament of parents must be considered not only as a constitutional peculiarity that may reappear in the child and possibly lead to illness, but also as an indication of the way in which the parents's behaviour influences the child. If as we assumed the siblings have an effect then also do the parents. A hypomanic mother in her vivacity may be so stimulating that her child could develop manic disease; the same thing may happen when there is too much stimulation from siblings. A dysthymic mother in a state of inhibition may not stimulate her child sufficiently; this is the same type of disadvantage seen when there was a lack of siblings. Thus an affective temperament on the part of the parents may contribute to the appearance of a pure phasic psychosis not only on a constitutional basis but also through its presence during the development of the child. In enlarging this supposition, that the illness itself, that is a pathologically manic or depressive behaviour of the mother, has a negative influence on the development of the child and may lead to the appearance of a psychosis, we learn that this supports the fact that we found many more affected parents than affected siblings. It is true that in other endogenous psychoses affected parents predominate, but in no case to such a degree. The question should be pursued whether patients in their early childhood development were exposed for a period of time to mothers going through a manic or a depressive phase. Unfortunately I have no data to answer this question.

Mothers rather than fathers were mentioned since among 8 affected parents listed in Table 33 seven mothers and only one father were noted. This too shows that it is not so much a matter of heredity as the illness of the mother. Against heredity there is also the fact that patients with affected parents have relatively more affected siblings than patients with healthy parents. The 8 patients with affected parents had 2 affected siblings, the 63 patients with healthy parents had only 3 affected siblings. There is no question of recessive heredity.

However in one special case, **pure melancholia,** heredity does play a role, for as shown in Table 51 there were 7.5% affected siblings. Here also was a great excess of affected parents (12.5%). I would not have taken into consideration these findings because of their small absolute numbers if I had not found something similar in earlier investigations. Another explanation is possible without pointing to heredity. In contrast to pure depressions pure melancholia shows inhibition of thinking and acting. This illness must therefore lead more strongly to a lack of stimulation for the children. Consequently such parents may contribute more to the illness of their child and often have an adverse influence on a second child as well. This may be the explanation for the increase in the number of affected parents and of affected siblings.

The question of the etiology of pure phasic psychoses is not hereby concluded, for a highly peculiar finding requires clarification. Among our **monozygotic twins** there were 6 cases of pure depression, of which 5 had the **agitated type of depression.** It was remarkable that in all 6 cases the **twin partner was also affected.** The rule according to which concordance in monozygotic twins is necessary for heredity is here totally broken. In motility psychosis where we saw something similar at least 2 pairs were discordant, whereas in agitated depression there was not a single one. Among dizygotic twins there were 2 pure phasic psychoses; in both cases the partner was healthy.

Two things need to be explained: first, why agitated depression which is only one of the quite rare pure depressions, should occur so frequently in monozygotic twins; second, why in all cases there was concordance even though the hereditary disposition in all pure forms of psychoses including agitated depression is particularly small. According to Table 33 among the siblings of pure phasic psychoses there were only 3.1% affected siblings.

My starting point is that in the case of agitated depression **instinctive feelings,** that is feelings arising in the instinctive sphere, are affected. I have described this in my "Biopsychologie der endogenen Psychosen" (LEONHARD 1970a) and proceed from there. The most important **instinctive feelings,** that of hunger, thirst, need for oxygen, and for proper body temperature are already developed at birth making life possible for the baby. If in the development of this instinctive sphere there is a disturbance this must happen before birth. Therefore, since a psychological explanation is no longer feasible, it is necessary to go back to physiology. With this presupposition I assume that interference with uterine nutrition may produce a disturbance in the instinctive sphere whereby a disposition towards agitated depression is created. The possibility of insufficient nourishment in the mother's womb may be a factor in the case of twins; for monozygotic twins this is especially the case. In dizygotic twins, having separate placentas, conditions for nourishment are more favorable; as is well known they have a lower mortality. In order to pursue my assumption further I checked the date of birth of the siblings with agitated depression. Two pairs were born in the German years of hunger 1918 and 1921, a third pair in 1929 during a time of great unemployment. The other monozygotic twins had no such temporal associations. It is easy to suppose that the mother of the patients with agitated depression in those days could not nourish the twins adequately in utero. The complete concordance is therefore explained because a uterine disturbance of nutrition affects both monozygotic twins in the same way. In dizygotic twins this is different, therefore there may be discordance.

Among dizygotic twins I found no agitated depression, but I did come across a case of **unproductive euphoria;** this, like agitated depression, shows a disturbance in the instinctive sphere, that is in its euphoric pole. Here we are dealing with an exceptionally rare disease, showing that the instinctive sphere even in the case of dizygotic twins may be vulnerable. It affected only one of the twins. Apart from this I could find only one dizygotic twin with pure phasic psychosis: this patient suffered from pure melancholia, the twin sister was healthy.

I cannot express an opinion with regard to twins suffering from pure depressions other than the agitated type as besides the 5 cases of the latter we found only one of **suspicious depression.** Even here there was concordance, thus it is possible that a disturbance in the instinctive sphere may have a bad influence on other layers of the emotions.

Generally it appears from its frequency in twins that agitated depression has different genetic prerequisites from the other pure forms. Whether in agitated depression the influences arising from the mother and older siblings are nevertheless of significance remains undecided. Here I do not have any indications. Perhaps the development of the instinctive sphere at the time of these influences has progressed so far that it no longer plays a role.

Having considered the pure phasic psychoses I pursued the question of the genesis of all endogenous psychoses. In conclusion I have to make the astonishing statement that **without exception in all forms the role of siblings** was very important. In my opinion this is an alarming finding. It shows that disharmonies in the relationship between children encroach deeply into biological events and may lead to endogenous psychoses. Of course it is known that children influence one another, but nobody thought that this influence could be so important. Not only in the field of endogenous psychoses but also in research into the ontogenetic and phylogenetic development of mankind unexpected points of view emerge.

The influence of children on one another begins quite early. At first however the influence of the mother is quite decisive, as we shall see in the following discussion of early childhood schizophrenia.

Early Childhood Catatonia

The special form of schizophrenia found in early childhood has not been described in previous editions of this book. The main reason for this is that in the course of my psychiatric activities I seldom saw severe cases of mental retardation from which I could obtain useful information. I did become acquainted with early childhood schizophrenia and as mentioned above have described it, but I thought that such cases were very rare, and fundamentally could not at all be separated from later childhood schizophrenia. I did not know that among the supposed severely retarded children there might be many early childhood schizophrenics. As I came across such cases I started to recognize their independent significance after the thought had occurred to me that I should be looking in departments of child psychiatry among mentally retarded patients for patients with schizophrenia, not of early childhood cases but throughout childhood. I thought I would discover at the most one or two cases, in fact I saw very few schizophrenics who had become ill in later childhood; in reality I found many cases which had begun in early childhood and thus got to know the great significance of this clinical picture.

Delineation of Early Childhood Catatonia

KRAEPELIN (1913, p. 912) once mentioned that some of the supposed cases of severe mental retardation had schizophrenia which began in early childhood, with the result that further psychic development had been arrested. He referred to patients who, according to him, showed certain mannerisms, some stereotypies of stance and movement or a "persistent negativistic insufficiency". WEYGANDT (1907) disagreed and believed that such symptoms of restlessness represented a syndrome of involution of the brain. MAYER-GROSS (1933) agreed with WEYGANDT. After that this question was no longer discussed, apparently KRAEPELIN'S opinion was not considered valid, as deduced from remarks by PENROSE (1969) and by HEALTON-WARD (1977). But more recently Elfriede ALBERT (1980) did agree with KRAEPELIN'S opinion, according to which people who developed schizophrenia in childhood showed not only symptoms of this psychosis, but also were retarded in their subsequent intellectual and general mental development. I myself go along with this opinion and go even further than KRAEPELIN, as there are still other syndromes not mentioned by him, indicating the presence of schizophrenia of early childhood.

I am referring to a large number of such cases, having undertaken investigations in units to which mentally retarded children had been admitted. Among 318 patients I found no less than 117 in which their severe mental retardation had a form of schizophrenia starting in early childhood. A first investigation gave 76 cases, a second 41.

Many of these children stayed in the units for long periods of time, thus I saw some who were already more than 30 years old. The average age was 22.6 years, the youngest child examined was 6 years old. For the proper diagnosis it was an advantage that the majority of patients had grown up, thus enabling lengthy observation after the beginning of the illness in early childhood. A disadvantage was however that I did not see these children during the first 3 years of their lives. But I was able to obtain further information through the opportunity I had to undertake investigations in homes for infants and small children. Here I found 7 children who could be considered to have early childhood schizophrenia. Four of them had not yet reached the age of 3 years, 2 not yet 4 years, and 1 not yet 6 years. I shall present these children below, and note that the diagnosis may be made at this early age. I shall point to differences in the clinical pictures. The symptoms of "automatic striking" and other automatic reactions that play an important a role later on are not found in the early years of life. If one holds out a hand to small children they don't grasp it or perhaps they assume this is a game and slap the hand.

The diagnosis is derived from the fact that 117 patients with childhood schizophrenia showed symptoms very similar to the ones that are seen in the systematic forms of adult schizophrenia. These are merely slight modifications due to the child's psyche. I shall describe this in detail. It is not true as claimed by many authors that schizophrenia in childhood is quite different from that in adulthood. The systematic forms of schizophrenia show fundamentally the same picture. It is these forms which are manifested in early childhood.

The Question of Organicity Underlying the Clinical Picture

WEYGANDT'S interpretation of KRAEPELIN'S cases was on an organic basis. In fact several patients to be described showed organic signs such as **enlargement of the ventricles** or **epileptic seizures.** Seventeen of the 117 patients had seizures at one time or another. Among these 5 patients had seizures only in childhood, probably in connection with infections. In 7 out of the other 12 cases the ventricles were somewhat enlarged. In no less than 19 other patients, that is in 26 cases, enlarged ventricles were present. In one case there was considerable enlargement of the posterior horn of the left ventricle; in other cases the enlargements were only slight. In these latter cases I do not wish to express a definite opinion. If a rather large amount of air was introduced when the air encephalogram was performed which happened often at that time, the ventricles of the child's brain which is still quite adaptable could easily be further enlarged. **Neurological signs** were described in only 5 cases: one with slight spasticity of the lower extremities, one with hypotonia and slight hyperreflexia, one with spontaneous Babinski and increased knee and ankle jerks, one with clonic knee jerk, one with divergent strabismus, right Babinski and left Oppenheim. It is remarkable that among all the 117 patients only 5 showed neurological changes, mostly of slight degree. In some cases it was noted that exact examination was difficult because of the child's restlessness, but usually a reliable neurological description was available. This shows that severe cerebral changes were absent. There were no cases of paresis, athetosis, or other extrapyramidal symptoms. The observed hydrocephalus therefore loses still more significance as an organic sign. But it should not be underestimated that indications of organic occurrence were found frequently. Changes in the electroencephalogram were not included. More gross changes were not found, the smaller ones in the case of these children were almost insignificant. In practically all cases it was mentioned that because of the child's restlessness the findings were not easily evaluated.

The presence of organic signs, as is quite well known, does not rule out childhood schizophrenia. This was found by many authors, for instance Kothe (1957), Cameron (1958), Villinger (1959), Stutte (1963), Eggers and Stutte (1969). It is to be kept in mind that external influences on the brain may bring about signs of systematic schizophrenia, which, considered as a system disease, is itself of an organic nature. But the connection may be seen in a different way. A disease of the brain, perhaps accompanied by a childhood disease, represents a severe condition which prevents the child from having contact with the environment for a period of time. We shall see later that lack of communication plays an essential role in the genesis of early childhood schizophrenia; lack of contact with the environment may also arise from the fact that a child may have a prolonged physical disability. This becomes even more clear when a child has to stay in hospital for a lengthy period of time. In this way sometimes it is not the physical disease but the lack of communication caused by the physical disease that produces psychic changes.

Since neurological signs were rarely present in these cases, the diagnosis of early childhood brain damage found in many cases of severe mental retardation stood on weak ground. When it has been established that a cerebral disease has run its course neurological changes are not lacking. Moreover if it is assumed that early childhood schizophrenia may occur in cases of hydrocephalus or epileptic seizures the diagnosis has to be made with great certainty. Therefore I would like to show at first that childhood schizophrenia – including that of later childhood – can be reliably diagnosed.

Diagnosis of Childhood Schizophrenia

During my clinical activity I noted every child presumably affected by an endogenous psychosis, and I have re-examined these patients when they were grown up. There were 41 children with different endogenous psychoses. One female patient was no longer in the German Democratic Republic at the time I wanted to re-examine her, but I was able to obtain the written communication that she was still affected. Similarly another female patient remained ill and died in a psychiatric hospital at the age of 20 years. One male patient I did not examine because the information obtained from his parents confirmed without doubt the continuation of the illness. I re-examined personally the remaining 38 patients. There were only **4 mistaken diagnoses.** Particularly among the endogenous psychoses there were some further corrections using my procedure for differentiated diagnosis. In the case of one female patient whom I had originally diagnosed as having a cycloid psychosis, on re-examination I changed the diagnosis to periodic catatonia. Conversely in two cases where I had originally diagnosed periodic catatonia, at the time of re-examination it was more likely that they had cycloid psychoses. In one case I assumed the patient had manic-depressive illness, but re-examination revealed the presence of monopolar depression.

Apart from these changes resulting from the use of my procedure for differentiated diagnosis, there were only 4 missed diagnoses, among the 41 cases there were two of psychopathy, one of a neurosis, and one of a puberty crisis in a mentally retarded patient. Thus I afffirm that endogenous psychoses of childhood may be accurately diagnosed.

Fifteen cases are of special interest – these belong to the **group of systematic schizophrenias,** since these patients who became ill in the early years of life have systematic schizophrenia. Here the diagnosis made at first was **confirmed without exception.** The **poor prognosis** which is to be generally assumed in the case of systematic schizophrenia was also confirmed. Among the 15 patients whom I re-examined 9 were permanently hospitalized in a psychiatric hospital, 3 were living an inactive life at home, a female patient with guidance was somewhat busy; two patients were occupied with simple tasks but continued to show catatonic stereotypies and rigid posture.

Table 53 gives an overview of the above patients, supplemented by cases which appeared later when in various hospitals we investigated adult schizophrenics whose illness had begun in childhood. In the last 15 years I have examined about 1500 patients with endogenous psychoses; among them there were 44 schizophrenics whose illness could be traced back into childhood. Since most of the 1500 patients were schizophrenic, this meant that in 2 to 3 patients out of 100 the illness began in childhood. This number is below the 6.2% of KRAEPELIN'S cases (1913, p. 910) and the 4.9% of Eugen BLEULER'S cases (quoted by LUTZ, 1970). From these figures it could be assumed that my figures were rather conservative. The 44 cases which were obtained retrospectively are added on Table 53 to the 37 cases which I had seen when they were children and whose diagnoses of endogenous psychoses I could confirm later. In addition I came across 6 cases which, as mentioned above, had been supposed to have severe mental retardation. These 6 patients did not become psychotic in early childhood as the others did but only in later childhood. Therefore they could not be included with the others.

As can be seen from Table 53 there were no cases of **paraphrenia** in childhood. Neither systematic paraphrenia nor affective paraphrenia was found

Table 53. Age of Children at Onset of Endogenous Psychoses

	Not yet 3 Years			Not yet 6 Years			Not yet 8 Years			Not yet 10 Years			Not yet 12 Years			Not yet 14 Years			Not yet 15 Years		
	♂	♀		♂	♀		♂	♀		♂	♀		♂	♀		♂	♀		♂	♀	
50 Systematic Catatonias	10	3	13	11	7	18	2	1	3	3	2	5	1	1	2	6	1	7	2	0	2
13 Periodic Catatonias	–	–	–	–	–	–	2	0	2	–	–	–	1	0	1	4	4	8	1	1	2
5 Cataphasias	–	–	–	–	–	–	–	–	–	1	0	1	–	–	–	3	0	3	1	0	1
11 Cycloid Psychoses	–	–	–	–	–	–	–	–	–	–	–	–	–	–	–	2	3	5	5	1	6
4 Manic-depressive Psychoses	–	–	–	–	–	–	–	–	–	0	1	1	–	–	–	1	0	1	1	1	2
4 Monopolar Depressions	–	–	–	–	–	–	–	–	–	–	–	–	–	–	–	0	1	1	3	0	3
87 Psychoses	10	3	13	11	7	18	4	1	5	4	3	7	2	1	3	16	9	25	13	3	16

in childhood. Where delusions are prominent the underlying illness may be periodic catatonia in which anxious ideas and hallucinations are common, or there may be an underlying cataphasia in which confabulatory ideas occur frequently. Likewise, anxiety-happiness psychosis is often characterized by delusions and hallucinations. Furthermore **hebephrenias** are not found in childhood; they appear only in the course of puberty. I must mention here that the concept of hebephrenia held by most psychiatrists is much wider than mine. Some cycloid psychoses appear in the patient's histories as hebephrenia.

The occurrence of catatonia as a form of systematic schizophrenia in childhood agrees with my earlier findings (LEONHARD 1960) as well as the observations of other authors (SPIEL 1961, WIECK 1965, SUCHAREWA 1967). However HERBAUER (1969) and LEMPP (1973) claim that catatonic symptoms do not belong to the cardinal symptoms of childhood schizophrenia. Perhaps these authors are thinking of cycloid psychoses which become very much more prominent in later childhood, as seen from Table 53. Moreover the different concept of these authors may be partly the result of the neglect of catatonic symptoms in modern psychiatric diagnosis. In former days catatonic symptoms were not overlooked, especially not in schizophrenias of childhood. KRAEPELIN described catatonias in childhood schizophrenia. RAECKE (1909) published an article "Catatonia in childhood", PÖNITZ (1913) published an article entitled "Contribution to the recognition of early catatonia". VOGT (1909) used the expression: "childhood insanity" ("Jugendirresein") in the title of his paper, but likewise mentioned that catatonic symptoms predominated.

The various forms of endogenous psychoses begin in different years of childhood, as shown in Table 53. In the first years of life only systematic catatonias are found. After 12 years of age cycloid psychoses become very frequent. After 6 years of age two of my patients became affected with periodic catatonia. I have to add here that I saw 3 other patients who had been ill since the first years of life and where periodic catatonia was suspected. Above all their aggressive impulsivity which frequently occurs in periodic catatonia in adulthood made me think of this diagnosis. One 15 years old imbecile boy with whom we began a conversation was asked to sit down. He did, but sat on a chair other than the one on which he was supposed to sit. When Frau VON TROSTORFF asked him in a friendly manner to change chairs he stood up, laughed, prepared to sit on the other chair, but interrupted this, stroked Frau VON TROSTORFF'S cheek, and then quite suddenly hit her in the face. At the same time he grabbed her glasses and pressed them so hard that they fell on the floor in many pieces. Subsequently he hit her again and pulled her hair. I put my arms around him and held his arms to his body. He then tried to reach my arms with his teeth and kicked my legs, fortunately he was wearing soft shoes. The nurse who was called spoke to him nicely and held him with both hands which he permitted. Later, when the medical director spoke with him, he was at first friendly but then suddenly bit him in the upper arm. We now were told that he often showed such dangerous behaviour. – A 16 year old girl, an imbecile, was similarly aggressive, hiting other children roughly, on one occasion even with an iron rod; another time she threw a large stone at the head of another child. This patient also tended to turn her aggression against herself, on one occasion stabbing herself with the points of a pair of scissors. – Another imbecile patient was quite anxious in the first years of life. He was afraid of other children, but could also be aggressive. After 9 years of age his aggressivity became stronger. He threatened to kill his grandmother and brother, and attacked with his fists his relatives and other people. During his admission into a psychiatric hospital at the age of 13 he was out of control, banged walls and cupboards with his head and fists and attacked his mother who was accompanying him. Now, at 27 years of age, because of continuing severe aggressiveness he could never again be discharged from the hospital. He suddenly hit others. A nurse, momentarily inattentive, was suddenly and without any apparent reason struck in the face.

All three patients were stiff in their movements and postures. The mother of the first patient after having been examined herself was shown to have periodic catatonia. It was through her that her son came to our notice. The third patient had a strong family history characteristic of periodic catatonia. His only brother was an imbecile like himself, but in addition, when we saw him at the age of 19 years, showed such severe restlessness of his whole body that I thought he had a hyperkinetic phase of periodic catatonia. The restless condition which had existed for just a few months, could well have been an acute catatonic phase. The patient's cousin likewise seemed imbecile, but his condition changed 6 months before being examined when he was 23 years old. He refused to work in the sheltered workshop, attacked his co-workers, and at home was aggressive towards family members and other people. He broke windows and attacked his mother. In hospital he was now partly friendly, partly irritated. His movements were stiff. Sometimes he turned his head repeatedly from side to side. He evidently was in a catatonic phase. There was no question of his having inherited imbecility. All four parents of the two brothers and of their cousins were intelligent people. Their common grandfather showed hereditary disposition to periodic catatonia, for he lacked contact, his posture was stiff, and his facial expression was discordant sometimes resembling grimacing.

If the three patients, as I presume, suffered from periodic catatonia, even though periodicity was hardly noticeable, it had to be assumed that this illness must have begun early in childhood and have led to retardation as well as to chronic irritability and impulsive aggressivity.

These patients might demonstrate that a very early onset of the illness leads to a particularly severe condition. But independent of these unusual cases it is possible to derive from the different **age of onset** a different **prognosis** of childhood schizophrenia. Systematic catatonias with their frequently very early onset generally lead to a severe defect; whereas cycloid psychoses with their later onset remit from each phase. Thus early onset does not necessarily mean a bad prognosis, it only indicates that the onset of the systematic schizophrenias with their unfavorable prognosis is especially early.

Distribution of Individual Forms of Early Childhood Catatonia

The rest of this chapter will be concerned with the systematic catatonias, as I was able to observe these with certainty in early childhood. I exclude 13 cases which are listed in Table 53 and I refer to 117 cases which I found in units supposedly for patients with severe mental retardation. In Table 54 these patients are listed with their more accurate diagnoses. Included are individual cases of systematic catatonia similar to the ones found in adulthood. The different frequency of the individual forms in Table 54 does not indicate a generally different incidence, but the difference in the need for treatment and care of patients with different symptoms. The negativistic children are the most difficult patients because of their aggression towards others and towards themselves. This explains the relatively large number found in these units. Manneristic catatonics which according to Table 54 are equally frequent cause great difficulty in their management because of their mannerisms of movement and neglectful behaviour.

There are **more male** than female patients. This I found to be similar to systematic catatonia of adulthood. An explanation might be that boys are more sensitive than girls because of isolation from friends and from their mother (see below). This difference between the sexes exists only with simple systematic catatonia; the combined forms occur in girls as often as in boys. When the illness begins in adulthood the number of men are closer to the number of women in the combined form but not to the same extent. In a series of investigations of 209 catatonic men the ratio of simple to combined cases was 100:47.1, in the case of 200 catatonic women it was 100:56.9. The figures 48 and 18 in Table 54 show that here male children have a ratio of 100 simple to 37.5 combined forms. Perhaps this could be explained by considering male children biologically more sensitive than female. They

Table 54. Specific Diagnoses of 117 Catatonics with Onset in the First Years of Life

Simple Systematic Catatonias	Male	Female	Both
Speech-Prompt Catatonia	3	6	9
Sluggish Catatonia	5	6	11
Proskinetic Catatonia	9	2	11
Negativistic Catatonia	14	6	20
Parakinetic Catatonia	7	3	10
Manneristic Catatonia	10	10	20
Subtotal	48	33	81
Combined Systematic Catatonias	18	18	36
Total	66	51	117

have as is known a higher mortality. This may lead to the following explanation: girls become afflicted relatively often only at a time when catatonia affects two systems, whereas boys die relatively frequently if two systems are affected at the same time. The danger to life should not be a surprise from the description of early childhood catatonia, especially the combined forms. The illness itself as well as frequent disturbances of nutrition may lead to death.

Distinguishing Early Childhood Catatonia from Mental Retardation

Before dealing with specific symptoms from which systematic catatonia of early childhood may be recognized with certainty some general remarks as to how the differentiation from severe mental retardation can be made are presented here. First should be mentioned the "praecox feeling", but on a somewhat more concrete basis. It used to be said that one always remained at a distance from schizophrenics "as through a pane of glass". In my opinion this so-called "feeling" arises through the **lack of natural movements of expression** which take place from person to person. This lack, as well as the stiffness characteristic of schizophrenics, becomes apparent as soon as one turns towards them. With schizophrenic children this "feeling" is present even more clearly than with adult patients. This is not the case with the mentally retarded. I can pointedly say: the mentally retarded behave quite normally in their primitive reactions which do not require intelligence, whereas schizophrenics do not. When talking to mentally retarded persons they turn their head and eyes towards you and their facial expression even if only rather bluntly shows an attentive attitude. If they are not profoundly retarded this attitude is all the more clear-cut as they are looking for contact. Asked for their names they try to pronounce it indistinctly. If they cannot do so, at least their face lightens up when they are called by their names. If this is done in a friendly manner they usually smile. If they happen to be unwilling or irritated one has to speak to them in a particularly friendly and encouraging tone of voice and in so doing they pay a certain amount of attention through which contact may be gained. Expressive gestures have a direct effect, a friendly look or a smile, even by a severely mentally retarded person touches you, thereby readily establishing closer emotional cntact.

In children with catatonia everything is quite different. Good contact is elusive; they remain distant, they may turn towards another person, but from their facial expression it cannot be ascertained whether it is an attempt to make contact or only a reflex or automatic movement. Sometimes they approach another person with head and eyes and even with the whole body, but what this means is an enigma, their facial expression reflecting nothing of significance. They may smile, their whole face may laugh, but you cannot tell what it means. Do they laugh because they are pleased, because they enjoy something, because they feel friendly, or only because it is an automatism or a stereotypy? Their expressive movements have nothing to say, their faces look like masks, the movements of the face like grimaces. From this lack of natural contact and expressive movements childhood catatonics appear like foreign bodies which do not fit into the community. Healthy children may attempt to seek contact, but nothing happens which they understand, thus they withdraw and avoid the sick child.

This schizophrenic strangeness felt by healthy people is somewhat different from the autistic withdrawal described by ASPERGER (1968) about his patients. Their behaviour is not incomprehensible to a normal person, for the latter under certain circumstances may also withdraw into himself. I agree with ASPERGER when he states: "It is quite generally a possibility that people may behave autistically." A healthy person understands the behaviour of an autistic person, but does not accept it as a permanent attitude and, while still a child, easily reacts with aggression. ASPERGER'S autistic patients were regularly exposed to the teasings of their contemporaries; they themselves understood why they were being teased, and reacted with angry outbursts or mean behaviour. This is a conflict between children of fundamentally different attitudes. With schizophrenic children such an exchange does not occur. Healthy children cannot go along with the behaviour of a schizophrenic child and therefore cannot react appropriately. Autistic children may be considered as foreign bodies in the community but never in such a peculiar manner. Put differently one can say that schizophrenic children have no feeling for healthy children, autistic children have a different feeling. The empathy for autism should not be confused with the inapproachability of schizophrenia.

My description of schizophrenic children probably applies to some of the autistic patients of KANNER (1958). If so the rest of the symptoms that show a schizophrenic disturbance will not be missed. Some of KANNER'S patients were, like mine, thought to be mentally retarded, whereas ASPERGER'S children were mostly above average intelligence.

The wrong diagnosis of mental retardation can moreover be avoided by paying more attention to the capacities which schizophrenic children still possess. These can be recognized through spending more time with the children and by knowing how to overcome the difficulties produced by the schizophrenic disturbance. It can then be found out that many abilities are present which patients with true mental retardation would certainly not have.

I have given a report on two girls who had been diagnosed as being mentally retarded (LEONHARD 1984), but who in reality had early childhood catatonia. One of them later on was able to work in a laundry, the other who did not speak and did not write, was able to learn through intensive intervention by her mother to typewrite. Of course the intellectual development of children who in the early years of their lives became catatonic lagged behind, due to lack of education rather than on a biological basis. It has to be kept in mind that a healthy child in the many years of going to school does not only acquire a great deal of knowledge but also is taught to think logically. Neither of the children whom I had observed was able to go to school because of their schizophrenic behaviour. One of my patients from an earlier observation, now 32 years old, succeeded with my and her parents' help to obtain individual instruction and thus gain a certain degree of education. Now she is able to work successfully in a sheltered workshop in spite of her severe catatonic disturbance. In addition to the lack of education of schizophrenic children the lack of interaction with other people, especially with their contemporaries, is such that their general knowledge remains quite insufficient.

Mentally retarded children often have a retarded development of the body and as grown-ups look like children. In this connection the development of catatonic children is similar as we shall see.

Clinical Pictures of Early Childhood Catatonia

Before dealing with the catatonic syndromes of early childhood as such I wish to mention a few **modifications** which follow from the psyche of the child. Together with general retardation the **development of speech** is held back. Consequently in syndromes where poverty of speech occurs in adulthood there is total mutism in childhood. If on the contrary a catatonic does have a heightened impulse to speech, language develops in spite of retardation, as seen in speech-prompt catatonia. The **greater mobility** of the child may be conducive to verbal expressions becoming more prominent even if these do not make any sense. The mobility of the child has another significance: ordinarily its extent may be recognized by observing how little children interact on the playground. The tendency to constantly turn towards events in the environment remains recognizable in the disorder. Whereas in adult catatonics there is an almost total turning away from outer events catatonic children do look around, turn the head when somebody enters through the door, look at the window if a sound comes from there. If however a syndrome contains the tendency to react to external events with automatic movements a child shows this peculiarity to a greater extent such as in seen in proskinetic catatonia. Another aspect of the child's greater mobility is the following: some catatonics tend towards compulsive movements; in children this tendency may become extremely pronounced so that from time to time, often for a long period of time, gross, hitting and turning movements are repeated over and over without a break. This is observed particularly in negativistic catatonia. The repeated rocking of the body is not to be included here as this also occurs frequently in mentally retarded children in so far as they are little stimulated. A constant turning and twisting of the head is more characteristic of catatonia. If restlessness of the whole body is part of the syndrome, as occurs in parakinetic catatonia, this restlessness occurs more prominently in a catatonic child.

These modifications are present when the child in the meantime has become older or has even reached adulthood. This is extremely important: it must be consistent with the general retardation. For this reason it may be deduced from the clinical picture that the illness had already begun in early childhood. If the onset is in later childhood these modifications, to be described in detail, are no longer found, with only a hint of them being present. Apparently with the full development of language the picture becomes more similar to that seen in adulthood. If the diagnosis of early childhood catatonia is made before the end of the **third year of life** this diagnosis correlates with the objective connections. I saw two catatonic children whose illness began at the age of 5 (one boy with speech-prompt catatonia and one girl with manneristic catatonia); both showed the clinical pictures to be found if the onset is in later childhood or in adulthood.

To the modifications at the beginning of the illness in early childhood must be added the **continued existence of early childhood emotional lability.** Although in catatonic children affective expressions lose their genuine nature they do arise quite strongly. If a catatonic syndrome in adulthood shows only an occasional smile, in childhood this may take the form of loud laughter which because of lack of genuine expression may appear as a form of grinning. A similar observation may be made vis-à-vis crying. Grown men who in their early years of life became catatonic may cry like children at the slightest provocation. This affective lability may also be expressed by anxiety or howling.

Next to be described are the characteristic syndromes of systematic catatonia and to show how they appear in children and adults in similar form but modified in the child.

Simple Systematic Catatonia of Early Childhood

In **speech-prompt catatonia** adults are found to give precipitate responses with a tendency to talk past the point ("Vorbeireden"). Another important symptom is that the patients look with empty expressionless face from which one cannot tell what their inner attitude is in connection with their spoken utterances. If the illness had already begun in early childhood the children not only answer but also speak spontaneously. The concept of speech-promptness ("Sprechbereitschaft") is thereby more clearly elicited than in adults. Speech-prompt catatonia is the **only systematic form** in which language in spite of onset in early childhood does not regress. The reason for this must be that the disorder itself promotes spoken utterances. In the other forms only **expressive language** is missing; the patients understand very well what is said to them.

A patient, now 25 years old, had always been regarded as severely mentally retarded and incapable of learning. During my examination he spoke with poor articulation, but one could understand him sufficiently. Most objects in the examination room he named correctly, thereby indicating that he was not severely retarded. Answers came more promptly than found in the early phases of the illness in adulthood. But beside the correct answers again and again there was **talking-past-the-point.** As is generally the case with these catatonics, if this happened once the patient stuck with his wrong wording, no matter how hard he was pushed. In this way he said that a picture on the wall was a book. In continuing the conversation and pointing to the picture he kept on saying: "A book, a book". Afterwards he still said when other objects were shown: "a book". Another question he answered by calling a ruler a "stick" in spite of it being flat, which could perhaps be understood, but then he even called a watering-can and a funnel "stick". When I asked him what his name was, which he gave correctly, he said shortly thereafter when I pointed to a pencil: "name". In addition the patient showed the characteristic **empty facial expression** typical of catatonics. As a childish peculiarity it may be stated that the patient regularly pointed with his finger at the objects which he named.

Quite apart from the fact that in the examination room the patient was able to name correctly objects which were not present in his every day life, a chance observation indicating that he was not an uneducable idiot was the following: towards the end of the examination the patient became increasingly tearful and eventually cried like a child. While I was wondering why, the nurse entered and asked if the examination was going to take much longer as the bus for an excursion was waiting outside. It turned out that the patient had cried because he was afraid that he would not be taken along on the excursion. He had correctly estimated the time of departure, and now he left, satisfied, with the nurse. The continuing emotional lability of the child was still succinctly recognizable in this man now 25 years old.

A 10 year old girl who likewise had been considered severely retarded and uneducable was able to talk with good articulation. With a rigid facial expression and pauses she spoke as follows:

"Today is a day – Uncle went home – Good bye – Today there is no pudding – The man doesn't come – Peggy (her name) is going home – Uncle goes home and the woman – Today there is stewed fruit – There is no meat ball – Uncle goes home – Today is a day – ..." Thus it went on. **Perseveration** seen in speech-prompt catatonia can be easily recognized here. **Talking-past-the-point** occurred after questions. After the patient had correctly named many objects in the examination room I pointed to the curtain, and she answered: "A house" (the question was repeated). "A sun". (The question was repeated forcibly) "A sun". (Pointing to the radiator) "A roof, Peggy goes home." (Clothes peg) "A stick, a sun, a roof, a sun." (Scarf) "A roof." (Are you a girl or a boy?) "Boy, the woman goes home" (Are you a boy or a girl?) "Boy, girl, good bye." (Are you a girl?) "Girl, boy, it is late on my watch, today is a day."

Just exactly as is found in adults the child could not be persuaded to answer a question correctly to which she had given a wrong answer before. Following the last question like a short circuit she responded to the last word. Similar answers may be obtained from children as well as from grown-ups. If asked: "Is today Wednesday or Tuesday?" they say "Tuesday". If subsequently asked: "Is today Tuesday or Wednesday?", they say "Wednesday".

The modification due to the child's psyche hardly changes the picture. To be added is that children **react impulsively.** Their increased mobility gives rise not only to increased talkativeness but also to increased activities. They respond to outer impressions; when naming objects they point to them, when angry they quickly become aggressive. The girl described above during the interview hit my associate Frau von Trostorff, apparently annoyed with her. The boy described above when angry tore clothes and often ran away. We are dealing here with something different from the much more impulsive aggressive and autoaggressive outbursts which we will get to know in discussing negativistic catatonia. Furthermore the autoaggressivity seen in negativistic catatonia is lacking.

Of particular interest was a **small child** seen by me in a home. The girl, 2 years and 2 months old, came into the home soon after her birth. When I saw her she was very restless, groped for everything she saw, handled it briefly, then turned towards the next object. She was constantly running in the room, giving the appearance of the restlessness of a little child; in between she could be quiet. Probably the impulsivity of speech-prompt catatonia had already become part of the picture. The little patient let herself be seated on a small chair, and when spoken to her facial expression showed no animation. In this manner she often looked at me for a long period of time with an almost challenging expression. Sometimes she came close to me with head and body and looked at me. Her facial expression never changed, the face remained quite empty. It looked like the turning to ("Zuwendung") of speech-prompt catatonics in which one never knew from their empty expression what it meant. The girl never answered any questions, she could hardly say anything. But in a striking manner she uttered a sound from time to time that sounded like "hoo". The nurse had already drawn this odd behaviour to our attention. Infrequently she made a longer sound which could have been a hint of an inarticulate word. – I think that this 2 year old child showed a readiness for speaking by means of this sound, as her language had not yet developed. It was a question not of an affective sound utterance such as occurs in catatonics of early childhood, but a simple stereotypically repetitive calling. The diagnosis was confirmed because the girl looked for no contact with other children as the nurse had already emphasized. Surely this is the autism typical of speech-prompt catatonia. It is highly interesting that speech-prompt catatonia can be recognized even if language has not yet developed.

In adult patients with **sluggish catatonia** the specific syndrome is characterized by the fact that the patients barely turn to one another, answer not at all or only with a few words, but are constantly preoccupied with their hallucinations. Children with this form of catatonia show other symptoms of restlessness not to be deemed reactions to hallucinations. The patients look at objects around them, apparently briefly; they may make a brief grasping movement towards the object. Childish mobility is no doubt responsible for this behaviour, as children normally show short circuit reactions to events in their environment.

A **16 year old boy** during examination showed the following: restlessly he looked around with head and eyes, raised his eyebrows, shook and nodded his head, his lips moved without sound, he smiled to himself, shrugged his shoulders, made some gesticulating movements, glanced at me fleetingly without fixing his gaze on me. When spoken to he did not reply, but turned to me quite briefly. Hallucinatory restlessness in this boy was clearly noticeable; it is seen in the beginning of the disorder in adulthood only if the patients are in a state of agitation. Moreover the boy showed movements not related to hallucinations but directed towards his environment. He glanced quickly at the medical record, even hit it on one occasion, then with his hands made indefinite movements and in-between touched his face.

It is worth noting that the catatonics who never say a word nevertheless do express themselves from within. The movements of their lips indicate a form of talking; but conversations with others do not take place As catatonia develops in early childhood lack of initiative resembles to a high degree that seen in adulthood. Rocking of the upper part of the body occurs especially frequently.

When seeing a **boy** only **2 years and 1 month old** in a home I could already diagnose sluggish catatonia. He did not turn towards me. I tried hard to get him to react, joked with him, laughed, gave him a challenging tap on his abdomen. All he did was to glance at me quite fleetingly. It was obvious that he had the schizophrenic incapacity for contact. Furthermore he was distracted from within. Without relation to his environment again and again small mimic movements appeared on his face, with raising and contracting his eyebrows. Sometimes he glanced around with a perplexed look without fixing it on anything. On one occasion he turned his head, opened his mouth, and smilingly emitted a sound. He kept on making small gesticulating movements. It was quite evident that the boy was distracted from within. He did react automatically when something happened in his environment. When the nurse passed through the room he followed her with his eyes; when there was a noise he looked in the direction from where the noise came. On one occasion he was busy with his legs. Such attention towards the outside occurred more frequently than in somewhat older children with sluggish catatonia. On one occasion he even ran towards me when I entered the room, but immediately turned around and went away. His stiff posture, the lack of reaction to being spoken to, and the constant inner distraction characterized sluggish catatonia.

Another boy aged 2 years and a few days showed a combined catatonia with a sluggish component. His clinical picture was typical of sluggish-proskinetic catatonia, to be described below. Proskinetic restlessness with turning and running towards objects was quite obvious. But in-between he stood still for a short time, glanced into empty space in a perplexed manner, paid no attention for the moment to what was going on around him, but appeared to be totally preoccupied with himself. Like the boy mentioned previously he showed mimic restlessness, contracted his eyebrows, as if he wanted to concentrate on something inside. He kept on looking obliquely upward with his head slightly inclined, his mouth open and giving the appearance of listening. Sometimes he smiled to himself. His inner distraction appeared to be confirmed by the nurse who said the boy was sometimes "as if he wasn't there", as if he heard nothing.

Both these small children showed inner distractibility but it is not clear if they heard voices. Probably this inner distractibility appears before linguistic function has developed.

In **proskinetic catatonia** adult patients speak with a low voice, which may turn to murmuring; in addition there may be odd fiddling with the hands and verbigeration. If the disorder begins in early childhood the clinical picture may at first appear to be quite different, as the patients do not talk and do not murmur. Rarely the uttering of a few indistinct sounds are heard. In adults the force of language is diminished only in so far as they speak less loudly.

Because of deficient speech development the diagnosis of proskinetic catatonia in childhood may be quite difficult, as murmuring and verbigeration are just about the most important symptoms of this disorder. But the diagnosis may be made more readily from the fact that restlessness with odd fiddling and grasping movements are especially noticeable during the onset of the illness in early childhood. Furthermore in childhood the underlying automatic movements predominate over these small movements of the hands. The patients not only move their hands oddly, but also bend forward in order to reach objects, they stand up and even run towards the objects. An **8 year old boy** kept on running around in the room during my examination and touched, knocked on, and scratched everything he could reach; he did this only for a moment, then he rapidly turned towards the next object. The symptom of automatic turning towards objects, which gave proskinetic catatonia its name, is therefore quite obtrusively present in early childhood and thus permits the recognition of catatonia relatively easily. "Automatic bumping" is found the same as in adulthood; as previously noted this may not be expected to occur when the children are still very small. The 2 years old boy described above, who had combined sluggish-proskinetic catatonia did not show it because he kept on turning from one object to another or ran towards them.

The exaggerated turning towards events in the environment seen in adult patients may also be observed in childhood. A boy who at the age of 16 was nearly grown up showed the following picture: Seeing on the table a dish filled with chocolates he grasped the dish and while we were about to intervene threw the chocolates quickly on the floor. Then he picked them up. Then he opened a cupboard in the room, closed it, opened a drawer, picked up a piece of wood (part of a toy), fingered it, then put it back, shook the night table, ran towards the examination table, picked up a bag which was lying there, put it down again, discovered a chocolate under the table, picked it up, and again took the piece of wood from the drawer, turning it this way and that. The piece of wood was taken away from him which he let happen without resistance. Likewise he could be interrupted in all his activities, even in his longing for the chocolates. He never resisted. In negativistic catatonia this is quite different. When patients with that disorder want to keep something, they strongly hold on to it. In proskinetic catatonia all this happens automatically.

Emotional expression is not lacking but is always without intensity. Anxiety, tearfulness, joy gives more life to the otherwise empty facial expression in the disorder when it begins in childhood than when it begins in adulthood. The outer expression of emotion may even occur in an exaggerated manner. This demonstrates the greater affectivity of children in general. There may even be laughter across the whole face. The 8 year old boy described above repeatedly howled with laughter. But this behaviour indicates more an outer manifestation of feelings than feelings themselves. Thus the propensity to automatism is manifested by means of expressive movements.

Flatness of affect may be observed in differentiating proskinesia of children from the **hyperactivity of excited imbeciles.** The turning towards objects by catatonics is automatic, whereas the hyperactivity of the mentally retarded is related to their emotions. The latter turn towards objects which they desire or in which they become interested. They often hurry towards anything that entices them with vivid joy in the face and cheerful sounds. If nothing is enticing or interesting their restlessness abates. A hyperactive mentally retarded patient in our observation could not be examined as long as the bag in which Frau VON TROSTORFF had chocolates was visible; almost forcefully he pushed himself towards the bag. After the bag was removed he calmed down and he did not turn towards other objects. Patients with proskinesia however do not care whether things are or are not interesting. When Frau VON TROSTORFF offered a chocolate they did grab it keenly, even rather quickly, but they constantly grabbed other objects for which they had no desire. The facial expression is different from that of the mentally retarded, there is always a stiffness, even vivid laughter appears mask-like. On the contrary expressive movements of excited imbeciles retain a natural character.

The picture of **negativistic catatonia** in childhood may be said to be even more typical than in adulthood. When I observed this type for the first time I described the symptoms somewhat forcefully. Later I moderated the description. But in negativistic catatonia of childhood the essential symptoms occur even more massively than I had previously described, that is to say, **negativism, negativistic excitement, impulsive longing,** and **aggressivity.** A child can also show **autoaggressivity.** I came across aggressivity during the investigations while walking through a unit and greeting patients fleetingly in order so see if there were any catatonics. Instead of replying to my friendly words one female patient slapped the glasses from my face. In negativistic catatonia such attacks appear so suddenly that they can hardly be prevented by supervision. They often have a violent and sometimes quite dangerous character.

A few days later I saw the girl who had knocked off my glasses in the examination room. Now she forcefully hit her face, with the right hand, with the left hand, at times with both hands simultaneously. Suddenly she struck me with her foot. When I defended myself with the sole of my shoe she struck it. Following this she produced a ribbon which she had brought with her, swinging it to and fro with a turning movement of her hand. At times she put the ribbon in her mouth, held it firmly with her teeth and tore off a piece of it. She swung the rest of it in a circle. I tried to take the ribbon away from her. She objected violently, hitting me. We shall come back to the repeated swinging of the ribbon. To start with she shows convincingly how much negativistic patients whose disorder began in early childhood tend to become aggressive. A patient whom I had observed had bitten off the tip of the finger of another patient. The patient was already 20 years old, which demonstrates that this childhood peculiarity continues into adulthood.

As I attempted to take away the ribbon the **negativistic excitement** typical for this disorder made its appearance.- With another patient I provoked it myself. This 12 year old boy in his negativism did not want to sit down; I tried to force him to do so. I pushed him against the chair, he defended himself and did this more violently as I tried to have my own way. He turned, moved his body sideways, tore himself away, slipped under my hands and almost fell on the floor. He moved his hands so quickly that I could not grasp them. I tried to lift him up, but with rapid movements he released himself and slipped down from the chair on which I had tried to place him. He successfully resisted me, although his physical growth had been greatly retarded and he had by no means the strength of a healthy 12 year old boy. – A female patient became negativistically excited when she wanted to leave the room and found that the door was locked. She started to shout very loudly and for such a long time that in-between she became quite breathless. With her fist and her legs she struck me, struck the door and pounded on the floor. I offered her a chocolate, she first looked away negativistically, but then took it and now alternated between shouting and chewing.

Such forms of excitement may appear without stimulus from outside and may be considered to be **affective discharges.** The patients may emit angry sounds, shout in an inarticulated manner, destroy objects, tear up clothes. There may be joyous outpourings with howling and loud laughter. The patient who had received chocolates from Frau VON TROSTORFF and expected to receive some more emitted howling sounds, jumped up and down on his chair and clapped his hands. Shortly before he had been very irritated and had stamped his feet on the floor. Moreover these patients tend to be very impulsive; greedily they demand food. We observed one patient while eating. He put so much food on his spoon and moved it so quickly that part of the food always fell off. In spite of this his plate was empty sooner than that of other patients. Then he grabbed the plate of the patient sitting beside him who however held on to it. Then with his spoon he went into the plate of another patient. When he did not succeed he ran after the nurse with his plate until he had received a second helping which he devoured as quickly as the first. The nurse reported about another patient: "He is always trying to find something to eat, knows exactly where the food is to be found in the kitchen and helps himself if he can; he even takes food from the garbage can." Likewise they are uninhibited in their sexual behaviour. A girl in erotic posture approached me and then made masturbatory movements between her legs. I did not observe sexual aggression by male patients towards the nurses. Apparently here the childish propensity to submit to a person in authority remains intact; for it was found again and again that otherwise aggressive patients obeyed the nurses. Male negativistic catatonics apparently masturbate excessively.

In the beginning of negativistic catatonia in early childhood aggressivity is associated with **autoaggressivity.** This is an important modification of the clinical picture, for in the beginning of the illness during adulthood this no longer happens. The compulsion may be so strong that the patient must have a nurse close by in order to prevent self-mutilation. The patients hit themselves in the face, knock their heads against hard objects, sometimes bite themselves. Some patients have scars on the face or elsewhere on the body from serious self-inflicted injuries. Autoaggressivity has of course the same basis as aggressivity. When the children want to release their inner tensions, because of their helplessness and fear of adults they turn their aggression inwards. There may be a transition between aggression and autoaggression when they tear their clothes. It is true that autoaggression also occurs to release tension in other catatonics and in mentally retarded persons, but never as strongly and as dangerously as in negativistic catatonia. Scars resulting from self-mutilation were seen only in negativistic or negativistic-combined children. If mentally retarded patients hit themselves or bite they do so only in situations of strong irritation. Sometimes it is quite evident that autoaggressive acts of catatonics have a compulsive character and are not really desired by the patients; for it happens that they hold out their hands to the nurses in order to be tied down. In examining a girl whose restraints had become loosened she helped me to tie up her hands again. Negativistic children do not tolerate anything that they do not like. As they get older autoaggression gradually recedes, although I did find it in some patients older than 20 years.

If at the moment there is no irritation these patients do not have to be grossly rejecting, but again and again their propensity thereto can be recognized. They do not like to make eye contact, they look downwards or sideways, they do not or only hesitatingly comply with requests, they pull back when one wants to shake hands with them. As with adults negativistic catatonics they may react with ambivalence when one wants to talk to them in an especially friendly manner. A hand which had just been pulled back is offered again, a smile may indicate an inner readiness.

Offering the hand in natural situations should not be confused with **counter-grasping,** which is based on **automatic forcible contact** ("Anstoßautomatie"). Patients with both adult and childhood negativistic catatonia when irritated jerkily pull back the hand one wants to shake, but when they are more calm they may keep on grasping a hand offered to them. More profound automatisms are at work here. In childhood automatic forcible contact occurs less frequently than when the disorder begins in adulthood. This is no doubt due to the fact that younger patients express their emotions more readily, especially when irritated. Negativistic refusal is behind the observation that these children never answer questions; in the case of adults when they are calm may give monosyllabic answers. But probably expressive speech does not develop at all when negativistic refusal begins to be a problem so early in life.

Negativism, ambivalence and **counter-grasping** are very closely associated with one another even more so in the child than in the adult. The **boy** whom I have just described because of his negativistic excitement exhibited the following: I had asked him to sit down and then indicated this with my hands. He stepped back against the chair, appeared to want to sit down, but then moved in the opposite direction. When I asked him again he repeated this behaviour. The third time he really sat down but immediately stood up. This ambivalence was very obvious as he was moving to and fro. Negativism followed when I asked the boy to sit down in a commanding tone of voice. Now he did not react at all. Following this my pushing and his negativistic attitude made him cry out loudly. When I stopped pushing and offered him my hand he stopped shouting and grasped my hand. He repeated this constantly as often as I offered him my hand even though his hand had become quite cramped.

The tendency to negativistic excitement is stronger in childhood because of the greater mobility of children. This accounts for another symptom seen in childhood negativistic catatonia. These children keep on **repeating certain movements.** The girl patient mentioned above was swinging a ribbon to and fro. Another patient constantly flung her hands to and fro. A male patient whom at the first examination had been quite uncooperative, when seen on another occasion in the dayroom, tore a bed sheet into one strip after another, then made a knot and threw that on the floor. The movements seen here are partly more complicated than in the more usual repetitive movements such as rocking the body. They probably represent discharges of tensions and become more repetitively uniform because they can think of no other forms of movement. There is perhaps a parallel to autoaggression, as the latter occurs most frequently in the form of a repetitive beating against the head. These repetitive movements interrupt the otherwise often lengthy periods of poverty of movement.

Discharges of nervous energy may also lead to the appearance of the phenomenon of **cramping** ("Verkrampfung"). Sometimes the whole body of the patient becomes rigid, the shoulders are pulled back, the arms are firmly pressed against the body, and the face is distorted into a grimace. After a few seconds these symptoms disappear, but may reappear several times during the examination.

When quite young children with negativistic catatonia do not show the same poverty of movement as seen when the illness develops in adulthood. As in sluggish catatonia the patients often turn towards objects in their environment and sometimes handle them. Such movements occur in normal tempo. In a state of emotional upset these movements in adults and even more in childhood become rapid and jerky. Movements are angular and abrupt, but are not clearly distorted. In one patient it could be observed particularly well how different his movements were when he was quiet and when he was excited. In keeping with his childhood mobility the 17 year old boy showed certain movements, such as turning towards objects lying in his field of vision. He grasped the water-tap, picked up the soap, licked it, then he rapped a radiator pipe, all this in quiet somewhat slow movements. But when I wanted to take away from him a spoon which he had seen on the table and had picked up, he prevented this with extraordinarily quick movements thus eluding me. When I finally succeeded in taking the spoon he suddenly grabbed it again and pulled it towards himself.

In states of excitement **expressive movements are** more prominent in these patients. While anger is not seen much with the patients' rigid faces a certain cramping may appear. At other times there may be ambivalent smiling which, when the patient is in a good mood, may turn into a grin all over the face.

I was able to observe in a home a girl only 3 1/2 years old with negativistic catatonia. When seen she held her head down and looked at me rather unwillingly when I was trying to cheer her up. I continued my joking, touched her on the nose, then on her belly. Eventually she raised her head and began to mimic my movements by touching her nose. I responded to this right away by showing her other movements: I opposed my index fingers, at first she did not react, but copied my movements with her own index fingers. She continued this movement for a while even though I had stopped it. In this way she repeated other movements which I made. Finally she folded her hands as I myself without being aware of this had folded my hands. For a while she moved the folded hands up and down. Then she hit her belly, also my hand. Then for a while she rubbed my hand. There followed further repetitive movements. Now she was smiling. Because of her unfriendly behaviour at the beginning, the subsequent ambivalent smiling and the tendency to repetitive movements I thought she had negativistic catatonia. The information provided by the nurse confirmed the diagnosis. She was often observed hitting her head against the wall, sometimes hitting her face. Three months ago she had torn out all her hair (her hair now was still quite sparse). Quite often she became aggressive and pulled the hair of other children. At a time when she was still younger contact could be established with her, but now no longer. The picture is therefore entirely in keeping with early childhood negativistic catatonia. Of particular interest is the echopractic behaviour typical of childhood, but which continued with the folding of her hands when I had stopped folding mine. I think that this was an example of early childhood automatic pushing, ordinarily the child would much rather copy others than counter-grasp.

Parakinetic catatonia is characterized in adulthood by motor restlessness with distortions of movement. Particularly affected are facial expression and gestures, furthermore both involuntary and voluntary movements are unharmonious. The patients may also act impulsively. If the disorder begins in childhood the same applies, yet restlessness is greater but more uniform. "Parakinetic manners", that is, stereotypically repetitive movements may also be present, but are more noticeable in the child. One patient in short intervals placed her hands in such a way that the backs of both hands touched one another, and then put them on the right side of her head, where they stayed for a short while during which the fingers were moving restlessly and her face was distorted more than usual. A parallel may be drawn to the fact that negativistic children make uniform movements often repetitively. The need to move when new movements are not available may be the basis for both phenomena. In the child to be described grasping for the nose and the mouth was such a stereotypy.

The **boy** was 7 years old and showed restlessness to a high degree. On entering the room he laughed excessively, looked at himself, then looked in another direction, climbed on the chair, sat down one way or another, kneeled on the chair, and dropped on the floor.

For a short time he sat down quietly, but then with his hands made turning movements into space. With rapid and distorted movements he groped for his nose, drew his hands over nose and mouth and at the same time made blowing and sniffing sounds. Then he turned one hand jerkily and laughed. Spreading his fingers he again made a distorted movement towards his nose and then brought them down to his mouth. Without apparent reason he again grinned. Grimacing movements of mouth and eyes were constantly present. Independent from laughter the boy occasionally made a noise that sounded like an "E" which no longer expressed anything in particular. Apart from distortion the movements had a somewhat jerky quality but they were not as sudden as patients with negativistic catatonia when irritated.

Another **patient** who became ill in early childhood even though he was now 26 years old showed still a great deal of restlessness. With rapid pace he practically ran into the room. His face was distorted into a grimace. In sitting down he twisted his hands forwards, put them together, then separated them, and at the same time made unnatural finger movements. In-between he pulled up his shoulders, distorted his face and made short noises. When requested to be quiet he interrupted his restlessness for a short while and held his hands stiffly on his thighs. Almost immediately his arms moved in a screwing manner. When standing his legs were equally restless, like a pitter-patter.

Patients whose disorder begins in adulthood do not show this kind of restlessness to the same degree. Sometimes it is not noticeable and is not mentioned in the case histories. Therefore the diagnosis is made more easily when the disorder begins early in life.

It is remarkable that patients with parakinetic catatonia developing in childhood **do not speak.** If certain sounds are made these are not sounds of speech, they are rather related to sounds of expression. Our patients did not even give their names. If the illness begins in later childhood or adulthood the characteristic "tangential remarks" are observed in which something is said about the situation or disconnected from the situation. It could have been expected that these children would show at least a badly articulated speech, as the speech-prompt catatonics at the beginning of their illness do speak even in the early years of life. Perhaps articulation itself which demands fine coordination is impeded by the parakinetic disturbance. A girl now 15 years old whom I saw apart from my investigative series could say words but these were so badly articulated that only the caregiver could understand them. The patient named every object in the room, but without help I could not understand a single word.

As with proskinetic catatonia it is necessary to differentiate parakinetic catatonia from the restlessness of **excited imbeciles.** When the latter are in a cheerful mood they often become very restless: they run around, shuffle their feet, make faces, gesticulate, and may make many vague shaking or sweeping movements with their hands in a jerky manner similar to those of patients with catatonic parakinesia. But in the latter the affect is missing. The movements of imbeciles are more closely related to choreatic than parakinetic movements. Likewise the constant restless movements of small children are more closely related to the choreatic ones. In profound mental retardation restlessness may appear even with slight irritation and without affect. This makes the differentiation more difficult. But then characteristically only the shaking and sweeping movements are found which in their speed may overflow into tremor. At the same time there is rigidity of the shoulders, as well as partial rigidity of the face. This symptom is mindful of what a healthy person may observe in himself. In the morning when it is time to get up he sometimes does not arise immediately but "stretches" at first, that is, he contracts body and face and sometimes shakes his arms. In a similar manner when yawning not only may the face contract but the body as well; subsequently the arms may shake. This is perhaps an atavistic sign originating in the extrapyramidal system. Parallel to "shaking palsy" it could be called a **"shaking cramp".** This syndrome can no longer be confused with a parakinetic disturbance. I do not know if it has been described in profound mental retardation; I have only recently come to know it.

Interestingly in a **5½ years old girl** seen in a home there was a combination of excessive and parakinetic restlessness. The girl was conspicuous as soon as I first entered the ward where she was. In observing the child more closely I noticed that parakinetic movements taking place with moderate speed frequently alternated with very rapid movements in which her arms were flung at times up and down, at times to and fro. These rapid movements did not look harmonic, but only the slow movements appeared distorted. Some of the latter were parakinetic stereotypies. The little girl kept on raising her twisted arm and put the back of her hand on her mouth or briefly rapped her face; at the same time she made croaking or grunting sounds. The rapid sweeping movements occurred in a state of excitement, for the distorted laughter which she showed otherwise turned into grinning over the whole face. Probably the girl was imbecile from early childhood; catatonia was superimposed later. This could be considered confirmed as the child, already over 5 years old, in spite of efforts by the nurses, "could not yet do anything". She was still incontinent and had learned to walk only the year before. She could not yet eat alone. Intelligent parakinetic catatonic children under guidance learn the simple daily routines quite quickly. Probably catatonia had developed very early, for at the age of 9 months the record showed "conspicuous turning movements with the hands and turning and swinging of the head". Perhaps catatonia began especially early because of the organic mental retardation. The combination of early childhood schizophrenia and organic brain disease was mentioned above.

Manneristic catatonia of adulthood may outwardly show quite different clinical pictures, depending on the way the mannerisms have developed. When sufficient stimulation is lacking the patient may show mannerisms of avoidance to a high degree. With therapeutic intervention the mannerisms however remain within tolerable limits. They may be transformed into an occupational activity of a manneristic-uniform character during the course of the day. Because of lack of expressivity the patient's postures and movements are rigid.

In children with manneristic catatonia the mannerisms are more uniform and appear more as **stereotypies.** This is probably due to the fact that firm patterns of activity have not yet been developed. Rocking of the body and the head occurs more regularly than in other catatonics. Several other stereotypies may be observed: one patient rocked his head not only in the usual slow manner but he intermittently shook it to and fro. Another patient from time to time pushed one of his arms to his side and moved his head in the same direction; it looked somewhat like a fencing posture. A female patient continued to make beckoning movements and did this even when a chocolate was offered to her. A male patient from time to time made a singing or smacking sound. Another patient repeatedly tapped on his knee. Mannerisms in the true sense of the word are not totally missing. A female patient refused to eat certain foods and from time to time lay down on the floor of the dayroom. A male patient used any opportunity to get out through the door or the window, not in order to leave without permission, but to walk up and down outside until he was brought in again.

A boy, not part of this series of examinations whose disorder began at the age of 10 years showed that the transition from mannerisms to stereotypies is related to the very early origin of the illness. He had many true mannerisms. When I called him into the room he paced up and down, turned his body to one side, walked forwards and backwards, without advancing. When I insisted that he come into the room and asked him to step over the threshold he raised and lowered his legs as if he was marking time in front of the door. When called more forcefully to come in he moved like this even more and in addition made small jumps. When called upon to open the door he grasped the outer end of the door handle with both hands and pushed it down. When I asked him to touch the door handle with one hand he reacted with a disconcerted look on his face. I took his hand in order to guide it to the door handle where upon he resisted so strongly that I did not succeed. At home one of his mannerisms was continually to bend down almost to the floor. Moreover he could not be touched; he immediately stepped aside. Such mannerisms, that is stereotypies related to certain situations, when the disorder begins in early childhood, are replaced by simple stereotypies not associated with the situations.

As a mannerism of avoidance all my patients exhibited **mutism.** Actually the patients can speak, unexpectedly they even say something. When I mispronounced the name of a patient the latter immediately interrupted me and gave the right name. But I could not get him to say anything else and I did not succeed in having him give his name once again. A patient who was usually mute suddenly during the examination said: "How much longer will it take?" An attendant informed me that a certain patient never said a word on the ward, but did talk with his visiting relatives. Clear-cut mannerisms of avoidance of a different kind I could not ascertain with these children; probably this was accounted for by their childhood mobility.

It is remarkable that in spite of the onset of the disorder in early childhood manneristic catatonics are able to speak, since in most catatonics with poverty of stimulation expressive speech hardly develops at all. But manneristic catatonics are different as they are influenced by their mannerisms of avoidance. Mannerisms of movement which are more prominent in childhood are even more pronounced. Only expressive movements are largely extinguished in manneristic catatonia.

Stereotypies of posture in manneristic catatonia of early childhood are more prominent than in adulthood. It may be said that in the form of a mannerism of movement certain postures are assumed and in the form of a mannerism of avoidance these posture are not changed. One patient constantly held his hands in front of his body; another patient contracted her hands into fists. A female patient held her fingers in an extremely contorted fashion, one finger placed obliquely over the others. A male patient walked heel to toe. Another male patient steadily twisted his mouth in such a way that his upper and lower teeth were wide apart while his lips were firmly pressed together; his face in this manner appeared to be remarkably elongated. Another male patient constantly held a small object, for instance a small piece of rubber, between thumb and index finger, and with all his strength refused to have the object taken away. Later I shall describe this patient together with his brother who was affected by a form of combined catatonia.

At the onset of the disorder in early childhood patients generally are more mobile; the childhood tendency to turn towards the environment is evident. Even if they do not react when spoken to they do observe events in their environment. Some of them during my investigations looked around so much that at first I had the suspicion that they were distracted by hallucinations. I also noticed that they reacted to the way I was dictating in their presence. When for instance I stated that the patient was continually turning his head hither and thither, he stopped this for a while. In adult age patients quite often smile when it seems that they are made to recognize that they could behave differently. In childhood this sometimes turns into a stiff grinning. The great mobility of the child is also evidenced by the fact that although the characteristic rigidity of posture and movement is present it is not so clearly marked as when the disorder begins later in life.

In a home I saw **a girl 1 year and 3 months old** whom I considered to have manneristic catatonia. All my efforts to stimulate her produced not a single reaction. She continued to sit rigidly on a small chair and held her arms sideways from her body. With slow movements she turned her head this way and that, and changed the direction of her look between me and Frau von Trostorff. Since the girl rarely looked in any another direction, this continuing movement had to be regarded as a form of stereotypy even if the tendency of the child to respond to outer events may have played a role. As I extended my arms in order to pick her up her arms did not move at all. Even on my arm she held both her arms away from her body. She did the same when picked up by the nurse. She completely lacked any response to encouragement; her stiff posture and her stereotyped body movements were evidence of manneristic catatonia which had already begun within the first 15 months of her life. When I asked the nurse whether the girl had peculiar habits the nurse replied that she had a propensity to allow her body to fall backwards when being lifted up. If picked up from the floor attention had to be paid to prevent her from hitting the floor with the back of her head. There was no evidence of muscular weakness – when sitting on a small chair or on my arm her attitude was no different from other children of the same age – a stereotyped behaviour had to be presumed.

A **boy** just 2 years old likewise probably had manneristic catatonia. He too during my attempts to cheer him up elicited no reaction whatsoever. His face was stiff, even rigid. His 1½ years older brother was also living in this home; he was a healthy vivid child. He began to harass his brother sitting beside him. He picked a stone from the pocket of his brother's trousers – probably beforehand he had inserted it –, placed it under his brother's cap and pulled the cap down on the face. Then he again looked for the stone, did not find it and then turned the head and body of his brother far forwards in order to look for the stone behind his back. When he could not find it he again put his hand under the cap and in so doing roughly turned the head of his brother this way and that. With all this the latter only made a very short defensive movement, otherwise he allowed everything to be done to him without reacting. In his stiff facial expression there was just a fleeting look of disapproval. Even when the nurse turned towards the boy on my request he remained motionless and expressionless. I could not elicit stereotypies; therefore I could not prove that my diagnosis was correct. But what else might have been wrong with him? A normal or pathological introversion could not extend to such a degree. Perhaps the disorder had begun only a short while ago so that the characteristic symptoms had not yet fully developed.

Combined Systematic Catatonia of Early Childhood

My opinions regarding combined catatonias are less binding since of the various combination I was able to observe only a few cases. But cases of the same combination resembled one another so clearly that the identified syndromes could be considered as sufficiently well established. Of course they were not discovered here for the first time, but had been observed before in adult age. Both descriptions complement one another. Some symptoms appeared in their characteristic form during the onset of the disorder in early childhood even more clearly, so that conversely the description which I had given for adult age was thereby confirmed.

When **speech-prompt** and **sluggish catatonia** were combined in adulthood there was an increase in the "perseveration" of speech-prompt catatonia, the patients continued repeating what they had said, and after inbetween questions they repeated it again. To this was added hallucinatory distractibility.

A **female patient** who had become affected around 3 years of age and was now 23 years old was constantly preoccupied during my examination and at other times when she was sitting in the dayroom. She whispered to herself, looked around, inbetween she gesticulated. Frequently she turned her head and eyes to one side appearing to be listening. This hallucination was confirmed by the fact that she often held her ears closed. In keeping with this behaviour the component of sluggish catatonia was evident. If this were present alone with the disorder originating in early childhood, no response would have been obtained. But the patient answered every question which I posed. Unfortunately since her speech was totally unarticulated nothing could be understood with certainty, but it seemed that she often repeated what I had asked. – A **male patient,** now 20 years old, at first also spoke quite indistinctly, but most of what he said could be understood. Echolalia here appeared clearly. His parents had found this out and said: "He repeats everything like a parrot." My questions produced the following: (Are we in Brandenburg?) "Brandenburg." (Or Berlin?) "Berlin." (Are you a boy?) "Boy." (Is this ward 11?) "11." (Or ward 12?) "12." When I said with a loud voice "window-frame", he repeated that word. In this way it continued as I was saying words without looking at the patient. When I said several words he repeated these too, but the last syllables were unarticulated, the same as the female patient mentioned above. When I said: "It's time to go home", the last word when he repeated this was an indistinct sound.

Echolalia we will find when considering other combinations of speech-prompt catatonia. This may represent a childish form of perseveration in that the patients do not later repeat the answer they had given, but repeat what they hear.

In cases where **speech-prompt catatonia** is combined with **proskinetic catatonia,** at the beginning of the disorder in adult age there is pressure of confused speech in which one can hardly find a sentence that makes sense. The first component can be recognized in that words that had been pronounced are later often repeated; the second component may result in several words being repeated in a form of verbigeration. Both these symptoms are more evident in children than in adults. Since there is little availability of thought content the pressure of speech consists only of perseverative and iterative words. Furthermore the clinical picture in the child is again modified because of the great readiness to pay attention to what is going on outside. In this way the conversation turns to what the patients see.

An 11 year old girl when entering the examination room immediately said: " Is this a book? – Is it a table? – We are fetching chocolate – What is that? (pointing at a folder) – Are you doing nothing? – We are fetching chocolate -Is uncle sick? -Are you looking for nothing? – Afraid – Pencil (pointing at the lady doctor who was writing). Is it a chair? (pointing at the armchair). Is it a chair, a chair? It is a chair – We are eating chocolate -Are you afraid? – The lamp is on, up there is a lamp – Are you doing nothing? – Are you doing nothing? – Who is that, that is an uncle. – Is it an aunt, is it an aunt? – A person who writes ("Rumschreiber") – "Arm of a watch", wrist-watch, arm of a watch – wrist-watch (This is a play on words in German: "Uhrarm", "Armbanduhr") – Are you doing nothing, are you afraid? – A girl (pointing at a picture with children) – A car (on the picture there was a car) – Closing the shoes (opens and closes the zippers of her shoes) – Uncle sick – Are you doing nothing – putting on watch" – (Pulls up her sleeve, not wearing a watch, apparently wants to have the watch of the lady doctor.) She then fingers an exercise book lying on the table and said: "Look at it, book, a hat, uncle is afraid, are you doing nothing?"

Here it can be clearly recognized how words which have been spoken at one time keep on appearing. Note also that some words are repeated; this occurred far more frequently than was recorded. A child with this condition turns much more to the examiner than an adult with the same condition, as observed from the fact that the girl kept on talking about what she could see.

Most of the objects in the room she named correctly, but besides that she showed **talking-past-the-point.** Looking at a picture in which there were no books she said "books". Looking at an empty page she said "Games", looking at a clay figure she said "rod". Though earlier she had correctly talked about books, she called a folder – as previously the picture – a "book". **Neologisms** also appeared: she had at first correctly named a paper-clip, later she said "Bindarm". Probably this sounded a bit like the word "Uhrarm" which previously she had called a watch ("Armbanduhr"). The expression "Rumschreiber" must have meant that with a ball-point pen one could write around ("herum"). There were apparently other neologisms, but designations not corresponding to objects shown to her were often difficult to understand, since the girl articulated badly and did not repeat such words. This girl in the way she talked resembled the girl described above under simple speech-prompt catatonia. That patient too pointed at objects which she named; but she did not turn towards these objects, she sat quietly, had a stiff face and appeared to be interested in nothing. This girl, however was constantly restless, pointed not only at the objects but bent down to get closer to them, and at times actually ran towards the objects. She also spoke much more haltingly. The other girl stopped talking for longer periods of time between her words. The presence of the proskinetic component was clear-cut in the one case, its absence in the other.

In spite of the characteristic picture in the case of the speech-prompt proskinetic girl I had at first some doubts with regard to the diagnosis, since the facial expression of the child was not coarsely stiff as is seen in speech-prompt catatonia and similarly in proskinetic catatonia of adults. But proskinetic children show, as mentioned above, more expressive movements than adults. Furthermore the facial expression of the girl just described was not mobile, but there was a practically constant smiling expression corresponding to her good mood. The face of another girl with the same catatonic combination was nearly always tearful during the examination.

I can show that the symptoms as formed in childhood continue into adulthood. In a previous paper I have briefly described a patient who developed systematic catatonia in the first years of life. At that time I could not as yet describe the combination precisely, in-between I had to make other observations. Now at the age of 30 years this patient on re-examination had the same picture as that described for the girl above. The symptoms from his childhood remained the same later on. The patient talked a lot earlier, and now again in the same manner, repeating words frequently. He now said: "But that takes a long time. Only old people get injections. How long is this going to take? This is only for old people. Do they get it more frequently, more frequently, more frequently? Three times and then no more? After 3 years have passed, have passed and then no more, not at all any more. How many more years is this going to take? Only old people get injections." – This patient turned to the person speaking to him with much more liveliness than is otherwise seen in adults with the combined catatonia. He came quite close to me, talking and occasionally tapping my shoulder.

In **speech-prompt-negativistic catatonia** patients of adult age still tend to talk-past-the-point, but often answers to questions are negative or the patients refuse to answer altogether. In addition there may be aggressive negativistic irritability. If the disorder begins in the first years of life patients tend to look around more and are inclined to become autoaggressive rather than aggressive against others. Otherwise the picture is the same. – A **female patient** now 24 years old partly looked around, partly looked at me with an empty face, at times with a hostile glance from the side. When questioned she sometimes gave answers, at other times she kept quiet in spite of my urgings. When I wanted to take away a wooden toy from her hand she resisted. Objects in the room were named correctly. To other questions she remained silent or talked-past-the-point. My usual questions were as follows: (Are you a boy or a girl?) "A boy, my name is Gitta." (Are you a boy or a girl?) "A girl." (Now what? A boy or a girl?) "A boy, a girl." After some other questions later she twice repeated: "A boy." – **Another patient,** blind from birth and now 18 years old, had an astonishing knowledge, even though he had been regarded as severely retarded. He correctly gave me the date of his birthday, likewise the address of his parents, also the date, and the day of the week on which he had been brought to the hospital the previous time. It is true that such answers were only given if he did not talk-past-the-point. We had the following conversation: (Where do you live?) "With Daddy." (In what place?) "With Daddy." (Here in Brandenburg?) "Yes." (Or are you from Potsdam?) "Yes." (Or from Berlin?) "Yes." (How old are you?) "10 years." (How long have you been here?) "10 years." (This question was emphatically repeated.) "I am in the children's unit." (Are you a boy or a girl?) "A boy and a girl." (You must be either a boy or a girl!) "A boy and a girl." As there was no success in obtaining a correction to a wrong answer it became evident that here was a symptom of speech-prompt catatonia. Yet the boy sometimes did not answer at all, but shouted with short interruptions, something that sounded like "Ah". Sometimes the pauses in-between were so brief that it sounded almost like a repetitive shouting. This looked like the negativistic symptom of iteration. Moreover the negativistic component was manifested in that this patient, if not mechanically restrained, bit into his upper arm until it bled. Probably there was a gnawing sensation on his arm such that even here iteration occurred. This special form of automutilation perhaps arose from the fact that because of his blindness this patient had to fall back on producing stimulation of the senses stronger from biting than from hitting himself. The shouting may have been a form of conation transferred from the optic to the acoustic area.

Where **speech-prompt** and **parakinetic catatonia** are combined in adult age pressure of speech may develop in which elements of the speech-prompt type are found. Besides that parakinetic restlessness becomes prominent. When the disorder arises in early childhood language is badly articulated. Furthermore there may be short-circuit answers and simple repetition of words.

A **patient** now 20 years old showed restlessness of the whole body and spoke a lot. But nothing could be understood because of the poor articulation. The nurse on his unit said that he repeated everything that was said to him. When I spoke some words to him he said words which could not be understood. A **female patient** showed less pressure of speech and in spite of poor articulation some of what she said could be understood. At the same time **talking-past-the-point** typical of speech-prompt catatonia was prominent. The patient, 35 years old, said she was 7 years old, she repeated this later although in-between she had mentioned other figures. Most of the objects shown to her she named correctly but also formed neologisms. For instance a cushion she named "chair cover" ("Stuhldecke"), a wooden ruler "wooden comb" ("Holzkamm"). There were also incomprehensible word forms such as "colored" ("Bunte") when shown a rubber stamp. A waste paper basket she called "filling" ("Füllung"), presumably indicating that the basket would be filled. Bad articulation is part of the picture of parakinetic catatonia, whereas the fact that speech is not absent indicates speech-prompt catatonia. This patient did not show echolalia, perhaps because the pressure of speech in her case was less than that of the patient described previously.

The complete lack of expression seen in speech-prompt catatonia may still be recognized in the combination, although there is constant facial grimacing of parakinetic catatonia. On the whole the restlessness of the face and also of the body is less marked than in simple parakinetic catatonia.

In the combination form of **speech-prompt-manneristic catatonia** of adult age talking-past-the-point appears to be more intentional than in the simple form. This is less often to be expected in a child who does not yet think like this. Moreover in adult age perseveration in general conversation is often repeated. In a child verbal stereotypies may arise. Furthermore there is a tendency to simply repeat questions. **Echolalia** here is even more prominent than in the other combinations described above. When I told a patient that I had already seen him in the dayroom he echoed: "dayroom". While I was dictating in his presence he kept on repeating a word from my dictation. But he did not repeat words pronounced to him directly. From that changeable behaviour the combination of echolalia with mannerisms can be recognized.

I was especially impressed when I encountered speech-prompt-manneristic catatonia of the childhood type in a **9 year old brother** of a female patient who herself was afflicted with manneristic catatonia. When a simple question was posed, for instance to name objects in the room, he answered correctly, but always at first repeated the question, for instance: (What is that?) "What is that, a tic-tac." (What is that called?) "What is that called, shoe." (Where do you live?) "Where do you live, Oranienburg." On one occasion when I did not understand him correctly the nurse asked him to speak clearly. He repeated: "Speak clearly, Ewald." Moreover the boy repeated stereotyped sentences which while he was still at home he had heard his father say and which he had repeated, for instance: "Ewald, what have you done, whether you want to or not, you are please from half eight!" ("ab halb achte!") This last sentence had been quoted in the medical record. The boy mentioned it to me less clearly than other sentences, probably because he had used it so frequently. Ordinarily he did not speak very clearly, but he could be understood. With sentences originating from his father he also copied the rhythm in which he had heard them. At home he had been looked after by his father, an invalid, while the mother had gone to work.

This boy, like another with the same combination, in-between briefly looked into space as if he was hallucinating. Actually hallucinatory distraction would not go against the presumption that the combination form was present, for adults with speech-prompt catatonia sometimes hallucinate.

Sluggish-proskinetic catatonia in adult age is characterized by hallucinatory distractibility, fingering movements, and murmurings. In addition there may be fantastic confabulations. The latter are present in adults and do not occur in early childhood, as neither sluggish nor proskinetic catatonics speak when the disorder originates in childhood. Murmuring too does not occur. However hallucinatory distractibility and proskinetic restlessness are more marked because of the child's mobility. A female patient was constantly restless with head and eyes and whispered to herself. Furthermore there was constant fingering, tapping and knocking the fingers against one another or against an object. In simple proskinetic catatonia of early childhood restlessness goes beyond fingering in that the patients turn towards more distant objects. In the combination inner distractibility inhibits turning towards the outside. Thus the female patient showed only small movements but these were much more vivid than those seen when the disorder begins in adult age. Another patient with the combined condition fingered her dress a little and constantly rubbed or tapped the fingers of the other hand or of the same hand. When spoken to these patients looked at you fleetingly and remained distracted by their hallucinations.

In the combination of **sluggish-negativistic catatonia** of adult age patients never answer questions, but they whisper to themselves and display hallucinatory distractibility. They do much fingering of their clothes or of the objects in front of them. If the disorder begins in early childhood hallucinatory restlessness is very evident. A female patient whispered not to herself but she constantly looked around, shook her head, made defensive movements with her hands and from time to time covered her ears. The negativistic component was characterized by the fact that when addressed she showed no sign of response. Chocolate given to her she did not put into her mouth but let it melt in her hand. When we wanted to take the chocolate away she objected strongly. Further the negativistic component was confirmed by the fact that she hit her own head. Aggression and autoaggression however in this patient and in another female patient with the combined form was less marked than in simple negativistic catatonia. Perhaps the inward distractibility had an attenuating effect. Fingering could be recognized from the fact that the patient often made grasping movements around herself and picked at her hair, but this was overshadowed by her strong hallucinatory restlessness. In between her face and body became cramped, similar to that observed in simple negativistic catatonia.

In **sluggish-parakinetic catatonia** in adult age we find parakinetic restlessness and talking with the voices. In addition there are fantastic confabulations, not found in children. Hallucinatory restlessness in adults is prominent in combination with parakinetic disturbances, in children this is even more so because of their mobility. A **female patient,** 28 years old, whose disorder began in early childhood, talked into the void with grimaces and parakinetic restlessness of the whole body. The patient looked up, sideways, sometimes also behind herself, and constantly spoke in a low voice. Because of bad articulation nobody could understand what she was saying. At times when she was more calm she was still moving her lips with a distracted facial expression. Her restlessness was partly hallucinatory but also parakinetic in nature. **Another patient** showing this combination with the same behaviour spoke inaudibly but was constantly moving his lips and whispering.

In the combination of **sluggish with manneristic catatonia** in adult age hallucinatory distractibility is accompanied by stereotyped movements. This is also true if the disorder begins in early childhood. A **female patient** with this combination when entering the examination room said something that sounded like "dödö" and "nana". In between she was constantly moving her lips and whispering; for a while she was looking up obliquely listening, and often made nodding movements. The restlessness with whispering, incomprehensible sounds, nodding, and listening postures were continued for a long time; when spoken to she did not respond. In addition there were **stereotypies** showing no connection with the hallucinations. Whisking her hand past her face could at first have been interpreted as a defense against hallucinations but the whisking continued indefinitively. Her face was stiff, lacking expressive movements. The latter is quite characteristic for the manneristic component, for in simple sluggish catatonia hallucinatory restlessness is combined with facial expressions. Stereotyped postures are also present in this combination. A female patient clenched her thumb pressing it against the other fingers and pulled her hand back when the examiner tried to open the hand. **Another female patient** showed a more impressive stereotypy of posture: during the examination most of the time she held her right hand against her face with index finger on her nose and thumb on her cheek bone.

In **proskinetic-negativistic catatonia** adult patients give reluctant and inconsequential answers. In these patients fine fingering is replaced by more gross movements with which they touch their bodies or objects. If the disorder begins in early childhood the combined form is more easily recognized because impulsive touching is more prominent. The patients' movements are also turned more frequently to the outside. Furthermore their negativistic urge to prevail appears more rapidly and strongly. However the correct diagnosis may be difficult as these patients do not reply to questions.

An **11 year old** boy who did not want to be interrupted in his compulsion to finger things maintained a negativistic attitude. Suddenly he grasped objects he saw, a pencil, a paper, books. With lightning speed he took a page of our notes crumpling it completely. Then he suddenly stood up and grabbed a briefcase lying at the other end of a long table. After that he suddenly ran to a water tap situated several meters behind him. While I held one of his hands he quickly tried to grab my glasses with the other hand. If the boy was prevented from acting out he made angry noises, stamped the floor with his feet, and tried to pull away his hands which had been restrained. On another occasion he was in a better mood. He now smiled at me, sounding more cheerful; several times he clapped his hands. This time he handled objects more quietly, rapped the table, fingered his clothes, shook the back of a chair. His face remained quite stiff, the occasional smile did not change anything; a natural expression never appeared.

The sister of the patient, now 19 years old, likewise was afflicted with proskinetic-negativistic catatonia starting in early childhood. While visiting her in her home I could not persuade her to come with us into a room. While talking with her parents I could see through the open door that she was continuously running to and fro in the corridor of the apartment and looking at us with a stiff face. When she was out of our field of vision she made a lot of noises: banging a door, hitting an object, or clattering a pot. She kept on shouting loudly. When she had calmed down I was looking for her and found her half lying on her bed holding a newspaper in front of her. I asked her in a friendly manner what was in the paper and wanted to take it in my hand to check what she was saying. She made a sudden dashing movement in-between, tore the paper towards herself, tried to hit me and kicked me with her feet. Following this her whole body was restless, she fingered her clothes, pulled her shirt out of her trousers, tugged the shirt, shook her hands and knocked repeatedly on her body. When friendly words were directed towards her she lost her irritability. Eventually she smiled at me and again and again automatically gave me her hand. When I held my hand 2 meters away she even stood up and grasped my hand. These were all characteristic symptoms of proskinetic-negativistic catatonia, including the fingering.

Autoaggressivity in spite of the negativistic component is not found here, aggressivity is found only in a small measure. This made it possible for the last named patient to stay at home. Another patient whom I saw with his parents and who had this combination of symptoms sometimes impulsively pulled the curtains down, but never attacked his family which included a baby.

In **proskinetic-parakinetic catatonia** both in adulthood and childhood the highest degrees of restlessness occurring in the systematic catatonias are found. To the distorted movements of the one form is added the automatic turning towards the environment of the other form. In adults tangential remarks typical of parakinetic catatonia and verbigeration seen in proskinetic catatonia may also be observed. When the disorder originates in early childhood verbal utterances are missing. However symptoms of restlessness are more prominent.

A 6 year old girl was constantly in motion during my examination, she climbed on a chair, hopped up and down, turned around, knocked the back of the chair, cried out, grasped objects which she saw, kneeled on the chair, stood on it, climbed down, tried to run away, made noises that sounded joyful. When prevented from running away she twisted her face, now clearly showing anxiety, but because of the stiffness of the face the type of expression could not be recognized. When grasping for objects movements were distorted, for instance one grasping arm could be twisted, the other overextended at the wrist. Frequently she made turning and shaking movements in front of her face without grasping. Noises could be heard repeatedly. For example she called “Mimimi”.

The fact that facial expression is not part of the general restlessness points to the proskinetic component in which the facial expression is empty. On the whole, increased turning towards the environment was always combined with distortion of the sequence of movements.

In **proskinetic-manneristic catatonia** adults show verbigeration and fingering movements. The latter occur in a stereotyped fashion in which at certain intervals the same form of fingering appears repeatedly. When the disorder arises in early childhood no verbal utterances are to be expected. In addition to fingering there may be more extensive movements which are also found in simple proskinetic catatonia of early childhood. These movements appear periodically in the same form and thus with the fingering movements are part of the stereotypy.

A **male patient** with this combination stared at me when I spoke to him but gave no answer. Instead he began to move one of his arms forwards and backwards and continued these movements for quite a while. The movements appeared as follows: he rubbed his little finger on his thumb, made movements of pronation and supination, at the same time quickly pushing his arm forwards and backwards. He repeated these movements without variation on that day and on other days. Sometimes the rubbing of his fingers occurred without the movement of his arm so that now only the fingering movements could be observed. During one examination his clothes were covered with wool fibres which he had plucked out. According to the caregiver he tore clothes, ripped bed sheets into small strips, tore up his own woollen cover and picked up the cover of another patient. This plucking and tearing corresponds to fingering movements and again is a part of stereotyped activity. – Another **male patient** stretched his arms to the side with finger movements and noddings of his head. – A **female patient** appeared to murmur verbigerations characteristic of proskinetic catatonia in emitting unarticulated sounds in a stereotyped fashion.

The **younger brother** of the first patient was suffering **from simple manneristic catatonia.** They were now 35 and 32 years old, respectively, but their illness began in their early years, they had always been considered severely retarded. They were on the same ward, but avoided one another. When I saw the younger one he was sitting down holding himself extremely rigidly with his arms held against his body and his fingers in different positions. While the little finger of his left hand was overextended in the proximal joint and the index finger of the right hand lifted away from the other fingers he held his hands firmly together. When I tried to take the hands apart he resisted so strongly that I could not do it. On another occasion he had a small object between the fingers (in the previous description of simple manneristic catatonia was mentioned a similar stereotypy of posture). He always had to hold on to something, either an object or one hand clasping the other. His face usually was completely stiff, nothing changed it, even though once he smiled and at another time grinned. It was interesting to note that the pictures of the two brothers were grossly different and yet one component of the disorder was present in both.

In cases where **negativistic** and **parakinetic catatonia** are combined in adult age the parakinetic restlessness is more uniform. In addition answers may be partly tangential as in the one form, partly unwilling as in the other form. As can be expected patients in whom the illness began in early childhood do not speak. In comparison with simple parakinetic children their restlessness is less marked. Aggressions and autoaggressions are less impulsive; they form part of the parakinetic movements in a playful manner.

A **9 year old boy** at first was sitting down quietly, but soon he became restless. He shook his head to and fro, held his head with arms twisted, made distorted movements with his fingers. He pulled up a leg and handled his foot. Sweeping movements of his arms followed. At the same time he was grimacing and laughing wryly. From time to time he let out a howl sounding like "Eia". In-between he beat his head with his fist. He continually tapped his shoulder or one of his thighs. On the ward he often attacked other children. This was not a case of simple parakinetic catatonia as there were agressive outbursts and iterations, although restlessness was not very marked. The howling was probably caused by both components of this form of catatonia.

A **half-sister,** 12 years old, showed simple negativisms and during my examination was greatly autoaggressive. I held her hands so that she could not beat herself, but she slipped forwards on the chair and hit her head against the back of the chair. I lifted her up so that her head would no longer hit the back of the chair; then she hit her head sideways against her shoulder. When I prevented this as well she succeeded temporarily in hitting her head against the wooden side-rest of the chair. I was hardly able to hold this frail girl so that she could not hurt herself. Her autoaggressivity was very marked when I had provoked a negativistic impulse by holding her. When I saw this girl on another day I was more gentle in my approach and tried to prevent her autoaggressivity by holding my arm between her hand and her head.

Negativistic-manneristic catatonics in adult age do not reply to questions and in general hardly react when spoken to. They do show restless movements of stereotyped or repetitive form. If the disorder begins in early childhood stereotyped and repetitive movements are more prominent. There is more turning towards others. In the examination room **a male patient** made repeated movements of his trunk sideways and at the same time turned his head. Then for a while he shook his hand this way and that and later repeated this again and again. He raised his legs from the floor and moved them forwards and backwards. In-between he sat down motionless holding his arms in front of his body. Such stiff postures have a character similar to stereotyped movements. When the latter stop the stiff posture is maintained for a while. Another **male patient** moved his hands jerkily towards his face, placed the finger tips of both hands together, and remained in this posture for some time. Then he put his hands on his cheek-bones. Sometimes his posture was grotesque in that he touched his face with his wrists and overextended his hands so that they spread out from his head like wings. Both patients showed the negativistic component in the form of aggressions and autoaggressions. The former patient tore clothes, hit himself and others. Once he hit another patient so hard that the latter had large haematomas on his head and knees. Last year on one occasion he pushed his head through a glass window suffering deep cuts in his face. He now has gross scars on nose and chin. Severe autoaggression is still present today even though he is now 25 years old. Such deliberate striving to self-mutilation indicates the presence not only of negativistic catatonia but also an intentional factor compatible with manneristic catatonia. A **female patient** showed a similar picture. When I saw her for the first time blood was running over her face as she had just hit her head hard against the edge of the bedstead resulting in several facial injuries. Negativistic tendency to self-mutilation is increased through the presence of the manneristic component, probably in connection with the fact that this tendency to autoaggressivity is enhanced and strengthened through the mannerisms.

Autoaggression differentiates children from patients whose illness began in adult age. Furthermore general mobility is better preserved. Repetitive movements are more changeable in their forms. Impulsive behaviour is more prominent. In this way we had some difficulty in diagnosing the first patient described above. As he had received chocolate after the first superficial examination, during the next examination he constantly reached for Frau VON TROSTORFF'S handbag containing chocolate. Before we had verified the combination we thought that this was a proskinetic type of behaviour.

In the combination of **parakinetic** and **manneristic catatonia** in adult age parakinetic restlessness is the same as in the simple form; movements are not only distorted but through the manneristic component appear stiff; the movements are also somewhat slowed down. Restlessness is rather more continuous than in simple parakinetic catatonia. It should be kept in mind that manneristic catatonics show mannerisms both of omission and movement. When the illness begins in early childhood restlessness is more marked. Otherwise the picture is the same. A **boy 16 years old** twisted and turned his head in different directions during the examination. He took my hand in an inharmonious movement, let it go and groped into space. Then he turned his face forwards and opened his mouth widely. Again he grasped into space and held his fingers unnaturally extended. At the same time he made a noise that sounded like "bömö". Other verbal utterances were missing. Often he smiled, sometimes he laughed loudly and sometimes he grimaced. His movements were very stiff but quick or jerky movements of simple parakinetic catatonia were absent. His restlessness was uniform, with some parakinetic stereotypies such as extending his arms forward, moving his head forward, and opening his mouth repeatedly. A **female patient** showed parakinetic stereotypies even more impressively. Apart from other uniformly repeated movements she showed the following: her left hand held rigidly in front of her face, fingers bent, thumb drawn in; then she knocked with her right hand on her left or brushed over it. The torsions in these two patients were not as evident as in simple parakinetic catatonic. The manneristic component apparently compensated somewhat for the rigidity.

Rarely **three catatonias** are combined. I have not mentioned such cases occurring in adult age, since it is difficult to analyse adequately the overall picture when the three syndromes are combined. However in early childhood certain symptoms seem to be more clear-cut, therefore I shall try to describe the following cases in which three components appeared to be present.

A **female patient** whose illness began in early childhood but who is now 25 years old showed striking disturbances which hitherto I had not seen. The patient in entering the examination room had a greatly distorted posture, head bent, shoulders pulled up, with her right hand held obliquely in front of her face. With contorted movements she shook her arms above her head and made similar movements repeatedly. In order to see her face I had to lift up her head and then I noticed constant grimacing movements that were without expression. At times she forcefully closed her left eye. In between she was sitting down in a cramped rigid position, squeezing her arms between her legs.

Features of parakinetic-manneristic catatonia were prominent. The cramped rigidity however was so strong that it could not be explained by just the two components. This rigidity of the whole body was like the one which often appears in negativistic catatonia. It may therefore be regarded as a third component. My impression was confirmed by the fact that earlier on she had been aggressive, had bitten other children, torn their hair and had sometimes pulled out her own hair. At the age of 10 she was described as follows: "Aggressive against other children, tried to choke them, was so quick that she had to be closely watched so that she would not cause any damage." The lightning speed of her movements indicates a negativistic component.

A **male patient** remained quiet as long as he was not stimulated; he gesticulated and whispered with his voices. When he was asked questions he perseverated and talked-past-the-point. I was wondering whether he had **speech-prompt-sluggish** catatonia, although the speech of other patients with this combination is unarticulated. But when tested for automatic touching, which should have been present, it was absent. Then I noticed that the face of the patient was not only stiff but quite rigid. Moreover in the case history mannerisms had been noted. For years it had been his habit to throw objects into the air and catch them. Moreover in a stereotyped fashion he kept on running away and remained for days in the neighbouring wood. I therefore had to assume the presence of a **manneristic** component which thus explained this comprehensible speech.

Etiology of Early Childhood Schizophrenia

Having demonstrated how catatonia arising in early childhood can be recognized and distinguished from mental retardation we should now examine its **etiology.**

Seeking for an explanation we decided to speak to the **mothers** in their homes; we asked about the development of the children in the first years of life and the care which they had received. Of the 117 patients I spoke with 86 mothers, 42 fathers, and several grandmothers for detailed information. My co-worker Frau VON TROSTORFF prepared the way. We were unable to contact 31 mothers, some had died, some had moved away. Fathers, grandmothers and patients' records helped us to obtain sufficient information.

Psychosocial Causes

Earlier investigations led me to the opinion that for the genesis of the illness in early childhood lack of communication was an important factor. Of the 11 children whom I knew at that time 7 at the beginning of the illness were in hospital or in a home. Two others were blind or deaf from birth. In only 2 cases nothing could be found which pointed to lack of communication. These observations brought to mind the findings of SPITZ (1945). This is even more relevant as SPITZ when examining babies and small children in a home found not only severe retardation but also catatonic symptoms. He mentioned "certain unusual postures of hands and fingers", "extremely bizarre hand and finger movements", and characterized the overall picture as "presenting a picture of stuporous catatonia". Unfortunately SPITZ'S cases were not re-examined later. I presume that many of them would become genuine catatonics. At first I thought of an overall lack of communication, not so much the absence of the mother, but in the first years of life this is obviously of great importance. It was not a question of a "schizophrenogenic mother"; a schizophrenic mother in most cases is better than no mother.

In the home of infants described by SPITZ the children were grossly neglected psychologically. This could no longer happen in modern homes, anyway not in the German Democratic Republic, where in the homes, operated partly by state authorities, partly by religious organizations, the children were well cared for. Infant mortality in the homes was very low. The model low infant mortality in the GDR was well known. However it is known that the mother can never be replaced by a home, and therefore in case of doubt it is better to leave the child with its mother and to support the latter. In some countries the authorities more quickly take away the child from a mother who apparently was a poor caregiver. This can be damaging to the child even if the home is excellent. This does not mean that a deficiency due to a negligent mother may not be greater than the unavoidable deficiency in a group home. Often children in their own home are bodily neglected so that because of imminent danger to life they have to be placed in another home.

I recently had the opportunity to review my former etiological observations on the **117 patients** mentioned above. Here was the result: 41 children grew up in a long-term home; 32 of these were admitted to that home soon after their birth, 9 somewhat later but before the end of the first year of life. Eight other children spent time in a long-term home during the first 3 years of life. Two children had to spend more than half of their first 3 years of life in a hospital. Eight children spent shorter periods of time in a hospital, but according to the reports they were afflicted with hospitalism. One child because of disease of the spine when 1 year old had to stay in bed in a plaster cast for 1½ years day and night. Ten children were totally or nearly totally blind from birth. Three were totally or nearly totally deaf. Two other children had bilateral radical operations for otitis media and suffered purulent discharges from their ears for months after their operations. In one case the child had to have a second operation and then stayed another 5 months in hospital, thereby suffering additional trauma. One child had parents who were deaf and dumb and was thereby was severely limited in communication. This accounts for 76 children. Many other children were reported to have had purulent ear infections in the early years of their lives; I am not counting them because we could not find out how severe the ear disease was and whether or not it would have produced hearing difficulties. Nor did I include a child born prematurely who had to spend four months in an "incubator". I believe that the 76 of the 117 children show clearly **that lack of communication is to be regarded a very essential cause of early childhood catatonia.**

Of special interest is the **blind patient** with speech-prompt-negativistic catatonia described above. He had a twin sister who stayed healthy. She too did not see well but as a small child was not greatly handicapped. Two operations improved her vision so that she could go to a school for the visually impaired and succeeded in passing ten grades. She is now being trained to become a physiotherapist. Her brother who also underwent treatment for his blindness which was not successful, suffered from severe lack of communication which led to early childhood catatonia, whereas the less severe visual impairment of his sister did not grossly prevent her communication with the environment and thus did not endanger her mental health.

In spite of the significance of the lack of communication unavoidable in a children's home it must be stated that most of the children in these homes remained healthy. Of the 41 children mentioned above only 2 were admitted to hospital every year at an average age of 22.6 years. Essentially at risk are only those children who grow up nowhere else but in the home. The fact is that apart from the 41 children staying in the home all the time only 8 who stayed there part of the time. This difference is multiplied if it is taken into account that the 41 children have much less contact with healthy children than the other 8 children who stayed in the home temporarily. In one children's home I counted how many children stayed in the home during their first 10 years of life continuously or only temporarily. The result was that only every 6th child grew up totally or almost totally in the home; for altogether there were 1031 children who stayed there temporarily and only 167 who stayed there all the time. In a smaller home only every 8th or 9th child was there constantly; here the figures were 506 and 58 respectively. Thus for children who only stay in a home[22] temporarily there is no serious danger.

[22] Translater's note: In German the word "Heim" may indicate the person's own home, or a foster home, or group home for children. In Leonhard's text the word "Heim" appears often without specifying whether it is a foster home or a group home. In most cases it seems to be the latter.

To be mentioned as well is that there is no danger for children staying in a home for some weeks. According to SCHMIDT-KOLMER (1960) they are slightly retarded when compared with children living with their families, but probably only on a psychological and not a biological basis, as less material is offered to them to work with. In my series of investigations of 5 early childhood schizophrenics from such homes 3 had stays in hospital, 1 had visual impairment; only in 1 case nothing could be found that could have reduced communication. Probably with such children there is a risk if even on weekends they have no mother to care for them.

In order to determine more exactly the danger existing when there is a very long stay in a home I added to my **retrospective** investigation a **prospective** one. With Frau VON TROSTORFF I visited several children's homes to look for the available information on children who had been admitted to the home in their first year of life and who were still there at the age of 3, in order to determine their further fate. At the beginning of this procedure I included children who stayed in the home temporarily; among the first 16 none were found with catatonia. The cases selected after that consisted of 130 children. In 21 cases Frau VON TROSTORFF could not find out where they were staying. Of the remaining 109 children (52 male, 57 female) I was able to determine whether after the end of their first years of life they were healthy or catatonic. The result was that 92 were normal or only had mental retardation, whereas 17 (7 male and 10 female) had become catatonic. These latter I examined personally; 12 had simple systematic catatonia, 5 had combined systematic catatonia. Of the children still in the home who did not have catatonia I examined 29 personally. However I did not count all the children who after staying in the home were placed back in their own or into another family. My idea was that the most severe early catatonics could hardly be treated at home. Exceptionally this can happen so that the number of catatonic children could be somewhat greater. But in general the **ratio of 92:17** corresponds to the objective data, that is for each case of catatonia there were 5 to 6 non-catatonic cases.

It should be mentioned that among the 17 catatonics the excess of male over female patients which I had found in 117 cases in the psychiatric hospital was not present (10 girls, 7 boys). This confirms the fact that boys are admitted to hospital relatively more frequently than girls because they are more difficult to manage. Of the 17 cases examined only 3 were in a psychiatric hospital, the others in a home, 1 even temporarily in his own home. Some of them when older and more difficult to handle probably will be admitted to hospital. Their average age was only 11.5 years.

I now return to the 117 patients and ask **what were the reasons for the fact that 41 of them had originally been admitted to a home.** It is possible that in some cases it happened because they already were abnormal so that their stay at a home could not be considered as a cause for the disturbance. Therefore I am stating the reasons for their admission to the home.

In not less than 22 cases single mothers returned to their jobs and could not keep their children with them. At the time when these children were born the regulation in the German Democratic Republic had not yet come into force according to which after childbirth the woman received maternity leave with pay for one year. In another case the mother soon after birth had to work because the father was drinking and did not care for the family. In 8 cases a mentally retarded mother had to look after several children and could not cope with yet another one. In 3 cases the mother gave her child away without an excuse having been recorded. In one of these cases soon after birth the parent moved away from the German Democratic Republic and left the child behind. One child was a result of incest; the mother and the father of the child were arrested. An epileptic mother as well as her husband wanted to go back to work after the birth of the child and therefore had the child placed in a home. In one case both parents were temporarily psychotic and were unable to take care of the child. In another case a schizophrenic mother declared – perhaps with justification – that she could not look after another child. In yet another case the father had open tuberculosis. Among these 39 cases there was none in which it could be stated that the child was already abnormal at the time of its admission to the home.

Only in 2 patients there may be a suspicion in this connection. After their birth they were admitted to hospital for a physical illness and then were transferred to a home. We were unable to find out why they were not returned to their parents. In one of these cases there must have been gross neglect by the mother, for she found out only 1½ years later where her child was. There is no evidence that an abnormality was responsible for the admission to the home. Thus I cannot see any other possibility than the various events as described above.

Let me report some other findings indicating that patients afflicted with early childhood schizophrenia suffered from lack of communication. Apart from the 22 illegitimate children admitted into the home there were 9 other illegitimate children. It has to be assumed that some of them did not receive enough care at home from their single mother or a substitute mother. Moreover 19 mothers were mentally retarded. Seven of these gave birth illegitimately to children who later became schizophrenic and are thus included among the 22 and 9 children just mentioned. Likewise 7 children of single mothers are included in the 41 children admitted to a home. Six others were neglected at home; they had problems due to poor nutrition and temporarily had to be treated in hospital. Should the thought arise that there might be a hereditary connection because many of the mothers were mentally retarded let me remark here that mental retardation has nothing to do with early childhood schizophrenia.

Frequently the neglect of a child was not due to the fact that the mother was mentally retarded, but because she had so many children that she could not cope with another one. A retarded mother whose husband likewise was retarded had 11 children; the affected child was the 10th. The other 10 children were said to have done all right in their lives. But usually some of the children of retarded mothers are retarded as well. One of the mothers had 7 children, 4 were older than the affected child; 4 of the siblings were retarded and received special education. – Another retarded mother had 10 children, 3 illegitimate before subsequent marriages, and 7 legitimate; the patient was the 7th child. Of his full siblings 4 did poorly in school. Examination of the 5th child led to the diagnosis of early childhood catatonia. Below I shall describe the development of the 2 affected children. – A retarded mother together with her retarded husband had 6 children. The first 3 were only 1 year apart in age; 4 children were retarded and had special education. The patient is the 3rd child. – A retarded mother from her first marriage had 2 children, then followed 2 illegitimate children which she neglected grossly. She married a second man and later lived with a third. Both illegitimate children developed early childhood catatonia. Their development will be described below. – The fact that frequently these mothers because of so many children were unable to give them proper care demonstrates that the affected child is often the youngest of the siblings. Of the 117 patients 34 were the youngest, only 15 the oldest.

It may be claimed that children of retarded mothers are more **physically than mentally neglected,** suffering from deficient nutrition and poor personal hygiene. But in order to demonstrate that the mental disorder cannot arise therefrom it should be kept in mind that children placed in homes become affected so frequently even though they are often well nourished and hygienically cared for. As stated previously the children described by Spitz had been bodily cared for without fault.

Still other observations show the importance of lack of communication in the development of early childhood schizophrenia. In 3 cases where no neglect could be demonstrated there was a **change of caregivers.** Perhaps the children had become so dependent on the first caregivers that they could not accept another one.

One of these children, a girl, soon after birth was given by her mother to a woman in the neighbourhood for private care. The mother seemed to have paid little attention to the child, as she could not say anything about how the baby developed with the caregiver. The latter had a child of the same age and took very good care of both of them. When the child was 1 year old it was placed in a crèche but did not adjust well and after a few weeks had to be taken out. The girl's own mother now stopped working and took her daughter home. The latter developed slowly and at the age of 3 years she came into a children's hospital where she was thought to have suffered early brain damage. After I had asked the mother to whom the girl was more attached, to her or to the caregiver, she replied that the girl was attached to neither one. I concluded that the girl did not accept her mother as a real mother. The lack of attachment of the child to her mother could not be understood otherwise. The caregiver must have been the person to whom the child had the best relation but the real mother was unable to say anything more precisely about this. The only thing she said was that the caregiver had looked after the child very well. I assume therefore that the change back to her mother destroyed the girl's communication which she had developed with the caregiver. It could also be assumed that the caregiver had given the child more emotional warmth than the mother who appeared to be rather cool.

A child until the age of 1½ years was with its mother. Then the mother had a second child and gave the first to the grandmother who was said to have spoiled the child very much. The mother gave examples of this so that we could suppose that the grandmother was the person to whom the child related the best. But the child had already become so firmly attached to its mother that when she lost her mother she could not become attached to her mother substitute.

Another child that had spent much time in hospitals during the first years of life and suffered therefrom, at home was cared for by its mother and even more so by its grandmother; to all appearances the grandmother was the more important person as was shown during our visit. The mother gave an objective account of her child without evidence of any emotional involvement. However the grandmother was deeply moved when telling us that she could suppress crying only with great effort; she declared that the child had been her favorite. The first 1½ years the mother and child were living with the grandmother. Then mother and child moved to another city. This meant a loss of the grandmother as the person to whom the child had related the best.

Another case was similar: A **boy** and **his sister** both of whom were catatonics – they have been described above – were strongly tied to their mother during the first 2 years of life. The mother felt much burdened by him as she had to look after his sister. Therefore at the beginning of his 3rd year she placed him in a crèche. Here the boy was restless, cried and ate little even though at home he had been eating well. After about 10 days he could no longer be managed in the crèche. His mother then placed him in a day centre where he had to stay in spite of his objection. After he had reached the age of 3 years, the upper age limit of the centre, the mother had to take the boy back home. However she did not keep him but placed him in a long-term home. As her reason for this she told us: "I had to sacrifice the boy on behalf of the family." It might have been assumed that the child only became ill in that home. But I could find no evidence that the lack of his mother was decisive after the 3rd year of life. The clinical picture that was revealed during my examination was that of children whose disorder had begun before the end of the 3rd year of life. The complete lack of verbal utterances, strong aggressivity and repetitive movements where factors in favor of such an early onset. My view of the connection was different: At the time when the mother forced the boy to stay in the day centre he refused to consider his mother as the best person to relate to. The mother too rejected the child. She had already been overburdened with the other child and was unable to give him motherly care.

Therefore many children were added to the 76 where lack of communication had to be considered an important factor. In fact of the 117 children there were not many where no such assumption could be made.

I must state here that **psychic hospitalism** plays a greater role than had been recognized so far. Only 8 children had been diagnosed with this condition while they temporarily stayed in hospital. Objectively there were perhaps more; later such a disturbance can often no longer be ascertained. In this connection I had the suspicion that the appearance of hospitalism is especially dangerous in cases where the attachment to the mother had previously been very close. One of the 8 children mentioned above had been cared for in the first years of life with great love by her mother. These days the latter still is very much attached to this child even though it is now 11 years old and because of her speech-prompt catatonia has become incapable of any meaningful contact. When her mother was asked how the child behaved during visits the mother said: "Quite strangely, as if we were foreigners." While saying this tears rolled down over her cheeks.

From the age of 3 months the child had to lie down for several months with both legs spread out and bandaged. The mother felt obliged to spend a lot of time with the girl. She developed well. But at the age of 2 years and 2 month she had to be admitted to hospital with pneumonia. At the time of her parent's visits they were only allowed to look at her through a pane of glass. She stood in her cot, rocked her body to and fro and did not react to her parents. The attending physician replied to their anxious questions that the child was suffering from hospitalism and he recommended that as early as 12 days after admission the girl be discharged with medical prescription. At home the child seemed to have forgotten everything that she had learned before, she stopped talking and became incontinent of urine again. Gradually her condition improved, the parents thought the child had become the same as she had been before, but soon after more serious disturbances appeared indicating the onset of catatonia. The girl could no longer find contact, she was rejected and abused by other children; she came home from the crèche with bite wounds and herself became aggressive. This speech-prompt-catatonic child with her peculiarities of speech ("Today is a day") was described above in detail.

The appearance of this severe hospitalism so quickly in the few days of the stay in hospital could be explained by the fact that previously the child had had a very close attachment to her emotionally warm mother. The contrast thereby was especially extreme and made deep inroads into the biological sphere. Apparently in spite of the short stay in hospital permanent damage was done which then led to early childhood schizophrenia.

I should like to append an **observation which I made in my own family.** At the time when my youngest daughter was 1 year and 2 months old my wife had to go on a trip in order to bring home things that had been taken away during the war. We placed the child in a good nearby private home where she was to stay about 10 days. When I visited the child after 3 days she made a good impression from a distance. On this warm summer day the children sat on blankets on a meadow and appeared to be playing. I thought when my daughter would see her father she would crawl or run towards to me and express her joy to see me. Instead as I approached her she glanced at me with an empty look and showed no reaction whatsoever to my greeting her. It seemed that she did not know me at all. I had an awful shock, took the child in my arms and took her immediately back into my office where I looked after her with the help of a lady doctor until my wife returned. She rapidly became friendly again – I think after she had slept a night in my room – , turned to me again, played again, but was anxious for a while. If I had not experienced this I could not have thought it possible that in only 3 days such a severe change could have occurred. Certainly the home into which she had been placed could not have been blamed, it had the best reputation; we ourselves had a very good impression of it at the time when we brought our daughter there. Hospitalism of this severity thus can appear in a home in the shortest period of time in spite of the best attention given. In the case of my own daughter what contributed to the situation – similar to that of the girl previously described – surely was the fact that she had been looked after by my wife with so much affection. She was a late child, born prematurely at the 8th month, for a long time quite delicate and therefore even objectively in great need of loving attention.

A person who has had such an experience will recognize very clearly how much the separation from the mother may affect the biological development of the little child. Thus I no longer consider it impossible that the severe hospitalism observed in the girl described above during her 12 days stay in the hospital led to the pathological development eventually resulting in catatonia.

With regard to my little daughter my observations showed the mother may not be the only person to whom the child is closely related but that the father too plays a role, either himself as a very important person or as a person incorporated in the role of the mother. I would not know what I would do today, at that time I did not pay that much attention to it, if the condition of our daughter had become completely normal again by the time when my wife returned, or if only her presence brought about the complete recovery. But in any case the severe hospitalism was rapidly interrupted when I took the child home with me. Having thus ascertained the significance of the father I may state that it is not a disadvantage for the child if besides the mother a grandmother may also be an important person. It might have been suspected that not only a change in the most important person but also the care by several other important persons might have been a disadvantage.

With the assumption that a child is particularly sensitive to the loss of motherly care if previously the mother had given it a very great amount of care 5 children are of great interest where I could not find any evidence of below average communication. In all cases the patient was a girl, the first born. The next child (4 times a girl, once a boy) was born 1 ½ to 2 ½ years later. The mothers confirmed that they had spent an especially large amount of time with their first born children. In all cases they had a lot of time for the child for none of them had a job outside the home. The children were tied particularly strongly to the mother and became especially sensitive to the decrease of attention following the birth of a sibling. In one case this became quite evident directly from the mother for she mentioned that after the birth of her little sister the girl had "withdrawn" and from then on had become more and more peculiar in her behaviour. Perhaps it is not a coincidence that all 5 patients were girls, for boys are stronger and are perhaps more adapt at making up for the diminishing care through their own activity or even to regain the care provided previously. It is well known that small children are competing for the care of their parents.

If the lack of communication with the mother or suitable substitute mother is regarded as the essential cause of early childhood catatonia a closer look should be taken at the **type of the maternal influence.** Becoming ill during this period of time should be considered even more from the **biological** point of view when a comparison is made with the origin of systematic schizophrenia in later childhood. A glance at the **animal kingdom** may be a helpful analogy. The new-born animal remains in the vicinity of the mother, is influenced primarily by the primitive forms of sensory impressions which are derived from bodily contact with its mother and her smell. While suckling gustatory stimuli are added. The higher senses, seeing and hearing, play a lesser role. Sounds of enticement or warning may be significant, but a young mammal, for instance a filly, constantly behind its mother, most of the time does not hear any sound from her and apparently does not look at her much either.

Primitive stimuli of smell, taste, and touch surely play a role in the influence which a human mother exercises on her child and thereby promotes the child's development. When the child is carried in the mother's arms it constantly senses the mother's body, it smells the skin and the breath of the mother. When breast-fed contact with the skin is especially close, the smell of the mother's skin comes from the immediate vicinity. Stimuli of taste and smell from the mother's milk are additional. Breast-fed babies thrive the best, that surely is not only due to the fact that they tolerate their mother's milk the best, but also to the accompanying stimuli of touch, smell, and taste. Therefore it is not a complete substitution when milk pumped from a woman is offered to the infant. And besides if a bottle-fed infant develops nutritional disturbances this is not always due to an error in the choice of nutrition, the disturbance may also derive from the fact that the nourishment is not offered in the manner provided by nature. If there is no or very little bodily contact with the mother or substitute mother in all probability there will be a nutritional disturbance, for the children described by Spitz all suffered bodily and many of them died in spite of excellent physical care.

Many of the schizophrenic children under my observation besides their mental illness had **deficient physical development.** Most of them looked younger than they were, many at the age of 20 hat not yet reached puberty, some adult women had never had a menstrual period, some men had no hair growth on their face and only little pubic hair. There may be extreme lack of maturation of the body. A female catatonic patient whose illness began in earliest childhood, at the age of 26 looked like a 15 year old, had never had a period and no growth of hair in the genital area. Her height was 133 cm and she was very thin, almost cachectic. A 12 year old female patient looked like a 7 year old, her height was 108 cm and her weight 22 kg.

It might be thought that inhibition in the area of the lower senses might produce a disturbance and a retardation of bodily development, thereby leading towards a tendency to develop early childhood schizophrenia. But against this is the fact that there are catatonics who although their illness began in early childhood had a completely normal development of the body, enter puberty at the usual age and later show all the signs of a mature man or woman. One of my male patients even had a premature and strong development of his body. At the age of 14 he was 175 cm tall and he appeared to be quite strong. It is true that his weight of 48 kg still corresponded to a childhood constitution. His genital organs were already developed and around the penis there was a lot of hair; growth of beard however there was not. I do not think that he had an abnormal growth pattern, as his overall constitution was harmonious. His height was in keeping with the fact that his father measured 183 cm and his mother was also quite tall. A strong bodily development therefore did not protect the boy from developing early childhood schizophrenia. Perhaps the **point of time of the damage** disturbing the subsequent physical development was earlier in infancy than the damage that produces early childhood schizophrenia. The above named patient during his first year of life was always with his mother and only in the second and third year of life repeatedly spent time in hospital because of nutritional disturbances and otitis. The critical phase for future physical development at that time was already behind him.

I suspect that the lower senses are significant only for the earliest physical development. Surely the higher senses, eyesight and hearing, are decisive in the lack of stimulation; for in systematic schizophrenias higher mental functions are affected. As a matter of fact we found many blind and deaf children among our patients. The inhibition in the lower senses and subsequently in the higher senses may in many cases overlap so that both the mental illness and physical underdevelopment occur at the same time.

Contact in the area of the higher senses is particularly human. The loving mother attending to the infant is constantly in contact with him optically and acoustically; she talks with the child, smiles at him, makes surprised, encouraging or other faces, nods or shakes the head. The child reacts to this, can already understand facial expressions, reacts differently to a cheerful, sad or angry face, does not yet understand language, but takes in the sounds which are constantly contained in the words of the mother, for the latter does not talk with an objective sound when for instance she says "my sweet child", "my darling", "you little fellow", but she speaks with a flattering, happy, encouraging, sometimes also sympathetic tone of voice which the infant already understands. Often she may call only "da, da, da" or similar senseless syllables. All this represents, as I believe, the **specifically human contact** which the child must not lack if it is to thrive mentally and not run the risk of developing early childhood catatonia.

Role of Heredity

In general hereditary disposition in the systematic schizophrenias is quite low. This is also true for the children under my observation. Among the 117 patients 10 (excluding the half-brother mentioned below) had a psychotic sibling. This comes to 4.4% of all the 228 siblings. Since many of the siblings have not yet become adults it still remains an open question whether or not some would later become schizophrenic. Eight of the children had a psychotic parent (6 mothers, 2 fathers). After subtracting 30 unknown fathers this comes to 3.9%. My later investigations showed that adult systematic catatonics had 2.7% affected siblings and 3.5% affected parents. There need not be any connection with heredity from this small increase of affected siblings of childhood catatonics; if a child was exposed to lack of communication, that also was frequently the case for the sibling. Similarly in 9 cases the sibling had early childhood catatonia; in one case a sister had anxiety-happiness psychosis. Among the 8 affected parents one mother had periodic depressions, the others were schizophrenic. The pathological behaviour of the affected mothers could not have contributed to the catatonia of their children because of other circumstances. Only one parakinetic mother had kept her child at home for the first year and a half before it was placed in a children's home. The child of the periodically depressed mother did stay with her during the first years of life, but her first phase appeared when the child was already 7 years old. The other patients had been removed from the schizophrenic mother shortly after their birth.

The parakinetic mother mentioned above also had a parakinetic child. Otherwise no similarity of the psychoses could be established. However this was the case in the siblings so that here heredity may be of significance. I have found homogeneity in the systematic schizophrenias in different members of the family even when the onset of the disorder occurred in adulthood, and have tried to explain above that though the disorder arises primarily from outer causes its formation is determined endogenously, depending on which "system" has a weakness. Since the etiological significance of the endogenous tendency is small it has to be assumed that the latter is found in most people at a time when outside causes are taking effect.

The **catatonic siblings** of our patients are presented here individually:

The brother of a **negativistic** patient had negativistic catatonia as well. – The sister of another **negativistic** patient had combined speech-prompt-negativistic catatonia. – The same combination was found in the sister of a girl with **negativistic** catatonia. – The sister of a **manneristic** female patient also had manneristic catatonia. – The brother of a **manneristic** male patient had proskinetic-manneristic catatonia. – The brother of a female **speech-prompt-proskinetic patient** had proskinetic catatonia. – The brother of a female **speech-prompt-negativistic** patient had speech-prompt-manneristic catatonia. – The sister of a **proskinetic-negativistic** male patient likewise was proskinetic-negativistic (I have described these two patients above). – The brother of a male **negativistic-manneristic** patient was a manneristic catatonic. – The half-sister of a male **negativistic-parakinetic** patient had negativistic catatonia (again relatively many siblings have the combined forms of the illness!).

As with systematic schizophrenia in adult age the siblings show in part the same subcategory of catatonia; in combination they have at least one form in common.

Should there be an inclination to ascribe to these findings a greater significance of heredity in the origin of early childhood catatonia, a counterargument comes from examining the **life history of the affected siblings.**

The first mentioned **negativistic** patient was admitted to a home at the age of 8 months and stayed there for 9 months. Before and after the stay in this home his mother had little time for him as she had to work hard on her own farm. The affected brother soon after his birth was admitted to a home because of below average weight and stayed there for 6 months. Then he too was poorly cared for by his mother. The patient had a twin sister who developed normally, did well in school, but at our visit behaved rather oddly. When we were speaking with her mother she sat beside her always looking away. We could not tell whether or not she was listening or why she did not go into another room as her sister had done. When I talked to her she answered with hesitation and shyness. At the same time her movements were unharmonic and quite angular. After her birth she was not in a home like her twin brother. It could be that because of this she remained healthy, even though a tendency towards catatonia was present when considering her peculiar behaviour and angular movements. – The **second negativistic catatonic** was one of 3 illegitimate and 7 legitimate siblings. He was neglected by the mentally retarded mother and besides that was hard of hearing for some time because of bilateral suppurative otitis. The affected sister was similarly neglected by her mother. Moreover in her second year of life she spent several months in a home because at that time her mother was unwell. – The **negativistic girl** as well as the sister gave no history of exogenous damage. However the following is worth noting: during the time between the patient's birth and the first years of her life 4 other children were present who had to be cared for by their mother. When the affected sister was born 5 older children were present. Since their mother had to work long hours on the farm it was unlikely that she could spend enough time with her children. – The **manneristic** female patient in the first years of her life was nearly always ill and frequently in hospital. Her sister too was ill for much of the time during the first years of her life. The mother had little time for them because of her job required her to be on night duty.- The **manneristic** male patient belonged to a family in which the mother had 2 legitimate children and 2 illegitimate children both of whom became catatonic, each having been severely neglected. When admitted to a psychiatric hospital the patient's brother seemed to have been even more grossly neglected. He was quite dirty. Both patients were described above. – The brother of the female **speech-prompt-proskinetic** patient also grew up in a children's home. – The **speech-prompt-negativistic** female patient belonged to a family in which a mentally retarded mother and a mentally retarded father had 9 children. After her birth the girl spent 4 months at home, then because of poor domestic conditions she was placed in a long-term home. Her affected brother was a late child. Since the mother was working he was cared for much of the time by his father who at the time of the child's birth was 60 years old. It has to be assumed that the retarded and elderly father was unable to replace the mother. This boy was described above with his marked echolalia. – The **proskinetic-negativistic** patient was the boy mentioned above whom his mother as she expressed herself had to "sacrifice". When the mother had to travel several weeks his sister, then in her third year, was placed in a home. Here she showed tearful resistance, related neither to nurses nor to other children, biting and scratching. In spite of dietary treatment she had diarrhoea. She suffered from hospitalism which perhaps contributed to the appearance of catatonia. – The **negativistic-manneristic** patient at the age of 1½ years was placed in a long-stay home after previously being cared for in a crèche. His mother had to work because the father was drinking and did not support the family. The affected brother was placed in a home in the first year of life for the same reasons. – Finally the half-siblings: the girl due to poor domestic circumstances was placed in a home shortly after birth. The boy for the same reasons stayed in a home for a whole year after his birth; then he was looked after with great neglect by his mentally retarded mother. The children were illegitimate with different fathers.

Therefore quite similar circumstances were found in both the affected siblings and the probands; this explains why the siblings became affected as well. Heredity imprinted the clinical pictures in their form, but had little to do with the onset of the illness.

During our visits to the homes we observed not only children but also **mothers** and, when present, **fathers.** In the cases where 2 children were affected neither parent had schizophrenia. Four mothers of these 10 children were mentally retarded, otherwise they showed nothing abnormal. Of the 3 fathers whom we saw, one of them, the father of the 2 manneristic siblings, spoke rapidly without modulation of the voice and with hardly any facial expression. Possibly he had a mild latent manneristic catatonia like his daughters.

The question remains what actually could be observed in the **86 mothers** and **42 fathers.**

A **father** resembled the father just mentioned. He showed a stiff almost rigid face. Most of the time he looked ahead without moving his head. His words came out impulsively and badly articulated. He too had a manneristic child. – The **father** of a proskinetic-parakinetic child showed practically no emotional expression. He spoke without modulation and at times his face twitched. – The **mother** of a negativistic child during examination was stiff in posture and expression. Never was there a friendly nor even a courteous smile in her face. She showed no sympathy towards her affected child; the last time she had seen the child was 6 years ago – at a time when the child was 5 years old. – The **mother** of a child with sluggish catatonia likewise was expressionless, giving the impression of being unfriendly. Sometimes it took a long time for her to answer questions and only then it could be recognized that she had remembered. – A **mother** had a very unharmonious motility reminiscent of the parakinetic catatonia of her son.

Peculiar behaviour which might be found in connection with schizophrenia was otherwise absent. Hereditary loading established in 8 affected parents was hardly any higher compared with healthy parents who had been looked at in more detail. Some other peculiarities could be observed which genetically had nothing to do with schizophrenia, but in some cases consisted even of accentuated joyousness during the interview. Some other peculiarities however might have contributed to neglect of the child. Mental retardation was mentioned above.

Four mothers were especially loquacious, two of them so intransigent that it was difficult to ask them questions. Three others in addition to their loquaciousness were quite vivacious. giving a hypomanic appearance. **Another mother** might also have been considered hypomanic, but exhibited such vivid emotional movements that I considered her to have an exuberant personality. The behaviour of **another mother** was similar but at the same time she showed paranoid trends and complained bitterly about the way physicians had treated her child. **Four mothers** showed deep emotional trends without being lively; I considered them to have emotional personalities. **Three mothers** were physically and mentally uncouth; one of them led an unstable life, the other 2 were impulsive in the way they reacted and during our conversation rapidly became resentful. All three might have been considered to have shown epileptoid trends. **Five mothers** were clearly hysterical, full of self-praise, even though they had in no way treated their children with care. One of them had an epileptoid trait; she was grossly asocial, went out with men while her children were locked up at home without food. Another of these 5 women was paranoid as well; the available case history confirmed this. A **mother** was a severe alcoholic and mentally depraved to such an extent that nothing more could be said about the structure of her personality.

A connection may be seen between the hysterical personalities and neglect of children, but a hereditary connection exists neither here nor with the other behaviours mentioned above. The mothers whom we had seen, apart from the ones mentioned, were inconspicuous, often quite open, that is to say not at all "schizoid". It was surprising that parents who showed so much pleasure in human contact had a catatonic child.

A feature to be commented on here is the **emotional coldness** of certain mothers. Again and again we noticed that mothers stopped visiting their affected children. Some of them did not even know where their children were until we told them. One mother had been informed that her child had died but later found out from us that this was not so. Another mother did not know that her child whom she had not seen since infancy was now blind. But from such observations it must not be assumed that all these mothers were emotionally cold. It is well known that the tendency to repress disagreeable situations which cannot be changed is a general human characteristic. The mothers had recognized that the child was incurable, they could not help by visiting, and they were reassured with the thought that the child was well cared for in the hospital. They continued with their every day duties. This attitude was facilitated by the schizophrenic behaviour of their children who, as has been described above, were no longer capable of direct emotional contact. The children became strangers to their mothers. Often the parents did not know whether or not their children recognized them when they visited. The children usually accepted food, fruit, sweets, otherwise they showed no emotional reactions. Some let their mothers sit alone as soon as they had taken the things to eat.

I think in spite of all this the behaviour of some mothers went beyond what could be understood psychologically. One mother had married again and then had some more children. The latter had no idea that they had an older sibling who was mentally ill. Perhaps the mother did not want to burden them with such thoughts, but she herself for years had stopped caring for her son now 21 years old; he was almost blind and almost deaf, suffering from negativistic catatonia and confined to a psychiatric hospital. During a long conversation with her about her affected son she remained cool and distant. – Another woman when visited by us was quite friendly and pleasant, she right away offered us coffee even though we came unannounced. However in the conversation concerning her affected daughter she showed no emotional expression. She did not even enquire about her daughter who had in the meantime reached the age of 26 years. Perhaps she could be excused because the child had been born illegitimately, but in the medical record it was noted reproachfully that she had "abandoned" the child, and in our conversation she had shown no emotional reaction. – Another mother had her child with her during the first years of life, and therefore could have become attached to it. But when we wanted to obtain information about its development she had almost nothing to say. She appeared to have stopped thinking about her child and for many years had not visited it in the hospital. This woman too lacked emotional expression during our conversation. – Still another mother, having received our questionnaire, came to see us in the Charité. She was afraid that her daughter, now 23 years old, would be discharged from the hospital. For her support she brought with her a healthy son. The daughter had been at home for 4 years, then was placed in a children's home. Since that time the mother had stopped caring for her daughter, never visited her in the home or in hospital, and never inquired about her condition. During our conversation she asked no questions at all about her daughter. This mother was living in good circumstances and with her husband had established a nursery for plants which she and her son continued to manage after her husband died. When we explained to her the reason for our investigation she was not unfriendly but objectively remained cool.

From these observations I would assume that several mothers of our patients were naturally unemotional. While they still had their children with them they showed them little motherly love so that the children suffered from this deprivation. But in none of these mothers did their cool attitude appear abnormal. As far as we could determine they showed nothing unusual in their behaviour and their life style.

From several statements by parents we could infer that right from the beginning the behaviour of the child itself led to the poor contact with its mother, for we heard repeatedly that **already before the onset of the illness the child had been very quiet.** We could not always assume that this "quietness" was already the beginning of a catatonia. Sometimes parents expressed the belief that this behaviour had started soon after the child's birth. The degree of pleasure from contact between people varies greatly, even infants may show this difference. When a child does not show much interest in the outside world the response of its mother may be diminished. Thus it could be that children in whose nature contact with others is weak, are more strongly exposed to the lack of communication which they need as much as other children. On the contrary perhaps a child constitutionally showing much pleasure in contact may still find enough communication even when the stimulation is lacking. As a example for this I may mention a 19 year old hypomanic female patient whose development was faultless even though she spent the first 5 years of her life in a home and was much neglected there. This girl learned to walk late and when she was received by foster parents at the age of 5 years she could hardly make herself understood. She had only a few words at her disposal, stammered these, and was unable to repeat a short sentence. The foster parents arranged for the child to receive speech therapy. She caught up rapidly, started school at the age of 7 years, passed 10 grades without difficulty, and is now taking a course in child psychology. With her hypomanic constitution she easily made happy contact with others and perhaps because of this remained healthy.

Prophylaxis of Early Childhood Catatonia

Certain possibilities of **prophylaxis** of this grave illness result from the findings which we have described. According to my conception many supposedly severely mentally retarded patients in fact have catatonia arising in early childhood. This means that we are faced with a big task. A great many children could be guarded against this adverse fate by protecting them during the first years of life from being isolated. Modern psychiatry independent from my views has already given the advice that a small child should not be separated from its mother for any long period of time, recommending that when the child is admitted to a hospital its mother participate in its care, and that in homes the nurses be trained to give individual care to the children so that they act as substitute mothers. Such interventions however must be stressed very much more in order to prevent the development of early childhood catatonia. The care establishments should always keep in mind that their task is not to give priority to physical care but that providing psychological care is very important. In having stated above that a mentally ill mother is better than no mother I would like to restate this as follows: A not too negligent mother is still better than no mother at all. This is not to say that neglect of a child in its own home should be ignored, this could lead to a gross deficiency and to illness even without stay in a hospital or in a home. Taking away the right of the mother to care for her child often cannot be avoided. But in drawing up a proper plan of action it should be kept in mind how the child can best be protected from damage and in every case this must be done with great care. Sometimes advice and warning of parents is sufficient to reach this objective. In the German Democratic Republic care by foster parents and subsequently through adoption was much promoted to the advantage of the child. As emphatic as my recommendations may be they should not lead to excessive solicitude during a stay in a home. It is only in homes for prolonged care that lack of communication is a danger; not however if the stay is for a short period of time. Of course there may be the danger developing of hospitalism from the transient separation of the mother, as may happen during a stay in a hospital. But a disadvantage from a stay in a short-term home could not be demonstrated. In considering over-anxious mothers who are afraid that they do not devote enough time to their child it should be emphasized that there is no evidence that a partial separation of the child from its mother through her job and the necessary care in a crèche would be disadvantageous. But precaution should be taken that the mother not accept duties outside the house that take her away from her child during the evening hours.

After what has been said the prophylaxis of early childhood catatonia is completely different from that of systematic schizophrenias in later childhood and adult age where contact with the mother is no longer decisive.

Therapeutic Feasibility in Early Childhood Catatonia

The therapy of endogenous psychoses is not dealt with in this book, but prophylaxis is. However in the case of early childhood catatonia I would like to say a few words concerning treatment, as certain valuable indications demonstrate. The first series of investigations having been concluded and for purposes of control subsequent investigations having been added, it was confirmed that of the children placed in homes several developed early childhood catatonia. According to the first observations for every 7 healthy such children there was one catatonic. Afterwards there were 7 patients among 62 healthy children in homes, that is approximately one in 9. That means insignificantly fewer patients. But a new finding was that of these 7 children 3 lived in the milieu of their home and attended a special school. Of the 14 previously examined children in homes only one boy, living in his own home, did not attend a special school. Of the 117 patients seen by me in the psychiatric hospital all without exception were considered incapable of going to school. The administration of the home in which our recent investigations were conducted was very keen on discharging the children at the latest in their third year of life, preferably into the milieu of their own home and not transferred to another home. In this way additional damage through hospitalization could be prevented, for the patients whom we had seen in their own homes had clearly milder clinical pictures than the children of our previous investigations. A parakinetic boy and a parakinetic-manneristic girl were not as restless as the earlier children with similar conditions. The childhood mobility which I had greatly stressed appears to have produced fewer disadvantages because it could be discharged in a more normal manner in an environment with more variety. Both of these children – different from the earlier patients – gave answers to simple questions. A speech-prompt girl did not incoherently join words and sentences, but gave evidence of the aforementioned mobility by her very rapid answers. The correct diagnosis was only obtained from the rest of the clinical picture of speech-prompt catatonia, that is, stiff posture, expressionless face, and autistic absence of relation to outer situations. Furthermore she showed automatic compulsive touching.

The opinion may be expressed that the course of the illness cannot be influenced by providing better external stimulation, yet here I should like to raise certain hopes deriving from my own observations. Although the catatonias of early childhood belong to the systematic forms which surely are of organic origin, theoretically it could be expected that as long as the nervous system has not yet matured it may be influenced. This has been considered in detail above in discussing the significance of communication. Here are my observations: Apart from the 7 children cared for in homes there were 3 whom I thought at the beginning of the investigation were catatonic, who however following further conversations turned to be healthy in the way they led and progressed in their lives. They made at first a crude impression, they shook the hand extended to them with a stiff movement, and showed not the slightest expression in their face. It was impossible to tell whether they were surprised, pleased, or annoyed about our arrival, or whether they had or had not been inwardly prepared for our visit about which they had been told. In the examination room they were sitting stiffly and gave monosyllabic answers to our questions without emotional expression. One of them sometimes smiled, but even this smile gave no clue as to what it signified. Speech was monotonous and the sound was of low volume. When I asked one of them to speak a little louder this did not happen at first but in between he suddenly pronounced a word unnaturally loudly. I could not tell why this change occurred. It might have been thought that all three were uncooperative, but this was contradicted by the fact that they responded willingly to questions and continued willingly to answer my many questions. Throughout the long exploration their stiff posture did not change. There were no gestures. Thus the three youths – they were 16, 18 and 21 years old – showed no mobility in their expression, as is so well known in catatonic patients.

But when I asked them about their lives up to now, nothing striking was revealed. The 16 year old boy had done well in school, for a time he was even the best in the class. Two years from now he will be taking the Abitur (the German school leaving examination). His plans for the future were rather unusual, as he himself admitted, for he was thinking about studying European sciences. But he had not made up his mind. He spent much time doing homework but had friends and went to discos. The 18 year old passed his 10 school grades without difficulty and was now training to be a mechanic. Here too he had no difficulties. Nothing abnormal could be sensed coming from himself and his parents. He had friends and a girl friend and belonged to a marine club as he intended to become a sailor in the navy. To the question as to whether he was a lone-wolf or not he and his parents answered in the negative.

The 21 year old man in addition to his stiff manner was markedly autistic. But here too there was nothing extraordinary in his vocational progress. In school at first he did not do well, but in the course of the years he caught up and passed the 10th grade successfully. He then took training as a brick layer, but did not stay with this occupation which he did not like; at present he was training to become a precision tool maker. During our visit we first met his mother who declared right away that her son was a lone-wolf, and was quite different from his 3 older siblings. He stayed by himself, had hardly any friends, never went to discos. But he was very industrious, and had already received several bonuses at work. At the moment he was not at home as a co-worker had asked him to help in moving to a new apartment. Otherwise in his free time he was at home. Later we spoke to the young man himself and thus confirmed his severe autism. As he said he had no interest in going around with other people, he looked forward to weekends when he could be by himself. Then he did chores, listened to the radio, looked at television or slept. He was interested in sports but, as he said himself, only passively, that is he looked at sports events on television. There was a time when he was interested in music, even learned to play an instrument, but had given it up. At no time in all these years had there been a change in his behaviour. There was no flaw in his personality. At work he was greatly appreciated. At home he had things in good order and recently repapered the walls in the apartment. He spoke all this without inward emotional expression; with his stiff face none could be recognized.

This might be a case of **autistic psychopathy** as described by ASPERGER, but quite apart from the fact that this supposition does not give any etiological clue, the autistic behaviour cannot be judged separately from the general stiffness and expressionlessness – this young man did have great rigidity of posture and expression.

Since these three who had been placed in homes as children so closely resembled catatonic patients, their expressionless behaviour precluded any direct emotional contact and clearly gave the impression of the so called "praecox-feeling". I assume that in their early childhood catatonia appeared but then was arrested before it could impair the shaping of their lives. On the basis of this assumption **therapeutic possibilities** may be considered. How was it possible that these three patients did not have a progressive illness? In catatonia of later childhood and of adulthood the **lack of siblings** represents a risk. Perhaps the progress of catatonia may be arrested therapeutically if to such a child many friends are supplied. These three following discharge from the home did have comrades. The last of the three had 3 siblings, one of them a vivacious man, 3 years older than the patient, whom we came to know ourselves. The 18 year old had no less than 5 siblings. The 16 year old had a brother 2 years older, who had many friends and took the patient into his circle. Thus I assume that a catatonia originated in their first year of life while the mother was missing but was later arrested when in the subsequent years many comrades were available who in that period of time were more important for the biological development than the mother. Should my assumption be correct the possibility of a genuine therapy would be opened up. If the onset of an early childhood catatonia cannot be prevented it could be arrested therapeutically. In general I would like to state here the following: Possibly the development of catatonia proceeds in 2 phases, in the first of which expressive movements are affected and in the second the rest of the motility. The final catatonia may arise only when the latter happens.

My remarks concerning the prophylaxis and perhaps the possible therapy of early childhood catatonia indicate once again how important fine nosological differentiation is in the field of the endogenous psychoses. Its theoretical significance I had begun to appreciate long ago; its practical application I have been able to demonstrate much more emphatically in this **new edition of the "Classification of Endogenous Psychoses and Their Differentiated Etiology" than in the earlier editions.**

Karl Leonhard's Life (1904–1988)

Karl LEONHARD was born on 21 March 1904 in Edelsfeld, Bavaria, son of a Protestant minister. He went to high school (Humanistisches Gymnasium) in Weiden, Oberpfalz, and then studied medicine in Erlangen, Berlin and Munich. After having finished his studies he became assistant physician for a short period of time at the Psychiatric Clinic of the University of Erlangen, under SPECHT. Then in 1931 he moved to the Psychiatric Hospital (Heil- und Pflegeanstalt) Gabersee, Upper Bavaria. A year later he was appointed senior physician and commenced pioneer investigations of the clinical pictures of defect-schizophrenics ("Defektschizophrene Krankheitsbilder"), attracting the attention of Karl KLEIST. In 1936 KLEIST called him to Frankfurt/ Main as senior physician, where a year later he was given an academic appointment in recognition of his work.

During the war years he was exempt from service at the front because of a recurring illness. In his position as clinical director he was able to save numerous patients from the so-called "euthanasia". In 1944 he was appointed extraordinary professor.

After World War II German psychiatry which had been discredited throughout the world turned almost entirely towards psychoanalysis and the anthropological and psychodynamic concepts of mental illnesses. Therefore there was no more opportunity for the scientific work and accomplishments of Karl KLEIST and Karl LEONHARD. Thus it was only in 1954 that the latter was called to become ordinary professor for psychiatry and neurology at the Medical Academy of Erfurt, situated at that time in the Soviet occupation zone. In 1957 he was appointed Director of the Department of Psychiatry of the Charité, the main teaching hospital of the Humboldt University in East Berlin, where he remained until he became 65 years of age.

In gratitude for declining the offer of an academic appointment in Frankfurt/Main, generous space and personnel were made available to him in Berlin enabling him to continue, even after having been appointed Professor Emeritus. He carried on his work with great energy right up to the time of his death. From Berlin he with Dr. Sieglinde VON TROSTORFF traveled frequently visiting various psychiatric hospitals for follow-up investigations; he personally examined large numbers of patients often obtaining their lifelong histories.

The tragic partition of Germany in 1961 to a great extent prevented him from influencing the development of psychiatry in the Western world. His books were printed sparingly and were rarely available in the West. Editors of

Western journals returned his papers because these were not in conformity with the standard practice of Anglo-American psychiatry and also because he pursued without compromise his own path derived from his findings. He always perceived this as rather painful. Evidence of his inner greatness is that he never allowed himself to be diverted from his scientific path by means of opportunistic conformity.

Nevertheless some of his books had several editions, and his "Classification of Endogenous Psychoses" was translated into English, Japanese, Italian, and Russian.

Karl LEONHARD died on 23 April 1988 shortly after his 84th birthday. During all his creative years he was supported by his wife Elfriede with great energy and devotion. She and his three children administer his scientific heritage.

References

Albert E (1980) Katamnesen frühkindlicher Schizophrenien. Psychiat Neurol Med Psychol Leipzig 32: 54–63

Albert E (1986) Über den Einfluß von neuroleptischer Langzeitmedikation auf den Verlauf von physischen und remittierenden Unterformen endogener Psychosen. In: Seidel K, Neumärker KJ, Schulze HAF (eds) Zur Klassifikation endogener Psychosen. Hirzel, Leipzig, S. 97–107

Angst J (1966) Zur Ätiologie und Nosologie endogener depressiver Psychosen. Springer, Berlin

Asperger H (1968) Zur Differentialdiagnose des kindlichen Autismus. Acta Paedopsychiat 35: 136–146

Bleuler M (1972) Die schizophrenen Geistesstörungen. Thieme, Stuttgart

Beck S, Lempp R (1965) Die Stellung in der Geschwisterreihe und ihre Bedeutung für das Auftreten psychoreaktiver Störungen. Dtsch Med J 16: 743–745

Cameron K (1958) A group of twenty-five psychotic children. Rev Psychiat Inf 25: 117–122

Eggers C, Stutte H (1969) Zur nosologischen Umgrenzung der kindlichen und präpubertalen Schizophrenie aus katamnestischer Sicht. Fortschr Neurol Psychiatr 37: 305–318

Essen-Möller E, Copenhagen L (1963) Über die Schizophreniehäufigkeit bei Müttern von Schizophrenen. Schweiz Arch Neurol Neurochir Psychiat 91: 260–266

Faust C (1941) Die paranoiden Schizophrenien auf Grund katamnestischer Untersuchung. Z Ges Neurol Psychiat 172: 308–393

Faust E (1953) Zur Frage der latenten Schizophrenien an den Sippen manifest Schizophrener. Mschr Psychiat 125: 65

Fleck U (1928) Über Beobachtungen bei alten Fällen von Schizophrenie. Arch Psychiat 85: 705–760

Fünfgeld E (1936) Die Motilitätspsychosen und Verwirrtheiten. Karger, Berlin

Harlow HF, Harlow MK (1965) Effects of various mother-infant relationships on rhesus monkey behaviour. In: Foss BM (ed) Determinants of infant behaviour, IV. Methuen, London, p. 15

Healton-Ward A (1977) Psychosis in mental handicap. Br J Psychiatry 130: 525–533

Herbauer (1969) Endogene Psychosen im Kindesalter. In: Huber G (ed) Schizophrenie und Zyklothymie. Thieme, Stuttgart

Heston DL (1966) Psychiatric disorders in foster home reared children of schizophrenic mothers. Br J Psychiat 112: 819–825

Jilek W (1974) In: Biological mechanisms of schizophrenia. Igaku Shoin Ltd, Tokyo

Kallmann FJ (1946) The genetic theory of schizophrenia. An analysis of 691 schizophrenic twins index families. Am J Psychiatry 103: 309–321

Kanner L (1958) The specificity of early infantile autism. Z Kinderpsychiatrie 25: 108–113

Kleist K (1928) Über zykloide, paranoide und epileptoide Psychosen und über die Frage der Degenerationspsychosen. Schweiz Arch Neurol Psychiat 23: 1

Kraepelin E (1913) Psychiatrie, 8. Aufl., Bd. III. Barth, Leipzig

Lempp R (1973) Psychosen im Kindes- und Jugendalter – eine Realitätsbezugsstörung. Eine Theorie der Schizophrenie. Huber, Bern
Leonhard K (1936) Die defektschizophrenen Krankheitsbilder. Thieme, Leipzig
Leonhard K (1937) Involutive und idiopathische Angstdepressionen in Klinik und Erblichkeit. Thieme, Leipzig
Leonhard K (1939) Das ängstlich-ekstatische Syndrom aus innerer Ursache (Angst-Eingebungspsychosen) und äußerer Ursache (Symptomatische Psychosen). Allg Z Psychiat 110: 101–142
Leonhard K (1939) Die Gesetze des normalen Träumens. Thieme, Leipzig
Leonhard K (1942) Zur Unterteilung und Erbbiologie der Schizophrenien. 2. Mitteilung: kombiniert-systematische und periodische Katatonien. Allg Z Psychiat 121: 1–35
Leonhard K (1948) Grundlagen der Psychiatrie. Enke, Stuttgart
Leonhard K (1949) Ausdruckssprache der Seele. Haug, Tübingen
Leonhard K (1950) Eine Sippe affektvoller Paraphrenie mit gehäuften Erkrankungen aus Verwandtenehen (zugleich ein Beitrag zur Frage der Paranoia). Arch Psychiat 184: 291–356
Leonhard K (1951) Gesetze und Sinn des Träumens, 2. Aufl. Thieme, Stuttgart
Leonhard K (1951) Grundlagen der Neurologie. Enke, Stuttgart
Leonhard K (1957) Aufteilung der endogenen Psychosen. Akademie, Berlin. 6. Aufl. 1986
Leonhard K (1959) Individualtherapie und Prophylaxe der hysterischen, anankastischen und sensohypochondrischen Psychosen. VEB Gustav Fischer, Jena
Leonhard K (1960) Über kindliche Katatonien. Psychiat Neurol Med Psychol 12: 1–2
Leonhard K (1961) Die Spielbreite der unsystematischen Schizophrenien, besonders der Kataphasie. Arch Psychiat Nervenkr 202: 513–526
Leonhard K (1963) Die Temperamente in den Familien der monopolaren euphorischen Psychosen. Psychiat Neurol Med Psychol 15: 203–206
Leonhard K (1963) Die präpsychotischen Temperamente bei den monopolaren und bipolaren phasischen Psychosen. Psychiat Neurol 146: 105–115
Leonhard K (1963) Kinderneurosen und Kinderpersönlichkeiten. VEB Volk und Gesundheit, Berlin. 4. Aufl. 1991 Verlag Gesundheit, Berlin
Leonhard K (1964) Prognostische Diagnostik der endogenen Psychosen (mit Sieglinde von Trostorff) Fischer, Stuttgart
Leonhard K (1964) Differenzierte Diagnostik der endogenen Psychosen, abnormen Persönlichkeitsstrukturen und neurotischen Entwicklungsstrukturen. VEB Volk und Gesundheit, Berlin. 4. Aufl. 1991 Ullstein Medical, Verlag Gesundheit, Berlin
Leonhard K (1964) Instinkte und Urinstinkte in der menschlichen Sexualität. Enke, Stuttgart
Leonhard K (1964) Normale und abnormale Persönlichkeiten. Akademie, Berlin
Leonhard K (1965) Die klinische Lokalisation der Hirntumoren in der Kritik der technischen, bioptischen und autoptischen Nachprüfung. Barth, Leipzig
Leonhard K (1966) Biologische Psychologie. Hirzel, Leipzig. 6. Auflage 1993 Hirzel, Stuttgart
Leonhard K (1969) Der menschliche Ausdruck. Barth, Leipzig. 3. Aufl. 1997 u.d.T. Der menschliche Ausdruck in Mimik, Gestik und Phonik. Würzburg
Leonhard K (1970) Biopsychologie der endogenen Psychosen. Hirzel, Leipzig
Leonhard K (1970) Akzentuierte Persönlichkeiten. VEB Volk und Gesundheit, Berlin. 2. Aufl. 1976
Leonhard K (1975) Ein dominanter und ein rezessiver Erbgang bei zwei verschiedenen Formen von Schizophrenie. Nervenarzt 46: 242–248

Leonhard K (1976) Zwillingsuntersuchungen mit einer differenzierten Diagnose der endogenen Psychosen. Psychiat Neurol Med Psychol 28: 78–88

Leonhard K (1981) Monopolar manias and euphorias. Rev Assoc Bras Psiq 3: 67–76

Leonhard K (1981) Individualtherapie der Neurosen, 3. Aufl. VEB Gustav Fischer, Jena

Leonhard K (1984) Als geistige Behinderung verkannte Kindheitsschizophrenie. In: Nissen G (ed) Psychiatrie des Schulalters. Huber, Bern (pp 28–46)

Leonhard K (1988) Bedeutende Persönlichkeiten in ihren psychischen Krankheiten. Akademie, Berlin. 2. Aufl. 1992 Ullstein Mosby, Berlin Wiesbaden

Leonhard K, v. Trostorff S (1964) Prognostische Diagnostik der endogenen Psychosen. VEB Gustav Fischer, Jena

Leonhard K, Korff I, Schulz H (1962) Die Temperamente in den Familien der monopolaren und bipolaren phasischen Psychosen. Psychiat Neurol 143: 416–434

Lutz J (1970) Zur Frage der Entstehung der Schizophrenie. In: Bleuler M, Angst J (eds) Die Entstehung der Schizophrenie. Huber, Bern

Luxemburger H (1936) Untersuchungen an schizophrenen Zwillingen und ihren Geschwistern zur Prüfung der Realität von Manifestationsschwankungen. Z Ges Neurol Psychiat 15: 351–394

Mayer-Gross W (1932) Die Klinik. In: Bumke (ed) Handbuch der Geisteskrankheiten, Bd. 9. Springer, Berlin

Mitsuda H (1967) Clinical genetics in psychiatry. Bunkosha, Kyoto

Neele E (1949) Die phasischen Psychosen. Barth, Leipzig

Otremba G (1963) Über das Krankheitsbild der Kataphasie. Psychiat Neurol Med Psychol 15: 61

Penrose LS (1966) The contribution of mental deficiency research to psychiatry. Br J Psychiat 112: 747–755

Perris C (1966) Study of bipolar (manic-depressive) and unipolar recurrent depressive psychoses. Acta Psychiat Scand 42, [Suppl], 194

Perris C (1974) A study of cycloid psychoses. Acta Psychiat Scand, [Suppl], 253

Pönitz K (1913) Beitrag zur Kenntnis der Frühkatatonie. Z Ges Neurol Psychiat 20: 343–357

Raecke J (1909) Katatonie im Kindesalter. Arch Psychiat Nervenkrankh 45: 245–279

Richter KH (1968) Über die Entstehung von Entfremdungserscheinungen bei der hypochondrischen Depression. Psychiatry Neurol Med Psychol 20: 181–184

Schmidt-Kolmer E (1960) Verhalten und Entwicklung des Kleinkinds, 2. Aufl. Akademie, Berlin

Schulz B, Leonhard K (1940) Erbbiologisch-klinische Untersuchungen an insgesamt 99 im Sinne Leonhards typischen bzw. atypischen Schizophrenien. Z Ges Neurol Psychiat 168: 587–613

Seitelberger F (1980) Das Gehirn und das Nervensystem im psychosomatischen Geschehen. Universitas 35: 33–40

Slater E (1953) Psychotic and neurotic illnesses in twins. Medical Research Council special report series No. 278. Her Majesty's Stationary Office, London

Specht G (1901) Über den pathologischen Affekt in der chronischen Paranoia. Deichertsche Verlagsbuchhandlung Nachf. G. Böhme, Erlangen

Spiel W (1961) Die endogenen Psychosen des Kindes- und Jugendalters. Karger, Basel

Spitz RA (1945) Hospitalism, an inquiry into the genesis of psychiatric conditions in early childhood. Psychoanal Study Child 1: 53–74

Stutte H (1963) Endogen-phasische Psychosen des Kindesalters. Acta Paedopsychiat 30: 34–42

Sucharewa GJ (1967) Die Bedeutung der vergleichenden Berücksichtigung des Lebensalters für die Untersuchung der Verlaufsgesetzmäßigkeiten der Schizophrenie bei Kindern und Jugendlichen. Acta Paedopsychiat 34: 307–320

Tienari P (1963) Psychiatric illness in identical twins. Munksgaard, Copenhagen
Toman W, Preiser S (1973) Familienkonstellation und ihre Störungen. Enke, Stuttgart
Trostorff Sv. (1966) Präpsychotische Temperamente bei den zykloiden Psychosen. Arch Psychiat 208: 61–90
Trostorff Sv. (1968) Über die hereditäre Belastung bei den bipolaren und den monopolaren phasischen Psychosen. Schweiz Arch Neurol Neurochir Psychiat 102: 235–243
Trostorff Sv. (1970) Extraversion und Introversion sowie Kontaktfreudigkeit und Kontaktarmut bei normalen und präpsychotischen Persönlichkeiten. VEB Gustav Fischer, Jena
Trostorff Sv. (1975) Verlauf und Psychosen in der Verwandtschaft bei den systematischen und unsystematischen Schizophrenien und den zykloiden Psychosen. Psychiat Neurol Med Psychol 27: 80–100
Trostorff Sv. (1981) Zur Frage eines dominanten Erbgangs bei der periodischen Katatonie. Psychiat Neurol Med Psychol 33: 158–166
Trostorff Sv. (1986) Rezessiver Erbgang bei affektvoller Paraphrenie. In: Seidel K, Neumärker KJ, Schulze HAF (eds) Zur Klassifikation endogener Psychosen. Hirzel, Leipzig, S. 108–115
Villinger W (1959) Symptomatologie der kindlich-jugendlichen Schizophrenien. Int Kongreßber 1: 345–350
Vogt H (1909) Über Fälle von Jugendirresein im Kindesalter. Allg Zschr Psychiat 66: 542–573
Wernicke C (1900) Grundriß der Psychiatrie. Thieme, Leipzig
Weygandt W (1908) Idiotie und dementia praecox. Z Erforsch Jugendl Schwachsinn 1: 311–332
Wieck C (1965) Schizophrenie im Kindesalter. Hirzel, Leipzig
Winokur G, Clayton P, Reich T (1969) Manic depressive illness. Mosby, St. Louis

Leonhard	*Stengel (1960)*	*Berman (1979)*	*Cahn (1999)*
Melancholie	Melancholia	Melancholy	Melancholia
Gehetzte Depression	Agitated Depression	Harried Depression	Agitated Depression
Selbstquälerische Depression	Self-Torturing Depression	Self-Torturing Depression	Self-Tortured Depression
Teilnehmsarme Depression	Apathetic Depression	Non-Participatory Depression	Apathetic Depression
Schwärmerische Euphorie	Exalted Euphoria	Enthusiastie Euphoria	Exalted Euphoria
Teilnahmsarme Euphorie	Indiferrent Euphoria	Non-Participatory euphoria	Indifferent Euphoria
Angst-Glück-Psychose	Anxiety-Bliss Psychosis	Anxiety-Happiness Psychosis	Anxiety-Happiness Psychosis
Erregt-gehemmte Verwirrtheits-psychose	Exicitation-Retardation Psychosis with Confusion	Excited-Inhibited Confusion Psychosis	Excited-Inhibited Confusion Psychosis
Hyperkinetisch-akinetische Motilitätspsychose	Hyperkinetic-Akinetic Psychoses	Hyperkinetic-Akinetic Motility Psychosis	Hyperkinetic-Akinetic Motility Psychosis
Affektvolle Paraphrenie	Affective Paraphrenia	Affect-Laden Paraphrenia	Affective Paraphrenia
Kataphasie (Schizophasie)	Schizophasia	Cataphasia (Schizophasia)	Cataphasia (Schizophasia)
Manierierte Katatonie	Manneristic Catatonia	Affected Catatonia	Manneristic Catatonia
Sprechbereite Katatonie	Talkative Catatonia	Voluble Catatonia	Speech-Prompt Catatonia
Sprachträge Katatonie	Untalkative Catatonia	Sluggish Catatonia	Sluggish Catatonia
Läppische Hebephrenie	Fatuous Hebephrenia	Silly Hebephrenia	Foolish Hebephrenia
Verschrobene Hebephrenie	Odd Hebephrenia	Eccentric Hebephrenia	Eccentric Hebephrenia
Flache Hebephrenie	Emotionally Flat Hebephrenia	Insipid Hebephrenia	Shallow Hebephrenia
Konfabulatorische Paraphrenie	Confusional Paraphrenia	Confabulatory Paraphrenia	Confabulatory Paraphrenie

Subject Index

Springer Medicine

Norbert Müller (ed.)

Psychiatry, Psychoimmunology, and Viruses

1999. VIII, 156 pages. 18 figures.
Softcover DM 128,–, öS 896,–
(recommended retail price)
ISBN 3-211-83249-1
Key Topics in Brain Research

The contributions in this book deal with basic research in psychoneuroimmunology, dysfunctions of the immune system in psychiatric disorders and the possible role of viruses in psychoneuroimmunological processes.

Growing knowledge about the reciprocal relationship between psychic processes and the immune system implicates more interest of clinical psychiatrists in this field because immunological dysfunction is found in depression and schizophrenia, but also other psychiatric syndromes.

Contents
- Viruses and the immune system in psychiatric disorders
- Serotonergic system, depression, and immune system
- Schizophrenia and the immune system
- Sleep, anxiety, headache, and immunity

Springer WienNewYork

Sachsenplatz 4–6, P.O.Box 89, A-1201 Wien, Fax +43-1-330 24 26, e-mail: books@springer.at, **Internet: http://www.springer.at**
New York, NY 10010, 175 Fifth Avenue • D-14197 Berlin, Heidelberger Platz 3 • Tokyo 113, 3–13, Hongo 3-chome, Bunkyo-ku

SpringerPsychiatry

Ernst Franzek, Helmut Beckmann

Psychoses of the Schizophrenic Spectrum in Twins

A Discussion on the Nature – Nurture Debate
in the Etiology of "Endogenous" Psychoses

1999. IX, 149 pages. 2 figures.
Softcover DM 78,–, öS 546,–
(recommended retail price)
ISBN 3-211-83298-X

The data of this polydiagnostic systematic twin study provide stro
evidence that the spectrum of schizophrenic psychoses is not a continu
of diseases. It consists of cycloid psychoses, unsystematic and systema
schizophrenias according to Karl Leonhard. The findings clearly indic
that in cycloid psychoses somatic and psycho-social factors are of ma
etiological importance, whereas genetic predisposition is the predon
nant cause of unsystematic schizophrenias.

Contents
- Introduction
- General remarks on twin origination
- The twins method in research
- Further questions
- Methodology of a systematic twin study
- Results
- Discussion
- Summary, Conclusion
- Appendix: Synoptic tables
- Bibliography, Index, Acknowledgements

SpringerWienNewYork

Sachsenplatz 4–6, P.O.Box 89, A-1201 Wien, Fax +43-1-330 24 26, e-mail: books@springer.at, **Internet: http://www.springer.at**
New York, NY 10010, 175 Fifth Avenue • D-14197 Berlin, Heidelberger Platz 3 • Tokyo 113, 3–13, Hongo 3-chome, Bunkyo-ku

Springer-Verlag and the Environment

WE AT SPRINGER-VERLAG FIRMLY BELIEVE THAT AN international science publisher has a special obligation to the environment, and our corporate policies consistently reflect this conviction.

WE ALSO EXPECT OUR BUSINESS PARTNERS – PRINTERS, paper mills, packaging manufacturers, etc. – to commit themselves to using environmentally friendly materials and production processes.

THE PAPER IN THIS BOOK IS MADE FROM NO-CHLORINE pulp and is acid free, in conformance with international standards for paper permanency.

GPSR Compliance

The European Union's (EU) General Product Safety Regulation (GPSR) is a set of rules that requires consumer products to be safe and our obligations to ensure this.

If you have any concerns about our products, you can contact us on ProductSafety@springernature.com

In case Publisher is established outside the EU, the EU authorized representative is:

Springer Nature Customer Service Center GmbH
Europaplatz 3
69115 Heidelberg, Germany

Batch number: 10732312

Printed by Printforce, the Netherlands